Accelerator Physics
Editors: F. Bonaudi
C.W. Fabjan

W. Blum   L. Rolandi

Particle Detection
with Drift Chambers

With 198 Figures and 44 Tables

Springer-Verlag

Berlin Heidelberg New York
London Paris Tokyo
Hong Kong Barcelona
Budapest

Authors
Dr. Walter Blum
Max-Planck-Institut für Physik
Werner-Heisenberg-Institut
Föhringer Ring 6
D-80805 München, Germany

Professor Dr. Luigi Rolandi
CERN, Div. PPE
CH-1211 Genève 23, Switzerland

Editors
Professor F. Bonaudi
Professor C.W. Fabjan
CERN, Div. PPE
CH-1211 Genève 23, Switzerland

The cover picture represents a computer reconstruction of an event measured in the drift chambers of the ALEPH experiment. One observes an event of electron-positron annihilation into two quarks which subsequently transform into several hadrons. Courtesy of H. Drevermann, CERN.

ISBN 3-540-56425-X Springer-Verlag Berlin Heidelberg New York
ISBN 0-387-56425-X Springer-Verlag New York Berlin Heidelberg

Library of Congress Cataloging-in-Publication Data. Blum, W. (Walter), 1937–   Particle detection with drift chambers/W. Blum, L. Rolandi.   p.   cm. Includes bibliographical references and index. ISBN 3-540-56425-X (Berlin). – ISBN 0-387-56425-X (New York) 1. Drift chambers. 2. Particles (Nuclear physics) – Measurement. I. Rolandi, L. (Luigi), 1953–   . II. Title. QC787.D74B58 1993 539.7'72 – dc20   93-13632

Typesetting: Macmillan India Ltd., Bangalore 25
55/3140/SPS-543210 – Printed on acid-free paper

Preface

A drift chamber is an apparatus for measuring the space coordinates of the trajectory of a charged particle. This is achieved by detecting the ionization electrons produced by the charged particle in the gas of the chamber and by measuring their drift times and arrival positions on sensitive electrodes.

When the multiwire proportional chamber, or 'Charpak chamber' as we used to call it, was introduced in 1968, its authors had already noted that the time of a signal could be useful for a coordinate determination, and first studies with a drift chamber were made by Bressani, Charpak, Rahm and Zupančič in 1969. When the first operational drift-chamber system with electric circuitry and readout was built by Walenta, Heintze and Schürlein in 1971, a new instrument for particle experiments had appeared. A broad study of the behaviour of drifting electrons in gases began in laboratories where there was interest in the detection of particles.

Diffusion and drift of electrons and ions in gases were at that time well-established subjects in their own right. The study of the influence of magnetic fields on these processes was completed in the 1930s and all fundamental equations were contained in the article by W.P. Allis in the *Encyclopedia of Physics* [ALL 56]. It did not take very long until the particle physicists learnt to apply the methods of the Maxwell–Boltzmann equations and of the electron-swarm experiments that had been developed for the study of atomic properties. The article by Palladino and Sadoulet [PAL 75] recorded some of these methods for use with particle-physics instruments.

F. Sauli gave an academic training course at CERN in 1975/76, in order to inform a growing number of users of the new devices. He published lecture notes [SAU 77], which were a major source of information for particle physicists who began to work with drift chambers.

When the authors of this book began to think about a large drift chamber for the ALEPH experiment, we realized that there was no single text to introduce us to those questions about drift chambers that would allow us to determine their ultimate limits of performance. We wanted to have a text, not on the technical details, but on the fundamental processes, so that a judgement about the various alternatives for building a drift chamber would be on solid ground. We needed some insight into the consequences of different geometries and how to distinguish between the behaviour of different gases, not so much a complete table of their properties. We wanted to understand on what trajectories the ionization electrons would drift to the proportional wires and to what extent the tracks would change their shape.

Paths to the literature were also required – just a few essential ones – so that an entry point to every important subject existed; they would not have to be a comprehensive review of 'everything'.

In some sense we have written the book that we wanted at that time. The text also contains a number of calculations that we made concerning the statistics of ionization and the fundamental limits of measuring accuracy that result from it, geometrical fits to curved tracks, and electrostatics of wire grids and field cages. Several experiments that we undertook during the construction time of the ALEPH experiment found their way into the book; they deal mainly with the drift and diffusion of electrons in gases under various field conditions, but also with the statistics of the ionization and amplification processes.

The book is nonetheless incomplete in some respects. We are aware that it lacks a chapter on electronic signal processing. Also some of the calculations are not yet backed up in detail by measurements as they will eventually have to be. Especially the parameter N_{eff} of the ionization process which governs the achievable accuracy should be accurately known and supported by measurements with interesting gases. We hope that workers in this field will direct their efforts to such questions. We would welcome comments about any other important omissions.

It was our intention to make the book readable for students who are interested in particle detectors. Therefore, we usually tried to explain in some detail the arguments that lead up to a final result. One may say that the book represents a cross between a monograph and an advanced textbook. Those who require a compendious catalogue of existing or proposed drift chambers may find useful the proceedings of the triannual Vienna Wire Chamber Conferences [VIE] or of the annual IEEE Symposia on Instrumentation for Nuclear Science [IEE].

Parts of the material have been presented in summer schools and guest lectures, and we thank H.D. Dahmen (Herbstschule Maria Laach), E. Fernandez (Universita Autonoma, Barcelona) and L. Bertocchi (ICTP, Trieste) for their hospitality.

We thank our colleagues from ALEPH TPC group, and especially J. May and F. Ragusa, for many stimulating discussions on the issues of this book. We are also obliged to H. Spitzer (Hamburg) who read and commented on an early version of the manuscript. Special thanks are extended to Mrs. Heininger in Munich who produced most of the drawings.

Geneva   *W. Blum*

1 April 1993   *L. Rolandi*

[ALL 56]   W.P. Allis, Motions of ions and electrons, in *Handbuch der Physik*, ed. by S. Flügge (Springer, Berlin 1956) Vol. XXI, p. 383

[IEE]   The symposia are usually held in the fall of every year and are published in consecutive volumes of the *IEEE Transactions in Nuclear Science* in the first issue of the following year

[PAL 75]   V. Palladino and B. Sadoulet, Application of classical theory of electrons in gases to drift proportional chambers, *Nucl. Instrum. Methods* **128**, 323 (1975)

[SAU 77]   F. Sauli, Principles of operation of multiwire proportional and drift chambers, Lectures given in the academic training programme of CERN 1975–76 (CERN 77-09, Geneva 1977), reprinted in *Experimental Techniques in High Energy Physics*, ed. by T. Ferbel (Addison-Wesley, Menlo Park 1987)

[VIE]   The Vienna Wire Chamber Conferences were held in February of the years 1978, 1980, 1983, 1986, 1989, and they were published the same years in *Nuclear Instruments and Methods* in the following volumes: **135, 176, 217,** A **252,** A **283**.

Contents

X   Contents

1. Gas Ionization by Charged Particles and by Laser Rays

Charged particles can be detected in drift chambers because they ionize the gas along their flight path. The energy required for them to do this is taken from their kinetic energy and is very small, typically a few keV per centimetre of gas in normal conditions.

The ionization electrons of every track segment are drifted through the gas and amplified at the wires in avalanches. Electrical signals that contain information about the original location and ionization density of the segment are recorded.

Our first task is to review how much ionization is created by a charged particle (Sects. 1.1 and 1.2). This will be done using the method of Allison and Cobb, but the historic method of Bethe and Bloch with the Sternheimer corrections is also discussed. Special emphasis is given to the fluctuation phenomena of ionization.

Pulsed UV lasers are sometimes used for the creation of straight ionization tracks in the gas of a drift chamber. Here the ionization mechanism is quite different from the one that is at work with charged particles, and we present an account of the two-photon rate equations as well as of some of the practical problems encountered when working with laser tracks (Sect. 1.3).

1.1 Gas Ionization by Fast Charged Particles

1.1.1 Ionizing Collisions

A charged particle that traverses the gas of a drift chamber leaves a track of ionization along its trajectory. The encounters with the gas atoms are purely random and are characterized by a mean free flight path λ between ionizing encounters given by the ionization cross-section per electron σ_I and the density N of electrons:

$$\lambda = 1/(N\sigma_I) . \tag{1.1}$$

Therefore, the number of encounters along any length L has a mean of L/λ, and the frequency distribution is the Poisson distribution

$$P(L/\lambda, k) = \frac{(L/\lambda)^k}{k!} \exp(- L/\lambda) . \tag{1.2}$$

It follows that the probability distribution $f(l)\,dl$ of the free flight paths l between encounters is an exponential, because the probability of finding zero encounters in the interval l times the probability of one encounter in dl is equal to

$$f(l)\,dl = P(1/\lambda, 0)P(dl/\lambda, 1)$$

$$= (1/\lambda)\exp(-l/\lambda)\,dl\,. \qquad (1.3)$$

From (1.2) we obtain the probability of having zero encounters along a track length L:

$$P(L/\lambda, 0) = \exp(-L/\lambda)\,. \qquad (1.4)$$

Equation (1.4) provides a method for measuring λ. If a gas counter with sensitive length L is set up so that the presence of even a single electron in L will always give a signal, then its inefficiency may be identified with expression (1.4), thus measuring λ. This method has been used with streamer, spark, and cloud chambers, as well as with proportional counters and Geiger–Müller tubes. A correction must be applied when a known fraction of single electrons remains below the threshold.

Table 1.1 shows a collection of measured values of $1/\lambda$ with fast particles whose relativistic velocity factor γ is quoted as well, because λ depends on the particle velocity (see Sect. 1.2.6); in fact, $1/\lambda$ goes through a minimum near $\gamma = 4$.

Table 1.1. Measured numbers of ionizing collisions per centimetre of track length in various gases at normal density [ERM 69]. The relativistic velocity factor γ is also indicated

Gas	$1\ \mathrm{cm}/\lambda$	γ
H_2	5.32 ± 0.06	4.0
	4.55 ± 0.35	3.2
	$5.1 \ \pm 0.8$	3.2
He	5.02 ± 0.06	4.0
	3.83 ± 0.11	3.4
	$3.5 \ \pm 0.2$[a]	3.6
Ne	$12.4 \ \pm 0.13$	4.0
	$11.6 \ \pm 0.3$[a]	3.6
Ar	$27.8 \ \pm 0.3$	4.0
	$28.6 \ \pm 0.5$	3.5
	$26.4 \ \pm 1.8$	3.5
Xe	44	4.0
N_2	19.3	4.9
O_2	$22.2 \ \pm 2.3$	4.3
Air	25.4	9.4
	$18.5 \ \pm 1.3$	3.5

[a] [SÖC 79].

Table 1.2. Minimal primary ionization cross-sections σ_p for charged particles in some gases, and relativistic velocity factor γ_{min} of the minimum, according to measurements done by Rieke and Prepejchal [RIE 72]

Gas	σ_p $(10^{-20}\,cm^2)$	γ_{min}	Gas	σ_p $(10^{-20}\,cm^2)$	γ_{min}
H_2	18.7	3.81	$i\text{-}C_4H_{10}$	333	3.56
He	18.6	3.68	$n\text{-}C_5H_{12}$	434	3.56
Ne	43.3	3.39	$neo\text{-}C_5H_{12}$	433	3.45
Ar	90.3	3.39	$n\text{-}C_6H_{14}$	526	3.51
Xe	172	3.39	C_2H_2	126	3.60
O_2	92.1	3.43	C_2H_4	161	3.58
CO_2	132	3.51	CH_3OH	155	3.65
C_2H_6	161	3.58	C_2H_5OH	230	3.51
C_3H_8	269	3.47	$(CH_3)_2CO$	277	3.54

In Table 1.2 we present additional measurements of a larger number of gases that are employed in drift chambers. These primary ionization cross-sections σ_p were measured by Rieke and Prepejchal [RIE 72] in the vicinity of the minimum at different values of γ and interpolated to the minimum σ_p^{min} at γ^{min}, using the parametrization of the Bethe–Bloch formula (see Sect. 1.2.7). The mean free path λ is related to σ_p by the number density N_m of molecules:

$$\lambda = 1/(N_m \sigma_p) \ .$$

The measurement errors are within $\pm 4\%$ (see the original paper for details). In comparison with the values presented in Table 1.1, the measurements are in rough agreement, except for argon.

1.1.2   Different Ionization Mechanisms

We distinguish between primary and secondary ionization. In primary ionization, one or sometimes two or three electrons are ejected from the atom A encountered by the fast particle, say a π meson:

$$\pi A \rightarrow \pi A^+ e^-, \pi A^{++} e^- e^-, \ldots \tag{1.5}$$

Most of the charge along a track is from secondary ionization where the electrons are ejected from atoms not encountered by the fast particle. This happens either in collisions of ionization electrons with atoms,

$$e^- A \rightarrow e^- A^+ e^-, e^- A^{++} e^- e^- \ , \tag{1.6}$$

or through intermediate excited states A*. An example is the following chain of reactions involving the collision of the excited state with a second species, B, of atoms or molecules that is present in the gas:

$$\pi A \rightarrow \pi A^* \tag{1.7a}$$

or

$$e^- A \to e^- A^*,$$ (1.7b)

$$A^*B \to AB^+ e^-.$$ (1.8)

Reaction (1.8) occurs if the excitation energy of A^* is above the ionization potential of B. In drift chambers, A^* is often the metastable state of a noble gas created in reaction (1.7b), and B is one of the molecular additives (quenchers) that are required for the stability of proportional wire operation; A^* may also be an optical excitation with a long lifetime due to resonance trapping. These effects are known under the names of *Penning effect* (involving metastables) and *Jesse effect* (involving optical excitations, also used more generally); obviously they depend very strongly on the gas composition and density.

Another example of secondary ionization through intermediate excitation has been observed in pure rare gases where an excited molecule A_2^* has a stable

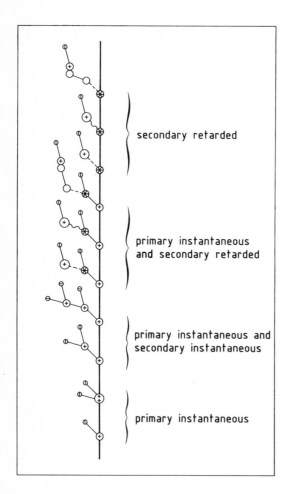

secondary retarded

primary instantaneous
and secondary retarded

primary instantaneous and
secondary instantaneous

primary instantaneous

Fig. 1.1. Pictorial classification of the ionization produced by a fast charged particle in a noble gas containing molecules with low ionization potential: $(-)$ electron; $(+)$ positive ion, single charge; $(++)$ positive ion, double charge; $(+)$ positive ion of the low-ionization species; $(*)$ state excited above the lower ionization potential of the other species; $( )(+)$ positive ion of noble gas molecule; $\sim$ photon transmission, $--$ collision

ionized ground state A_2^+:

$$A^*A \rightarrow A_2^* \rightarrow A_2^+ e^- \ . \tag{1.9}$$

The different contributions of processes (1.6–9) are in most cases unknown. For further references, we recommend the proceedings of the conferences dedicated to these phenomena, for example the Symposium on the Jesse Effect and Related Phenomena [PRO 74].

A pictorial summary of the processes discussed is given in Fig. 1.1.

1.1.3 Average Energy Required to Produce One Ion Pair

Only a certain fraction of all the energy lost by the fast particle is spent in ionization. The total amount of ionization from all processes is characterized by the energy W that is spent, on the average, on the creation of one free electron. We write

$$W\langle N_{\mathrm{I}} \rangle = L \left\langle \frac{\mathrm{d}E}{\mathrm{d}x} \right\rangle , \tag{1.10}$$

where $\langle N_{\mathrm{I}} \rangle$ is the average number of ionization electrons created along a trajectory of length L, and $\langle \mathrm{d}E/\mathrm{d}x \rangle$ is the average total energy loss per unit path length of the fast particle; W must be measured for every gas mixture.

Many measurements of W have been performed since the advent of radioactivity, using radioactive and artificial sources of radiation. The amount of ionization produced by particles that lose all their energy in the gas is measured by ionization chambers or proportional counters. The value of W is the ratio of the initial energy to the number of ion pairs. The energy W depends on the gas – its composition and density – and on the nature of the particle. Experimentally it is found that W is independent of the initial energy above a few keV for electrons and a few MeV for alpha-particles, which is a remarkable fact.

When a relativistic particle traverses a layer of gas, the energy deposit is such a small fraction of its total energy that it cannot be measured as the difference between initial and final energy. Therefore, there is no direct determination of the appropriate value of W, and we have to rely on extrapolations from fully stopped electrons.

A critical review of the average energy required to produce an ion pair is given in a report of the International Commission on Radiation Units and Measurements [INT 79]. A treatment in a wider context is provided by the book of Christophorou [CHR 71] and by the review by Inokuti [INO 75]; see also the references quoted in these three works. For pure noble gases, W varies between 46 eV for He and 22 eV for Xe; for pure organic vapours the range between 23 and 30 eV is typical. Ionization potentials are smaller by factors that are typically between 1.5 and 3. Table 1.3 contains a small selection of W-values for various gases.

Table 1.3. Energy W spent, on the average, for the creation of one ionization electron in various gases and gas mixtures [CHR 71]; W_α and W_β are from measurements using α or β sources, respectively. The lowest ionization potential is also indicated

Gas	W_α (eV)	W_β (eV)	I (eV)	Gas mixture[a]	W_α (eV)
H_2	36.4	36.3	15.43	Ar (96.5%) + C_2H_6 (3.5%)	24.4
He	46.0	42.3	24.58	Ar (99.6%) + C_2H_2 (0.4%)	20.4
Ne	36.6	36.4	21.56	Ar (97%) + CH_4 (3%)	26.0
Ar	26.4	26.3	15.76	Ar (98%) + C_3H_8 (2%)	23.5
Kr	24.0	24.05	14.00	Ar (99.9%) + C_6H_6 (0.1%)	22.4
Xe	21.7	21.9	12.13	Ar (98.8%) + C_3H_6 (1.2%)	23.8
CO_2	34.3	32.8	13.81	Kr (99.5%) + C_4H_8-2 (0.5%)	22.5
CH_4	29.1	27.1	12.99	Kr (93.2%) + C_2H_2 (6.8%)	23.2
C_2H_6	26.6	24.4	11.65	Kr (99%) + C_3H_6 (1%)	22.8
C_2H_2	27.5	25.8	11.40		
Air	35.0	33.8	12.15		
H_2O	30.5	29.9	12.60		

[a] The quoted concentration is the one that gave the smallest W.

Values of W measured with photons and with electrons are the same. Values of W measured with α-sources are similar to those measured with β-sources: W_α/W_β is 1 for noble gases but can reach 1.15 for some organic vapours [CHR 71]. In pure argon, the value of W for very slow electrons such as the ones emitted in primary ionization processes (1.6) is greater than for fast electrons. Figure 1.2 contains measurements reported by Combecher [COM 77].

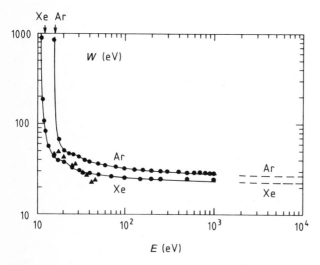

Fig. 1.2. Average energy W spent for the creation of one ionization electron in pure argon and in pure xenon as a function of the energy E of the ionizing particle, which is an electron fully stopped [COM 77]. The *dashed lines* represent the values W_β from Table 1.2 measured at larger E

From Table 1.3 it is apparent that the total ionization in a noble gas can be increased by adding a small concentration of molecules with low ionization potential. The extra ionization comes *later*, depending on the de-excitation rate of the A* involved. For example, a contamination of 3×10^{-4} nitrogen in neon–helium gas has caused secondary retarded ionization that amounted to 60% of the primary, with mean retardation times around 1 μs [BLU 74].

Whether the number of primary encounters that lead to ionization can be increased in the same manner is not so clear; the fast particle would have to excite the state A* in a primary collision (reaction (1.7a)). A metastable state does not have a dipole transition to the ground state and is therefore not easily excited by the fast particle.

1.1.4   The Range of Primary Electrons

Primary electrons are emitted almost perpendicular to the track. They lose their kinetic energy E in collisions with the gas molecules, scattering almost randomly and producing secondary electrons, until they have lost their kinetic energy. A practical range R can be defined as the thickness of the layer of material they cross before being stopped. The empirical relation

$$R(E) = AE\left(1 - \frac{B}{1 + CE}\right) \tag{1.11}$$

with $A = 5.37 \times 10^{-4}$ g cm^{-2} keV^{-1}, $B = 0.9815$, and $C = 3.1230 \times 10^{-3}$ keV^{-1} is shown in Fig. 1.3 and compared with experimental data in the range between 300 eV and 20 MeV. It is shown in [KOB 68] that the parametrization (1.11) is applicable to all materials with low and intermediate atomic number Z. In the absence of low-energy data this curve may be taken as a basis for an extrapolation to lower E. Below 1 keV the relation reads

$$R(E) = 9.93\left(\frac{\mu g}{cm^2\, keV}\right)E \;.$$

In argon (N.T.P.) an electron of 1 keV is stopped in about 30 μm, and one of 10 keV in about 1.5 mm. Only 0.05% of the collisions between the particle and the gas molecules produce primary electrons with kinetic energy larger than 10 keV. We will compute these probabilities in Sect. 1.2.4.

1.1.5   The Differential Cross-section dσ/dE

Every act of ionization is a quantum mechanical transition initiated by the field of the fast particle and the field created indirectly by the neighbouring polarizable atoms. A complete calculation would involve the transition amplitudes of all the atomic states and does not exist.

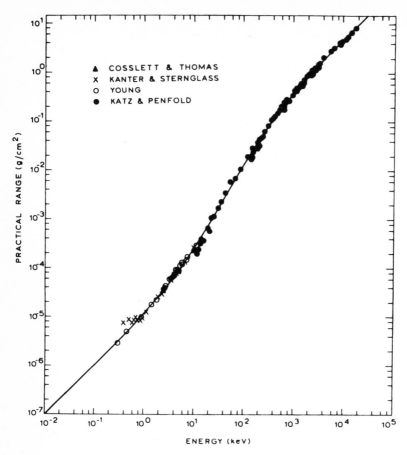

Fig. 1.3. Practical range versus kinetic energy for electrons in aluminium. The data points are from various authors, which can be traced back consulting [KOB 68]. The curve is a parametrization according to (1.11)

For the detection of particles, what we have to know in the first place is the amount of ionization along the track and the associated fluctuation phenomena. For this it is sufficient to determine for an act of primary ionization with what probabilities it will result in a total ionization of $1, 2, \ldots,$ or n electrons. The total ionization over a piece of track and its frequency distribution can then be determined by summing over all the primary encounters in that piece.

Using the concept of the average energy W required to produce one ion pair – be it a constant as in (1.10) or a function of the energy of the primary electron as in Fig. 1.2 – the problem is reduced to finding the energy spectrum $F(E)\,\mathrm{d}E$ of the primary electrons, or, equivalently, the corresponding differential cross-section $\mathrm{d}\sigma/\mathrm{d}E$. Once this is known, we obtain λ and $F(E)$ with the relations

$$1/\lambda = \int N \frac{\mathrm{d}\sigma}{\mathrm{d}E}\,\mathrm{d}E$$

and

$$F(E) = \frac{N(\mathrm{d}\sigma/\mathrm{d}E)}{1/\lambda} \, ,$$

where N is the electron number density in the gas and $\mathrm{d}\sigma/\mathrm{d}E$ the differential cross-section per electron to produce a primary electron with an energy between E and $E + \mathrm{d}E$.

1.2   Calculation of Energy Loss

In order to calculate $\mathrm{d}\sigma/\mathrm{d}E$ we will begin by investigating the average total energy loss per unit distance, $\langle \mathrm{d}E/\mathrm{d}x \rangle$, of a moving charged particle in a polarizable medium. Here a classical calculation is appropriate in which the medium is treated as a continuum characterized by a complex dielectric constant $\varepsilon = \varepsilon_1 + \mathrm{i}\varepsilon_2$. Later on we will interpret the resulting integral over the lost energy in a quantum mechanical sense.

1.2.1   Force on a Charge Travelling Through a Polarizable Medium

It is the field E_{long} opposite to its direction of motion, created by the moving particle in the medium at its own space point, which produces the force equal to the energy loss per unit distance

$$\langle \mathrm{d}E/\mathrm{d}x \rangle = eE_{\mathrm{long}} \, ,$$

where e is the charge of the moving particle. We follow the method of Landau and Lifshitz [LAN 60] in the form used by Allison and Cobb [ALL 80].

Maxwell's equations in an isotropic, homogeneous, non-magnetic medium are

$$\mathbf{V} \cdot \mathbf{H} = 0 \, ,$$
$$\mathbf{V} \times \mathbf{E} = -\frac{1}{c}\frac{\partial \mathbf{H}}{\partial t} \, ,$$
$$\mathbf{V} \cdot (\varepsilon \mathbf{E}) = 4\pi\varrho \, ,$$
$$\mathbf{V} \times \mathbf{H} = \frac{1}{c}\frac{\partial(\varepsilon \mathbf{E})}{\partial t} + \frac{4\pi}{c}\mathbf{j} \, .$$

(1.12)

Since there will be no confusion between the electric field vector E and the energy lost, E, we will use the customary symbols.

The charge density and the flux are given by the particle moving with velocity βc:

$$\varrho = e\delta^3(\mathbf{r} - \boldsymbol{\beta} ct), \qquad \mathbf{j} = \boldsymbol{\beta} c\varrho \, .$$

(1.13)

We work in the Coulomb gauge and introduce the potentials ϕ and A:

$$H = \nabla \times A \ ,$$

$$\nabla \cdot A = 0 \ , \tag{1.14}$$

$$E = -\frac{1}{c}\frac{\partial A}{\partial t} - \nabla\phi \ .$$

Equations (1.12), expressed in terms of the potentials, are

$$\nabla \cdot (\varepsilon\nabla\phi) = -4\pi e\delta^3(r - \beta ct) \ ,$$

$$-\nabla^2 A = -\frac{1}{c^2}\frac{\partial}{\partial t}\left(\varepsilon\frac{\partial A}{\partial t}\right)\frac{1}{c}\frac{\partial}{\partial t}(\varepsilon\nabla\phi) + 4\pi e\beta\delta^3(r - \beta ct) \ . \tag{1.15}$$

The solutions can be found in terms of the Fourier transforms $\phi(k, \omega)$ and $A(k, \omega)$ of the potentials.

The Fourier transform $F(k, \omega)$ of a vector field $F(r, t)$ is given by

$$F(k, \omega) = \frac{1}{(2\pi)^2}\int d^3r\, dt\, F(r, t)\exp(ik \cdot r - i\omega t) \ ,$$

$$F(r, t) = \frac{1}{(2\pi)^2}\int d^3k\, d\omega\, F(k, \omega)\exp(ik \cdot r - i\omega t) \ . \tag{1.16}$$

The solutions of (1.15) are

$$\phi(k, \omega) = 2e\delta(\omega - k \cdot \beta c)/k^2\varepsilon \ ,$$

$$A(k, \omega) = 2e\frac{\omega k/k^2c - \beta}{(-k^2 + \varepsilon\omega^2/c^2)}\delta(\omega - k \cdot \beta c) \ . \tag{1.17}$$

Using the third of (1.14), the electric field for every point is calculated according to

$$E(r, t) = \frac{1}{(2\pi)^2}\int \frac{i\omega}{c}\{A(k, \omega) - ik\phi(k, \omega)\exp[i(k \cdot r - \omega t)]\}\, d^3k\, d\omega \ , \tag{1.18}$$

and the energy loss per unit length is

$$\langle dE/dx\rangle = eE(\beta ct, t) \cdot \beta/\beta \ , \tag{1.19}$$

which is independent of t because the field created in the medium is travelling with the particle. This may be seen by inserting (1.17) into (1.18).

In the evaluation of (1.18) with the help of (1.17) we integrate over the two directions of k using the fact that the isotropic medium has a scalar ε. We further use $\varepsilon(-\omega) = \varepsilon^*(\omega)$ and obtain finally

$$\left\langle\frac{dE}{dx}\right\rangle = \frac{2e^2}{\beta^2\pi}\int\limits_0^\infty d\omega \int\limits_{\omega/\beta c}^\infty dk\left[\omega k\left(\beta^2 - \frac{\omega^2}{k^2c^2}\right)\right.$$

$$\left.\times \operatorname{Im}\frac{1}{-k^2c^2 + \varepsilon\omega^2} - \frac{\omega}{kc^2}\operatorname{Im}\left(\frac{1}{\varepsilon}\right)\right] \ . \tag{1.20}$$

The lower limit of the integral over k depends on the particle velocity βc and is explained later.

The energy loss is determined by the manner in which the complex dielectric constant ε depends on the wave number k and the frequency ω. Once $\varepsilon(k, \omega)$ is specified, $\langle dE/dx \rangle$ can be calculated for every β; $\varepsilon(k, \omega)$ is, in principle, given by the structure of the atoms of the medium. In practice, a simplifying model is sufficient.

1.2.2 The Photo-Absorption Ionization Model

Allison and Cobb [ALL 80] have made a model of $\varepsilon(k, \omega)$ based on the measured photo-absorption cross-section $\sigma_\gamma(\omega)$. A plane light-wave travelling along x is attenuated in the medium if the imaginary part of the dielectric constant is larger than zero. The wave number k is related to the frequency ω by

$$k = \sqrt{\varepsilon}\,\omega/c \quad (\varepsilon = \varepsilon_1 + i\varepsilon_2) . \tag{1.21}$$

It causes a damping factor $e^{-\alpha x/2}$ in the propagation function and $e^{-\alpha x}$ in the intensity with

$$\alpha = 2(\omega/c)\,\mathrm{Im}\,\sqrt{\varepsilon} \approx (\omega/c)\varepsilon_2 . \tag{1.22}$$

The second equality holds if $\varepsilon_1 - 1$, $\varepsilon_2 \ll 1$ such as for gases.

In terms of free photons traversing a medium that has electron density N and atomic charge Z, the attenuation is given by the photo-absorption cross-section $\sigma_\gamma(\omega)$:

$$\sigma_\gamma(\omega) = \frac{Z}{N}\alpha \approx \frac{Z}{N}\frac{\omega}{c}\varepsilon_2(\omega) . \tag{1.23}$$

The cross-section $\sigma_\gamma(\omega)$, and therefore $\varepsilon_2(\omega)$, is known, for many gases, from measurements using synchrotron radiation. The real part of ε is then derived from the dispersion relation

$$\varepsilon_1(\omega) - 1 = \frac{2}{\pi}P\int_0^\infty \frac{x\varepsilon_2(x)}{x^2 - \omega^2}dx \tag{1.24}$$

(P = principal value). Figures 1.4 and 1.5 show graphs of σ_γ and ε_1 for argon.

In the quantum picture, the (ω, k) plane appears as the kinematic domain of energy $E = \hbar\omega$ and momentum $p = \hbar k$ exchanged between the moving particle and the atoms and electrons of the medium. The exchanged photons are not free, not 'on the mass shell', i.e. they have a relation between E and p that is different from $E = pc/\sqrt{\varepsilon}$ implied by (1.21). Their relationship may be understood in terms of the kinematic constraints: for example, the photons exchanged with free electrons at rest have $E = p^2/2m$, the photons exchanged with bound electrons (binding energy E_1, approximate momentum q) have $E \approx E_1 + (p + q)^2/2m$. The minimum momentum transfer at each energy E depends on the velocity β of

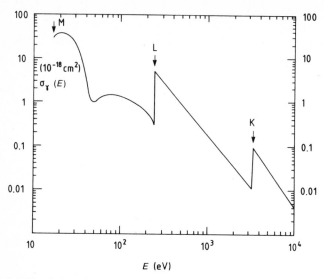

Fig. 1.4. Total photo-ionization cross-section of Ar as a function of the photon energy, as compiled by Marr and West [MAR 76]. The imaginary part of the dielectric constant is calculated from this curve using (1.23)

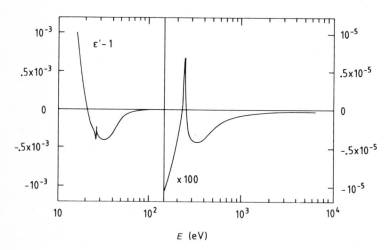

Fig. 1.5. The real part of ε as a function of E, calculated from Fig. 1.4 using (1.24) [LAP 80]

the moving particle and is equal to $p_{min} = E/\beta c$. It delimits the kinematic domain and is the lower limit of integration in the integral (1.20).

Figure 1.5 presents a picture of the (E, p) plane. The solution of the integral (1.20) requires a knowledge of $\varepsilon(k, \omega)$ over the kinematic domain. From Eqs. (1.23) and (1.24) it is known only for free photons, i.e. outside, along the line $p_{f\gamma}$ in Fig. 1.6.

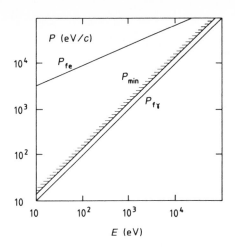

Fig. 1.6. Kinematic domain of (ω, k) or (E, p) of the electromagnetic radiation exchanged between the fast particle (β) and the medium. The minimum momentum exchanged is $p = E/\beta c$; the momentum exchanged to a free electron is $p_{\text{fe}} = \sqrt{2mE/c^2}$. The momenta exchanged with bound electrons are smeared out around the free-electron line; p_{min} delimits the physical domain, $p_{\text{f}\gamma}$ is the free photon line

The model of Allison and Cobb (photo-absorption ionization, or PAI model) consists in extending the knowledge of ε into the kinematic domain. For small k in the resonance region below the free-electron line, they take

$$\varepsilon(k, \omega) = \varepsilon(\omega) , \tag{1.25}$$

independent of k and equal to the value derived by (1.23) and (1.24). On the free-electron line, they take a δ function to represent point-like scattering in the absorption:

$$\varepsilon_2(k, \omega) = C\delta(\omega - k^2/2m) . \tag{1.26}$$

The normalizing constant C is determined in such a way that the total coupling strength satisfies the Bethe sum rule for each k:

$$\int \omega \varepsilon_2(k, \omega) \, d\omega = \frac{2\pi^2 N e^2}{m} . \tag{1.27}$$

For a justification of this procedure and a more detailed discussion, the reader is referred to the original article [ALL 80].

Using (1.23–26), we are now able to integrate expression (1.20) for $\langle dE/dx \rangle$ over k. What remains is an integral over ω:

$$\left\langle \frac{dE}{dx} \right\rangle = \int\limits_0^\infty d\omega \frac{e^2}{\beta^2 c^2 \pi} \left[\frac{Nc}{Z} \sigma_\gamma(\omega) \ln \frac{2mc^2 \beta^2}{\hbar\omega[(1 - \beta^2\varepsilon_1)^2 + \beta^4\varepsilon_2^2]^{1/2}} \right.$$
$$\left. + \omega\left(\beta^2 - \frac{\varepsilon_1}{|\varepsilon|^2} \right)\theta + \frac{1}{Z\omega} \int\limits_0^\omega \sigma_\gamma(\omega') \, d\omega' \right] . \tag{1.28}$$

Here we have obtained the energy loss per unit path length in the framework of the electrodynamics of a continuous medium, although our knowledge of $\varepsilon(k, \omega)$ was actually inspired by a picture of photon absorption and collision.

At this point we leave the frame given by the classical theory and recognize the energy loss as being caused by a number of discrete collisions per unit length, each with an energy transfer $E = \hbar\omega$. Therefore, we reinterpret the integrand of (1.28) to mean E times a probability of energy transfer per unit path per unit interval of E. Now a collision probability per unit path is a cross-section times the density N. Therefore, we write the integral (1.28) in the form

$$\left\langle \frac{dE}{dx} \right\rangle = \int_0^\infty EN \frac{d\sigma}{dE} \hbar \, d\omega \ . \tag{1.29}$$

This gives us the differential energy transfer cross-section per electron:

$$\frac{d\sigma}{dE} = \frac{\alpha}{\beta^2 \pi} \frac{\sigma_\gamma(E)}{EZ} \ln \frac{2mc^2\beta^2}{E[(1 - \beta^2\varepsilon_1)^2 + \beta^4\varepsilon_2^2]^{1/2}}$$
$$+ \frac{\alpha}{\beta^2 \pi} \left[\frac{Z}{N\hbar c} \left(\beta^2 - \frac{\varepsilon_1}{|\varepsilon|^2} \right) \theta + \frac{1}{ZE^2} \int_0^E \sigma_\gamma(E') \, dE' \right] . \tag{1.30}$$

The spectrum of energy transfer is determined by expression (1.30). The normalized differential probability per unit energy is

$$F(E) = \frac{N(d\sigma/dE)}{\int N(d\sigma/dE) \, dE}, \tag{1.31}$$

where N is the number of electrons per unit volume; $\varepsilon = \varepsilon_1 + i\varepsilon_1$ is to be obtained using (1.23) and (1.24); $\theta = \arg(1 - \varepsilon_1\beta^2 + i\varepsilon_2\beta^2)$; and α is the fine-structure constant.

The number of primary encounters per unit length is given by

$$1/\lambda = \int N \frac{d\sigma}{dE} \, dE \ . \tag{1.32}$$

In the gas of a drift chamber the largest possible energy transfers cause secondary electron tracks which do not belong to the primary track. Depending on the exact method of observation, there is always an effective cut-off E_{max} for the observable energy transfer, which is independent of β. For simplicity, we keep the '∞' as the upper limit of integration. Figure 1.7 shows the energy spectrum for argon according to a numerical calculation along this line by Lapique and Piuz [LAP 80]; they have evaluated the model of Chechin et al. [CHE 72, ERM 77], which is very similar to the PAI model. The second peak beyond $E = 240 \, \text{eV}$, which is due to the contribution of the L shell, is easily visible.

1.2.3 Behaviour for Large E

For energies above the highest atomic binding energy E_K, the fast particle undergoes elastic scattering on the atomic electrons as if they were free, and (1.30) becomes the differential cross-section for Rutherford scattering on one

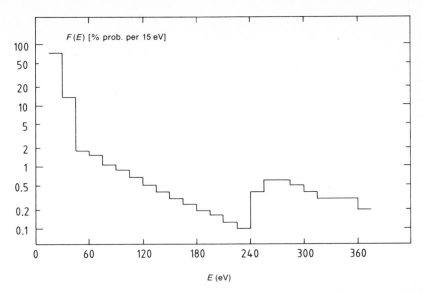

Fig. 1.7. Distribution of energy transfer calculated by Lapique and Piuz for argon at $\gamma = 1000$, based on the formula by Chechin et al. [CHE 76], similar to our (1.30). Adapted from Table 1 of [LAP 80]

electron. Using (1.27) and (1.28) we find

$$\frac{d\sigma}{dE} \to \frac{2\pi r_e^2 \, mc^2}{\beta^2 \, E^2} \quad (E \gg E_K) , \tag{1.33}$$

where r_e is the classical electron radius, equal to $e^2/mc^2 = 2.82 \times 10^{-13}$ cm. This happens because the third term in (1.30) is the only one surviving at large E, where $\sigma_\gamma(E)$ vanishes quickly so that the sum rule (1.27) applies, together with (1.23).

This behaviour at large E means that the energy spectrum $F(E)$ has an extremely long tail. Although $\int F(E)\,dE$ converges, the mean transferred energy per collision, $\langle E \rangle$, has a logarithmic divergence,

$$\langle E \rangle = \int EF(E)\,dE = \int dE/E \propto \log E . \tag{1.34}$$

1.2.4   Cluster-Size Distribution

An effective description of the ionization left by the particle along its trajectory is provided by a probability distribution of the number of electrons liberated directly or indirectly with each primary encounter. It is known under the name cluster-size distribution, because the secondary electrons are usually created in the immediate vicinity of the primary encounter and, together with the primary electrons, form clusters of one or several – sometimes many – electrons.

Although the secondary electrons are not always so well localized, we will use this name.

In order to calculate the cluster-size distribution $P(k)$ we need to know the spectrum of energy loss, $F(E)dE$, and, for each E, the probability $p(E, k)$ of producing exactly k ionization electrons. The cluster-size distribution is obtained by integration over the energy:

$$P(k) = \int F(E)p(E, k)\,dE . \tag{1.35}$$

We may also form the integrated probability $Q(j)$ that a cluster has more than j electrons:

$$Q(j) = 1 - \sum_{k=1}^{j} P(k) . \tag{1.36}$$

The quantity $p(E, k)$ contains the details of the various ionization mechanisms described in Sect. 1.1 and is generally not known. Lapique and Piuz [LAP 80] have made a computer model of the atomic processes involved in pure argon and have thus been able to calculate a cluster-size distribution. It is presented in the integrated form in Fig. 1.8.

Apart from early cloud chamber studies, there is now one careful experimental determination of cluster-size distributions from the Heidelberg group

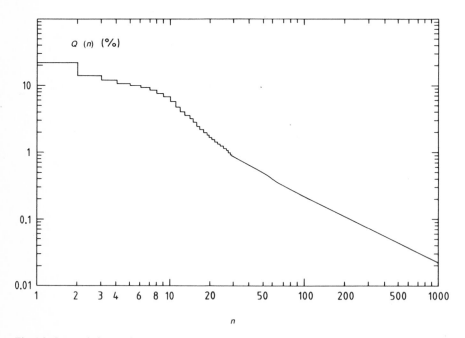

Fig. 1.8. Integral cluster-size distribution for fast particles ($\gamma = 1000$) in argon; $Q(n)$ is the probability that the cluster has more than n electrons. Calculated from [LAP 80] and [ERM 77]. For large n, $Q(n) \approx 0.2/n$

Table 1.4. Experimental cluster-size distributions $P(k)$ in per cent by Fischle et al. [FIS 91]. Whereas the measured data and their errors are shown in Fig. 1.9, this table contains interpolated values for numerical applications. Values in brackets are extrapolations according to the expectation of free electrons

k	CH_4	Ar	He	CO_2
1	78.6	65.6	76.60	72.50
2	12.0	15.0	12.50	14.00
3	3.4	6.4	4.60	4.20
4	1.6	3.5	2.0	2.20
5	0.95	2.25	1.2	1.40
6	0.60	1.55	0.75	1.00
7	0.44	1.05	0.50	0.75
8	0.34	0.81	0.36	0.55
9	0.27	0.61	0.25	0.46
10	0.21	0.49	0.19	0.38
11	0.17	0.39	0.14	0.34
12	0.13	0.30	0.10	0.28
13	0.10	0.25	0.08	0.24
14	0.08	0.20	0.06	0.20
15	0.06	0.16	0.048	0.16
16	(0.050)	0.12	(0.043)	0.12
17	(0.042)	0.095	(0.038)	0.09
18	(0.037)	0.075	(0.034)	(0.064)
19	(0.033)	(0.063)	(0.030)	(0.048)
≥ 20	$(11.9/k^2)$	$(21.6/k^2)$	$(10.9/k^2)$	$(14.9/k^2)$

[FIS 91]. It covers the range up to approximately 15 electrons in argon, helium, methane and several hydrocarbons. We reproduce their measurements in Fig. 1.9. For numerical applications Table 1.4 contains best estimates for the probabilities $P(k)$, based on the hand-drawn lines of Figs. 1.9 and their extrapolations according to the $1/n^2$ behaviour expected for unbound electrons.

If one compares these measurements with the calculation of Lapique and Piuz, then one observes that there is somewhat less structure than they expected in the function $P(k)$. The Heidelberg group suggest that the absorption of free photons with its strong energy variation, which is the basis for the model of Chechin et al. and the PAI model, may not be directly applicable to the calculation of $d\sigma/dE$.

1.2.5 Ionization Distribution on a Given Track Length

The importance of the cluster-size distribution lies in the fact that, once it is known, the ionization distribution $G(x, n)$ on the track length x is calculated simply by summing the cluster size as many times as there are primary encounters in the track length; $G(n)$ depends only on the cluster-size distribution and

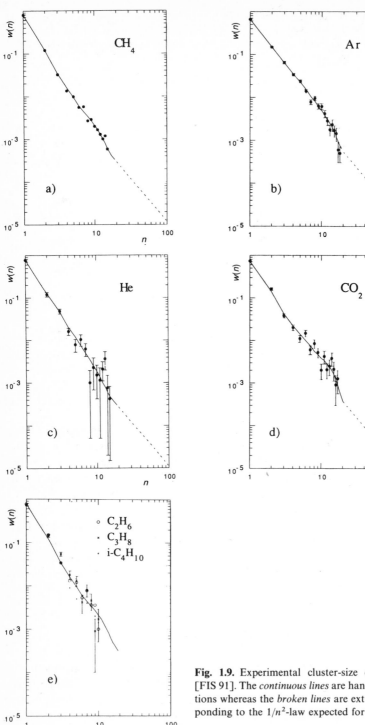

Fig. 1.9. Experimental cluster-size distributions from [FIS 91]. The *continuous lines* are hand-drawn interpolations whereas the *broken lines* are extrapolations corresponding to the $1/n^2$-law expected for large n

the number of clusters. The most practical way to achieve the summation is by the Monte Carlo method on a computer, especially when the cluster-size distribution exists only in the form of a table.

In this case, we proceed in two steps. First, we make a random choice from the Poisson distribution (1.2) of a number m of encounters in the track length x: $\langle m \rangle = x/\lambda$. Second, we make m random choices from the cluster-size distribution $P(k)$ of m cluster sizes $k_1, k_2, \ldots, k_m$. The number

$$n = \sum_{i=1}^{m} k_i \tag{1.37}$$

is one entry in a frequency distribution $G(x, n)$ of the number of ionization electrons in x.

Figure 1.10 contains ionization distributions for several track lengths (mean numbers $\langle m \rangle$ of clusters) in argon that were computed in this way using the integrated probability Q depicted in Fig. 1.8. The ionization distributions develop a peak that defines *the most probable ionization* I_{mp}. They also have a *full width at half maximum* W, although the mean and the root-mean-square deviation exist only if an upper cut-off is introduced. The value of I_{mp} is not proportional to the mean number of clusters but rises from $\langle m \rangle$ to $\sim 3 \langle m \rangle$ as $\langle m \rangle$ increases from 5 to 1000. Figure 1.11 shows this increase. It does not obey a simple law because it is influenced by the atomic structure of argon. Neglecting the atomic structure, it approaches a straight line; compare the remarks made on Δ_{mp} after (1.50). The distributions of Fig. 1.10 also become more peaked, so that the ratio W/I_{mp} decreases from ~ 1.3 to ~ 0.3 in this range of $\langle m \rangle$.

Relative widths W/I_{mp} of measured pulse-height distributions quoted by Walenta [WAL 81] are shown in Fig. 1.12 as a function of the gas sample thickness pL. The ratio decreases with increasing pL. If we parametrize the decrease in the form of a power law,

$$\left(\frac{W}{I_{\mathrm{mp}}} \right)_1 : \left(\frac{W}{I_{\mathrm{mp}}} \right)_2 = [(pL)_1 : (pL)_2]^k , \tag{1.38}$$

then from Fig. 1.12 we get k between -0.2 and -0.4. Since the ionization distribution depends only on the cluster-size distribution and the number of clusters, it depends on the sample length L and the gas pressure p through the product pL.

For practical purposes, we introduce an upper cut-off E_{95} in the transferred energy at the level of 95% of the integrated probability distribution $F(E)$:

$$\int_0^{E_{95}} F(E)\,dE = 0.95 .$$

This allows us to compute the *average ionization* $\langle I \rangle_{95}$ from the distributions of Fig. 1.10. It is plotted in Fig. 1.13 as a function of $\langle m \rangle$.

The role of the rare cases of large cluster sizes can be appreciated by looking at Fig. 1.18. Since 1% of the clusters are larger than 30 electrons, it takes of the order of 100 primary interactions (3 cm of argon NTP, $\gamma = 1000$) to have one

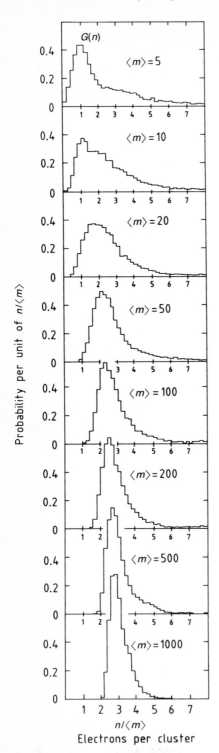

Fig. 1.10. Ionization distribution obtained by summing m times the cluster-size distribution of Fig. 1.7, using the method described in Sect. 1.2.3. On 1 cm of Ar in normal conditions there are, on the average, $\langle m \rangle = 35$ clusters ($\gamma = 1000$); the eight distributions then correspond to track lengths of 0.14, 0.29, 0.57, 1.4, 2.9, 5.7, 14, and 29 cm

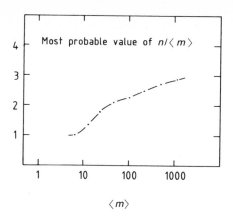

Fig. 1.11. Values of the most probable number of electrons (expressed in units of $\langle m \rangle$) as a function of the mean number $\langle m \rangle$ of clusters in Ar, obtained from the data of Fig. 1.8

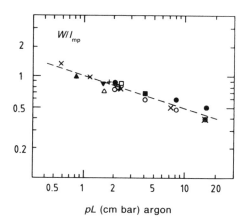

Fig. 1.12. Measured relative widths of pulse-height distributions as collected by Walenta [WAL 81]. The plot shows ten different experiments with the gas sample thickness pL varying between 0.6 and 15 cm bar. The line is drawn to guide the eye

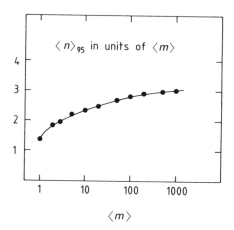

Fig. 1.13. Average ionization after cutting the upper 5% of the probability distribution, calculated as a function of the average number $\langle m \rangle$ of clusters

such cluster among them. It will contribute about 10% to the total ionization. The larger the number m of clusters one needs to sum, the smaller will be the probabilities $Q(n)$ that have to be taken into account. Let us compute the probability P that the very rare event of a large cluster above n_0 occurs at least once in a number of m clusters:

$$P = 1 - [1 - Q(n_0)]^m \approx 1 - e^{-mQ(n_0)} . \tag{1.39}$$

It can be shown that (1.39) is correct within 10% as long as $Q(n_0) < 1/\sqrt{(5m)}$.

If the cluster-size distribution is not known, we first calculate the energy loss Δ on a given track length x and its probability distribution $F(x, \Delta)$. This is then converted into the ionization distribution by dividing Δ by the appropriate value of W. The energy transfer spectrum is independent of any other collision, and we have to sum the contributions from as many collisions $\langle m \rangle = x/\lambda$ as there are on the length x, using expression (1.32).

These days this is best achieved by the Monte Carlo method using a computer. Working with analytical methods, one may also perform a stepwise convolution. For example,

$$F(2\lambda, \Delta) = \int_0^\infty F(\Delta - E)F(E)\,dE ,$$
$$\tag{1.40}$$
$$F(4\lambda, \Delta) = \int_0^\infty F(2\lambda, \Delta - E)F(2\lambda, E)\,dE ,$$

etc.

The review of Bichsel and Saxon [BIC 75] contains more details about how to build up $F(x, \Delta)$.

Another way of constructing $F(x, \Delta)$ from $F(E)$ was invented by Landau in 1944 [LAN 44]. He expressed the change of $F(x, \Delta)$ along a length dx by the difference in the number of particles which, because of ionization losses along dx, acquire a given energy E and the number of particles which leave the given energy interval near Δ:

$$\frac{\partial}{\partial x} F(x, \Delta) = \int_0^\infty F(E)[F(x, \Delta - E) - F(x, \Delta)]\,dE . \tag{1.41}$$

(For the upper limit of integration, one may write ∞ because $F(x, \Delta) = 0$ for $\Delta < 0$.) The solution of (1.41) is found with the help of the Laplace transform $\bar{F}(x, p)$, which is related to the energy loss distribution by

$$\bar{F}(x, p) = \int_0^\infty F(x, \Delta)e^{-p\Delta}\,d\Delta , \tag{1.42}$$

$$F(x, \Delta) = \frac{1}{2\pi i} \int_{-i\infty + \sigma}^{+i\infty + \sigma} e^{p\Delta} \bar{F}(x, p)\,dp . \tag{1.43}$$

Here the integration is to the right ($\sigma > 0$) of the imaginary axis of p. Multiplying both sides of (1.41) by a $e^{-p\Delta}$ and integrating over Δ, we get

$$\frac{\partial}{\partial x}\bar{F}(x, p) = -\bar{F}(x, p)\int_0^\infty F(E)(1 - e^{-pE})\,dE \;, \tag{1.44}$$

which integrates to

$$\bar{F}(x, p) = \exp\left[-x\int_0^\infty \bar{F}(E)(1 - e^{-pE})\,dE\right], \tag{1.45}$$

because the boundary condition is $F(0, \Delta) = \delta(\Delta)$ or $F(0, p) = 1$. Inserting (1.42) into (1.45), Landau obtained the general expression for the energy loss distribution, valid for any $F(E)$:

$$F(x, \Delta) = \frac{1}{2\pi i}\int_{-i\infty+\sigma}^{+i\infty+\sigma} \exp\left[p\Delta - x\int_0^\infty F(E)(1 - e^{-pE})\,dE\right]dp \;. \tag{1.46}$$

This relation was evaluated by Landau for a simplified form of $F(E)$ that is applicable at energies far above the atomic binding energies where the scattering cross-section is determined by Rutherford scattering and where the atomic structure can be ignored (see (1.33)). Inserting

$$F(E) = \frac{2\pi r_e^2 \, mc^2}{\beta^2}\frac{}{E^2}\,N \tag{1.47}$$

into (1.46), Landau was able to show that the probability distribution was given by a universal function $\phi(\lambda)$:

$$F(x, \Delta)\,d\Delta = \phi\left(\frac{\Delta - \Delta_{mp}}{\xi}\right)d\left(\frac{\Delta - \Delta_{mp}}{\xi}\right). \tag{1.48}$$

Here Δ_{mp} is the most probable energy loss, and ξ is a scaling factor for the energy loss, proportional to x:

$$\xi = x2\pi r_e^2 \frac{mc^2}{\beta^2}\,N \;. \tag{1.49}$$

The function $\phi(\lambda)$ and its integral $\psi(\lambda)$ are given as a graph in Fig. 1.14. Computer programs exist for the calculation of ϕ and for the random generation of Landau-distributed numbers: see Kölbig and Schorr [KÖL 84]. The integral probability for an energy loss exceeding Δ is

$$\int_\Delta^\infty F(x, \Delta')\,d\Delta' = \psi\left(\frac{\Delta - \Delta_{mp}}{\xi}\right). \tag{1.50}$$

We notice that 10% of the cases lie above the value of Δ that is three times the FWHM above the most probable value. For large positive values of the argument, $\phi(\lambda)$ tends to $1/\lambda^2$, $\psi(\lambda)$ tends to $1/\lambda$. The assumption (1.47) makes the Landau curve a valid description of the energy loss fluctuations only in a regime of large Δ (corresponding to $x \approx 170$ cm in normal argon gas, according to an analysis of [CHE 76] [see also Fig. 1.21]). We skip a discussion of Landau's

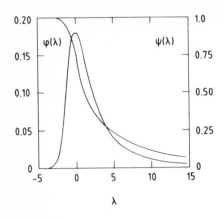

Fig. 1.14. Landau's function $\phi(\lambda)$ and its integral $\psi(\lambda)$. The scale on the left refers to ϕ, the one on the right to ψ

expression for Δ_{mp} and of the normalization of his $F(E)$. Let us remark, however, that, as a function of the length x, Δ_{mp} is proportional to $x \log x$. In practice, the Landau curve is often used to parametrize energy loss distributions with a two-parameter fit of ξ and Δ_{mp}, without reference to the theoretical expressions for them.

Generalizations of the Landau theory have been given by Blunck and Leisegang [BLU 50], Vavilov [VAV 57], and others. The interested reader is referred to the monograph by Bugadov, Merson, Sitar and Chechin [BUG 88] for a comparison of these theories of energy loss.

When the summation of the energy lost over the length x has been achieved with any of the methods mentioned above, the energy loss distribution must be converted into an ionization distribution. We have to make the assumption that to every energy loss Δ there corresponds, on the average, a number n of ion pairs according to the relation

$$\Delta = nW , \tag{1.51}$$

where W is the average energy for producing an ion pair (Sect. 1.1.3). Expression (1.51) is to hold independently of the size or the composition of Δ (whether there are one large or many small transfers), and W is to be the constant measured with fully stopped electrons. It is hard to ascertain the error that we introduce with this assumption. The W measured with fully stopped electrons is known to increase for energy transfers below ~ 1 keV (Fig. 1.2).

Using (1.51), we obtain the probability distribution $G(x, n)$ of the number n of ionization electrons produced on the track length x:

$$G(x, n) = F(x, nW) W . \tag{1.52}$$

In Fig. 1.15 we show two examples of ionization distributions calculated in this way by Allison and Cobb [ALL 80], compared with measured pulse heights from Harris et al. [HAR 73] on argon samples of 1.5 cm thickness. Although there is a small systematic shift and an excess of data at small n it is a remarkable fact that the pulse-height distribution can be predicted so well using the theory

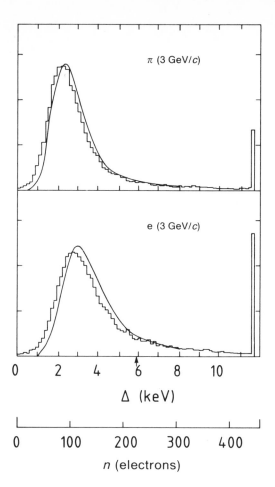

Fig. 1.15. Measured pulse-height distributions from [HAR 73], compared with the predictions of the photoabsorption model of [ALL 80]. The experimental overflow is collected in the last bin. The horizontal scale is normalized to the peak of the ^{55}Fe spectrum (5.9 keV = 223 electrons)

described above. For a comparison with the predictions of other models for calculating $F(x, \Delta)$, see Allison and Cobb [ALL 80], and Ermilova, Kotenko and Merzon [ERM 77].

1.2.6   Velocity Dependence of the Energy Loss

Let us go back to (1.30). The term proportional to θ is connected with Cerenkov radiation of frequency ω. It makes only a small contribution to the cross-section but has a very characteristic velocity dependence. As soon as ε_2 vanishes, this radiation will be emitted into the medium above a threshold given by

$$\beta_0^2 = \frac{1}{\varepsilon_1},\tag{1.53}$$

where θ jumps approximately from 0 to π as β increases. The emitted intensity in photons per unit path length per unit of photon energy interval is then

$$N\left(\frac{d\sigma}{dE}\right)_{Ce} = \frac{\alpha}{c}\left(1 - \frac{1}{\beta^2\varepsilon_1}\right) = \frac{\alpha}{c\gamma_0^2}\frac{\gamma^2 - \gamma_0^2}{\gamma^2 - 1} . \tag{1.54}$$

The behaviour of this intensity as a function of $\gamma = 1/(1 - \beta^2)^{1/2}$ is plotted in Fig. 1.16. The threshold can be expressed by

$$\gamma_0^2 = \varepsilon_1/(\varepsilon_1 - 1) . \tag{1.55}$$

We now discuss the velocity dependence of the remaining terms in (1.30). They give the main contribution to the cross-section. The equation may be written in the form

$$\frac{d\sigma}{dE} = \frac{a}{\beta^2}\left[b + \ln\frac{\beta^2}{[(1 - \beta^2\varepsilon_1)^2 + \beta^4\varepsilon_2^2]^{1/2}}\right], \tag{1.56}$$

where a and b depend on E and $\sigma_y(E)$ but are independent of β.

There is an overall factor of $1/\beta^2$ which dominates at small β. For β near 1, the behaviour is determined by the logarithmic term which first rises and then remains constant as the relativistic velocity increases:

$$\ln\frac{\beta^2}{[(1 - \beta^2\varepsilon_1)^2 + \beta^4\varepsilon_2^2]^{1/2}} = \ln\frac{\gamma^2 - 1}{\{[\gamma^2(1 - \varepsilon_1) + \varepsilon_1]^2 + (\gamma^2 - 1)^2\varepsilon_2^2\}^{1/2}} \tag{1.57}$$

$$\rightarrow \ln\frac{\gamma^2 - 1}{\varepsilon_1} \quad \text{for } \gamma^2 \ll 1/|1 - \varepsilon| , \tag{1.58}$$

$$\rightarrow \ln\frac{1}{|1 - \varepsilon|} \quad \text{for } \gamma^2 \gg 1/|1 - \varepsilon| . \tag{1.59}$$

The region (1.58) is called the relativistic rise and the region (1.59) the plateau of the energy loss (see Fig. 1.17).

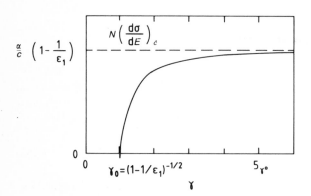

Fig. 1.16. Intensity of Cerenkov radiation as a function of γ, according to (1.54)

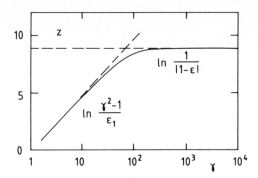

Fig. 1.17. Behaviour as a function of γ of the logarithmic term (1.57):

$$Z = \ln \frac{\gamma^2 - 1}{\{[\gamma^2(1 - \varepsilon_1) + \varepsilon_1]^2 + (\gamma^2 - 1)\varepsilon_2^2\}^{1/2}} ;$$

ε was put equal to $1 - 10^{-4} + 10^{-4}i$

The beginning of the plateau is characterized by a relativistic velocity factor γ^*, such that

$$\frac{\gamma^{*2} - 1}{\varepsilon_1} = \frac{1}{|1 - \varepsilon|} ; \quad \gamma^* \approx 1\sqrt{|1 - \varepsilon|} . \tag{1.60}$$

As long as ε_1 is larger than 1 – as is usually the case for visible light – there is always some γ_0 to fulfill the Cerenkov condition (1.55), and Cerenkov radiation is emitted into the medium if it is transparent ($\varepsilon_2 = 0$). Under these conditions, the beginning of the plateau region is at the same velocity: $\gamma^* \approx \gamma^0$. But if ε_1 is smaller than 1, as happens for energies E above the highest resonances of the atoms of the medium, then there is no Cerenkov light to correspond to the onset of the plateau.

In the theory of optical dispersion, the behaviour of $\varepsilon(\omega)$ for $\omega = E/\hbar$ far above the resonances is described using the concept of plasma frequency ω_p, which is given by the charge e, the mass m and the density N of the electrons in the gas (e.g. [JAC 75]):

$$\omega_p^2 = 4\pi N e^2/m . \tag{1.61}$$

It is shown that there

$$\varepsilon_1(\omega) \approx 1 - \omega_p^2/\omega^2 , \tag{1.62}$$

$$\varepsilon_2(\omega) \ll 1 - \varepsilon_1(\omega) . \tag{1.63}$$

In connection with (1.60), the beginning of the plateau is given by the plasma frequency for every value E of the energy loss:

$$\gamma^* \approx \omega/\omega_p = \left(\frac{m}{4\pi N c^2}\right)^{1/2} \frac{E}{\hbar} . \tag{1.64}$$

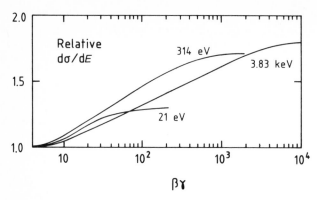

Fig. 1.18. Energy dependence of dE/dx: variation with $\beta\gamma$ of the energy loss cross-section for Ar at normal density, normalized to $\beta\gamma = 4$, according to a calculation by Allison and Cobb [ALL 80]. The curves are for three values of the energy loss E, corresponding to M-, L-, and K-shell ionization

It turns out that this relation (which was derived for a free-electron gas) already holds very well for all E above the M-shell resonances in argon.

The existence of a plateau is caused by the density of the medium. Both $(1 - \varepsilon_1)$ and ε_2 are proportional to the electron density N and go to zero when N does. We can see from (1.60) and also from (1.30) that the relativistic rise of the energy loss will continue for all γ in the limit of vanishing N, and there will be no plateau. The medium with vanishing density behaves like a single atom, which has no plateau for the collision cross-section.

Figure 1.18 shows the energy loss cross-section in argon as a function of $\beta\gamma$ for three representative values of E, calculated by Allison and Cobb from (1.30) without the small Cerenkov term. The variable $\beta\gamma$ instead of γ is chosen because it is more appropriate at small β, and also it is the variable in which energy loss had been discussed before using the Bethe–Bloch formula (Sect. 1.2.7). The three curves are normalized at their minimum approximately at $\gamma = 4$. The onset of the plateau is at values γ^* proportional to E, as described by (1.64). From the point of view of particle identification, the large E are the best, but they are also very rare (see Figs. 1.7 and 1.8).

After integration over E, the calculated most probable energy loss had the form depicted in Fig. 1.19, where it is compared with measurements (see Lehraus et al. [LEH 82] and references quoted therein). Both the measured and the calculated values are normalized at $\gamma = 4$. The agreement is remarkable and indicates that the theory gives an adequate explanation of the velocity dependence of gas ionization.

The effect of the gas density on the relativistic rise is visible in the data of Fig. 1.20. There is also a small dependence on the gas sample thickness, for which we refer the reader to an article by Walenta [WAL 79] and to Fig. 9.5.

There is a velocity dependence not only of the most probable value but also of the whole shape of the ionization distribution. This dependence is particularly pronounced in very small gas samples, i.e. with small numbers of primary

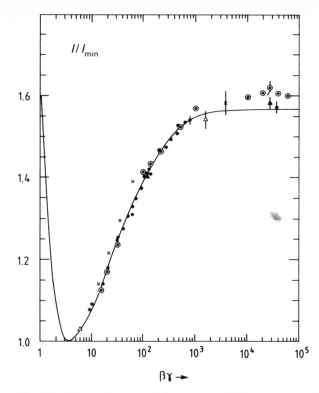

$\beta\gamma \rightarrow$

Fig. 1.19. Measured values of the most probable energy loss in an Ar + methane mixture as a function of $\beta\gamma$, compared with the photo-absorption model of Allison and Cobb for Ar. For details, see [ALL 80] and [LEH 82] and references quoted therein

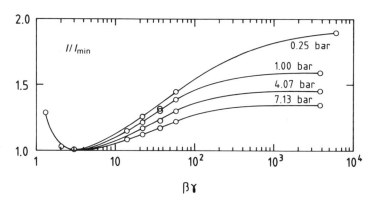

$\beta\gamma$

Fig. 1.20. Density dependence of dE/dx: variation with $\beta\gamma$ of the most probable value of the pulse-height distribution, normalized to the value at the minimum, according to measurements by Walenta et al. [WAL 79] in Ar (90%) + CH_4 (10%) over a length of 2.3 cm. The measurements are at four values of the gas pressure. The interpolating curves were calculated according to the theory of Sternheimer

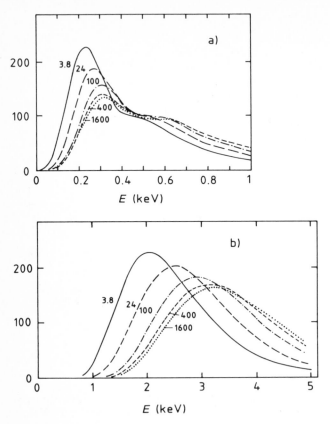

Fig. 1.21a, b. dE/dx-distributions calculated with the PAI model in argon [ALL 81] for different values of γ (**a**) sample length 0.3 cm; (**b**) 1.5 cm, both at normal density

encounters, where the change in statistics – a consequence of the velocity dependence of the primary ionization – is relatively large. We see in Fig. 1.21 the energy-loss distributions calculated with the PAI model for thin ($x = 1.5$ cm) and very thin ($x = 0.3$ cm) argon samples, at five different velocities. The curious disappearance of the sharp peak in the very thin sample is caused by the shell structure of the argon atom.

1.2.7 The Bethe–Bloch Formula

Historically, Bethe was the first to calculate, in 1930, the average energy loss with a quantum theory of collision between the travelling particle and a single atom. After adding the energy lost to all the atoms in the vicinity of the particle, the energy loss per unit of pathlength is given by

$$\frac{dE}{dx} = \frac{4\pi Ne^4}{mc^2\beta^2}z^2\left(\ln\frac{2mc^2\beta^2\gamma^2}{I} - \beta^2\right) \tag{1.65}$$

([BET 33], Eq. 6.16). In this equation mc^2 is the rest energy of the electron, z the charge of the travelling particle, N the number density of electrons in the matter traversed, e the elementary charge, β the velocity of the travelling particle in terms of the velocity c of light, and $\gamma^2 = 1/(1 - \beta^2)$. The symbol I denotes the mean excitation energy of the atom. Bloch calculated values of I using the Thomas–Fermi theory of the atom. Equation (1.65) is also called the Bethe–Bloch formula. It describes the integral over all the energies lost to the individual atoms of the medium. This integral extends up to the maximum of the transferrable energy, and it is for this reason that (1.65) is only valid for travelling particles heavier than electrons. These have different kinematic limits because of their small mass and because they are identical with their collision partner; also their spectrum of transferred energy is different. A modification of (1.65) will make it valid for electrons and heavier particles alike above some value of γ; see Sect. 1.2.8.

The factor at the front of (1.65) can be brought into a different form by expressing N through Avogadro's number N_0, the gas density ρ and the ratio of atomic number Z and atomic weight A of the medium: $N = N_0(Z/A)\rho$. Also, e^2/mc^2 is equal to the classical electron radius $r_e = 2.82$ fm. Therefore, (1.65) can be written as

$$\frac{dE}{dx} = 4\pi N_0 r_e^2 mc^2 \frac{Z}{A} \rho \frac{1}{\beta^2} z^2 \left(\ln \frac{2mc^2}{I} \beta^2 \gamma^2 - \beta^2 \right)$$

$$= 0.3071 \left(\frac{\text{MeV}}{\text{g/cm}^2} \right) \frac{Z}{A} \rho \frac{1}{\beta^2} z^2 \left(\ln \frac{2mc^2}{I} \beta^2 \gamma^2 - \beta^2 \right) \qquad (1.65')$$

As $\beta\gamma$ is increased dE/dx falls at first with $1/\beta^2$, then goes through a minimum and rises again for larger values of $\beta\gamma$. The logarithmic term describes the relativistic rise. Its strength is given by the mean excitation energy I. There are two physical origins of this effect. One is in the time behaviour of the electromagnetic field carried by the travelling particle. As seen from the atom, the field components of short duration become stronger as γ goes up, increasing the cross-section for the excitation and ionization. The other reason is kinematic. As $\beta\gamma$ is increased the maximal possible energy that can be conveyed from the particle to the atom goes up too and makes the average energy increase with $\beta\gamma$. For example, a pion of 1 GeV can transfer 50 MeV, and a pion of 10 GeV can transfer 3.3 GeV to an electron at rest. Such large energy transfers are not applicable to tracks in drift chambers; this is another reason for a necessary modification of (1.65); see Sect. 1.3.

The mean ionization energy I is calculable for simple atoms, but it has most often been considered a parameter to be fitted from the measurements of the ionization energy loss near the minimum. A critical collection of such determinations of I is contained in an article by Seltzer and Berger [SEL 82], a selection for relevant drift chamber gases is presented in our Table 1.5. The mean excitation energy for the chemical elements is found to increase with the atomic number Z and follows approximately

$$I = AZ$$

Table 1.5. Mean excitation energies and specific ionization in the minimum for various gases. Values of I from [SEL 82], values of $(dE/dx)_{min}/\rho$ calculated from (1.65')

Gas		Z	A	Density ρ (N.T.P.) (g/l)	Mean excitation energy I (eV)	$(dE/dx)_{min}/\rho$ (MeV/g cm^{-2})
Hydrogen	H_2	2	2.016	0.090	19.2	4.11
Helium	He	2	4.003	0.178	41.8	1.94
Neon	Ne	10	20.18	0.90	137	1.73
Argon	Ar	18	39.95	1.78	188	1.52
Krypton	Kr	36	83.8	3.74	352	1.36
Xenon	Xe	54	131.3	5.89	482	1.26
Oxygen	O_2	16	32.00	1.43	95	1.81
Nitrogen	N_2	14	28.01	1.25	82	1.83
Methane	CH_4	10	16.04	0.72	41.7	2.43
Ethane	C_2H_6	18	30.07	1.25	45.4	2.31
Propane	C_3H_8	26	44.11	1.88	47.1	2.27
Isobutane	C_4H_{10}	34	58.12	2.67	48.3	2.25
Ethylene	C_2H_4	20	28.05	1.18	51	2.73
Carbon dioxide	CO_2	22	44.00	1.98	85	1.83

with A decreasing from 20 eV (H) to 13 eV (C) and roughly 10 eV (Ar and higher). Values of I for compounds and mixtures can be calculated in a first approximation according to Bragg's additivity rule, which states that

$$\ln I_{compound} = \sum n_i \ln I_i / \sum n_i \,,$$

where the n_i and I_i are the electron densities and the mean excitation energies belonging to element i.

It was discovered later that the relativistic rise would not continue to indefinitely large values of γ. In 1939, Fermi calculated the 'density effect' as the coherent effect of the surrounding polarizable atoms, which shield the field of the travelling particle [FER 40]. A corresponding correction term $\delta(\beta)$ is introduced into (1.65), which reads in its conventional form

$$\frac{dE}{dx} = \frac{4\pi N e^4}{mc^2} \frac{1}{\beta^2} z^2 \left(\ln \frac{2mc^2}{I} \beta^2 \gamma^2 - \beta^2 - \frac{\delta(\beta)}{2} \right). \tag{1.66}$$

The exact behaviour of $\delta(\beta)$ obviously has to depend on the substance and its state of aggregation. Sternheimer and others have made parameter descriptions of $\delta(\beta)$ which consist of piecewise power-law fits with coefficients derived from the known oscillator strengths of the relevant substances. Such parameters are tabulated in [STE 84], and the literature can be traced back from there. Near the minimum, $\delta = 0$, and in the limit of $\beta \to 1$, the density correction approaches

$$\delta \to \ln \frac{\hbar^2 \omega_p^2}{I^2} \gamma^2 - 1 \tag{1.67}$$

([FAN 63], Eq. 50), where $\hbar \omega_p$ is the quantum energy of the plasma oscillation of the medium (see (1.61)). The correction term $\delta(\beta)/2$ in (1.66), being a linear function of $\ln \gamma$ cancels some but not all of the relativistic rise – the dynamic part has disappeared, and the kinematic part still makes the total energy loss increase, albeit with a smaller rate than before.

1.2.8 Energy Deposited on a Track – Restricted Energy Loss

The track ionization for drift chambers cannot be calculated on the basis of the total energy loss described in (1.65). This formula includes all the high-energy transfers that do not contribute to a track, although they are kinematically possible. Above a certain energy, an electron knocked out of a gas atom will form a second track, a δ electron, and will not contribute to the first one any more. Above what energy the new track is recognized and its ionization no longer attributed to the first track depends on the circumstances. Among these circumstances there is the *range* (i.e. the length of δ ray until it is stopped) in the particular gas. If the range is below the typical size of the pick-up electrodes, there is no separation yet. If the range is large compared to this size, then the second track may be separated by the electronics and the pattern recognition program. If there is a *magnetic field* and the track of the δ electron is curved, a similar idea applies to the radius of curvature.

There are also statistical circumstances that produce an effective cut-off: in any finite number of measurements there is always one largest energy transfer. A value of energy transfer so improbable that no track of a given series of measurements, or only an insignificant part of them, contains one is also an effective cut-off. We have calculated some examples of cut-off energies in Table 1.6 which result from the different effects mentioned above.

It appears from Table 1.6 that in a typical argon chamber without magnetic field a cut-off E_{max} somewhere between 30 and 250 keV is at work on account of the range. Inside a magnetic field the situation is more complicated because some δ rays will curl up and stay with the primary track, even up to 1 or several MeV; but this happens only at the 1% probability level per bar m of tracklength. In summary: there is a cut-off with a value somewhere between 30 keV and 1 MeV (for a typical argon chamber), depending on the apparatus. It has to replace the kinematic limit if it is smaller.

The modified Bethe–Bloch formula for the energy loss restricted in this way reads as follows ([FAN 63], Eq. 88):

$$\left(\frac{dE}{dx}\right)_{restricted} = \frac{4\pi Ne^4}{mc^2} \frac{1}{\beta^2} z^2 \left[\ln \frac{\sqrt{2mc^2 E_{max}} \beta\gamma}{I} - \frac{\beta^2}{2} - \frac{\delta(\beta)}{2} \right]. \tag{1.68}$$

Equation (1.68) holds for the range of $\beta\gamma$ where E_{max} is smaller than the kinematic limit, and for $\gamma^2 \gg E_{max}/mc^2$. In comparison to (1.66) the value at the minimum as well as the relativistic rise have become smaller. Formulae (1.65) and (1.66) were valid for particles heavier than electrons, because electrons have different kinematic limits and a different spectrum. The restricted energy loss

Table 1.6. Some energy transfers E relevant for the introduction of an upper cut-off in the definition of a track of ionization in argon gas

Range of δ electron (bar cm)	E (keV)
1	30
3	60
10	120
30	250

Radius of curvature of δ-electron track in different magnetic fields (cm)	E (MeV)
0.5 in 0.4 T	0.3
0.5 in 1.5 T	1.8
5 in 0.4 T	5.5
5 in 1.5 T	22

Probability of finding one interaction with minimum deposit of E on 1 bar m of tracklength	E (MeV)
0.1%	12
1%	1.2
10%	0.12

Kinematic limits of maximum transferrable energy (MeV)

$\beta\gamma$	e	π	p
0.1	0.001	0.010	0.010
1	0.106	1.01	1.02
4	0.80	15.9	16.3
10	2.31	95	101
100	25.3	5.8×10^3	9.2×10^3
1000	255	121×10^3	490×10^3

(1.68) is also applicable for electrons because the different kinematic limits have been replaced by the common cut-off E_{max}. It can be shown ([BET 33], Eq. 55.8) that electrons travelling with $\gamma^2 \gg E_{max}/mc^2$ produce essentially the same spectrum as heavier particles with the same γ. The universal validity of (1.68), which implies that the restricted energy loss is a function only of the particle velocity and not of the mass or the energy separately, is a consequence of the introduction of the cut-off energy E_{max}.

In the limit $\beta \to 1$ we now have complete cancellation of the γ dependence, and the restricted energy loss reaches the 'Fermi plateau'; insertion of (1.67) into (1.68) yields

$$\left(\frac{dE}{dx}\right)_{\text{restr } \beta = 1} = \frac{4\pi N e^4}{mc^2} z^2 \ln \frac{\sqrt{2mc^2 E_{\text{max}}}}{\hbar \omega_p} . \tag{1.69}$$

The ratio between the values of the restricted energy loss on the plateau and in the minimum is given by

$$R(E_{\text{max}}) = \frac{(dE/dx)_{\text{restricted}}^{\text{plateau}}}{(dE/dx)_{\text{restricted}}^{\text{minimum}}}$$

$$= \frac{\ln \dfrac{\sqrt{2mc^2 E_{\text{max}}}}{\hbar \omega_p}}{\ln \sqrt{\dfrac{2mc^2 E_{\text{max}}}{I}} (\beta \gamma)_{\text{min}} - \dfrac{\beta_{\text{min}}^2}{2}} . \tag{1.70}$$

If we insert the values of ω_p and I for argon gas at N.T.P., using (1.61) and Table 1.5, we obtain $R = 1.61$, 1.55 and 1.49 for $E_{\text{max}} = 30$ keV, 150 keV and 1 MeV, respectively. A precise evaluation of R requires better knowledge of E_{max}. When varying the gas and its density, E_{max} will also change in most applications.

The interest in particle identification by ionization measurement brings into focus the *accuracy* and hence the statistical aspect. The *average* energy loss of the Bethe–Bloch formulation has large fluctuations whereas the *most probable* energy loss fluctuates less.

The analytic method of Landau for the determination of the most probable energy loss cannot be a basis for the calculation of the energy dependence of ionization in drift chambers – their gas layers are too thin, and the cut-off required by the definition of a track is not part of Landau's method.

A statistical formulation of the problem, treated with Monte–Carlo methods along the lines of Ermilova, Kotenko and Merzon [ERM 77] or of Allison and Cobb [ALL 80], is better suited to describing the velocity dependence of the most probable energy loss. The properties of the medium enter into these models through the frequency dependence of the dielectric constant, which in turn is derivable from experimental photoabsorption coefficients. We have presented the photoabsorption ionization model [PAI] of Allison and Cobb in Sect. 1.2.2. Some more information including numerical values of the energy loss from the PAI model on the Fermi plateau is contained in Chap. 9. A step-by-step comparison of the method of Bethe–Bloch–Sternheimer with the PAI model does not exist, to our knowledge.

The reader who is interested in more historical details may find useful the famous old book by Rossi [ROS 52]. An in-depth discussion of ionization is contained in the monograph by Bugadov, Merson, Sitar and Chechin [BUG 88] as well as in the article by Fano [FAN 63].

1.2.9   Localization of Charge Along the Track

As we have seen in Sects. 1.1.1 and 1.2.2, the charge along a track is created in discrete clusters that vary greatly in size. This has some consequence for the coordinate measurement of the track. The precise way in which the discrete nature of ionization determines fundamental limits of accuracy depends on the measurement method in question and will be dealt with in Chap. 6. In preparation, we discuss here the following question: Given a certain track length L – for example, the part of the track collected onto one wire – with what precision is the centre of gravity of the discretely deposited ionization located along the track?

If the coordinate along the track is x, and in the ith cluster we have n_i electrons, then the centre of gravity is at

$$\bar{x} = \sum_{i=1}^{m} (x_i n_i) \Bigg/ \sum_{i=1}^{m} n_i \,. \tag{1.71}$$

Let the track extend around 0 from $-L/2$ to $+L/2$. The probability distribution for every x_i is

$$f(x_i)\,\mathrm{d}x_i = (1/L)\,\mathrm{d}x_i \,. \tag{1.72}$$

Averaging over all x_i, we have for every fixed set of n_i

$$\langle \bar{x} \rangle = 0$$

and

$$\langle \bar{x}^2 \rangle = \left\langle \sum_{i=1}^{m} x_i^2 n_i^2 \right\rangle \Bigg/ \left(\sum_{i=1}^{m} n_i \right)^2$$

$$= \frac{L^2}{12} \sum_{i=1}^{m} n_i^2 \Bigg/ \left(\sum_{i=1}^{m} n_i \right)^2 = \frac{L^2}{12} \frac{1}{N_{\mathrm{eff}}} \,. \tag{1.73}$$

This is because the x_i are created independently of each other so that the mixed terms $x_i x_j$ in the square drop out. The statistical factor that multiplies $L^2/12$ has been denoted $1/N_{\mathrm{eff}}$ in (1.73); its value depends on the partitioning of the charge over the m clusters. If all clusters were equally large, then N_{eff} would take on the value m. It is equal to the equivalent number of independently fluctuating entities of equal sizes that create a fluctuation that is the same as the one created by the clustered charges. As these are different in size, N_{eff} has to be smaller than m.

For a calculation of N_{eff}, which governs the localizability of any cluster ionization charge, we have to integrate over the cluster-size distribution. Starting from the one plotted in Fig. 1.7 we calculate, with the Monte Carlo method, the values listed in Table 1.7.

There are two properties of N_{eff} that are remarkable: (i) it is quite small compared with the average number $\langle m \rangle$ of clusters (let alone the total ionization); (ii) it does not grow as fast as $\langle m \rangle$ (see the graph in Fig. 1.22). The reason

Table 1.7. Values of N_{eff} calculated for various average numbers $\langle m \rangle$ of clusters, using the cluster-size distribution of Fig. 1.7

Average number $\langle m \rangle$	1	5	10	20	50	100	200	500	1000
N_{eff}	1	2.4	3.7	5.2	8.0	12	15	26	41

for the latter is that the relative contribution of the rare large clusters grows with the increase of the number of clusters as the tail of the cluster-size distribution is sampled more often.

The variation of N_{eff} with $\langle m \rangle$ seen in Fig. 1.22 may be described roughly as

$$N_{eff} = \langle m \rangle^{0.54} . \tag{1.74}$$

There is a statistical correlation between N_{eff} and the total charge deposited in L. It can be used to increase N_{eff} by cutting on the pulse height, but the gain is not overwhelmingly large.

If diffusion is taken into account, N_{eff} may become considerably larger in time, owing to the declustering effect. This is treated in Sect. 6.2.

1.2.10   A Measurement of N_{eff}

The effective number of independently fluctuating charges along the length L of a track can be measured. The r.m.s. fluctuation of the centre of gravity of the charge deposited along L is, by definition of N_{eff}, equal to

$$\delta y_c = \frac{L}{\sqrt{12 N_{eff}}} .$$

A particle track whose ionization is sampled over a length L by n wires with pulse heights P_i $(i = 1, \ldots, n)$ has the centre of its charge at

$$y_c = \sum P_i y_i / \sum P_i ,$$

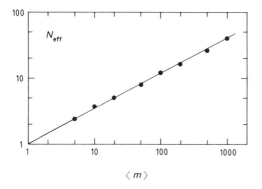

Fig. 1.22. Dependence of N_{eff}, the effective number of independently fluctuating entities of charge, as a function of the mean number $\langle m \rangle$ of clusters. Result of a calculation using the cluster-size distribution in Ar

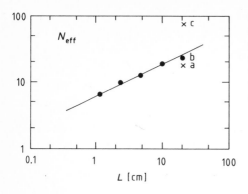

Fig. 1.23. N_{eff} measured with cosmic rays as a function of the track length L. The last data point has the largest dependence on the treatment of the overflow bin (see text). The three values correspond to extrapolation according to (a) a $1/PH^2$ law, (b) the Landau distribution or (c) no extrapolation, respectively

where the y_i are the wire positions along the track. A measurement of y_c of many tracks allows a determination of δy_c, and hence of N_{eff}, to be made.

In an experiment using a model of the ALEPH TPC, cosmic-ray tracks were collected in a small interval of momentum and angle of incidence. The chamber contained an Ar (90%) + CH_4(10%) gas mixture at atmospheric pressure and had a sense-wire spacing of 0.4 cm. Track lengths L between 0.8 and 20 cm were specified and N_{eff} was calculated for every L. The result is shown in Fig. 1.23. The choice of L is limited on the low side by the wire spacing and on the high side by the dynamic range of the pulse-height recording. In the case under consideration, the overflow bin contained 1.5% of the pulses. For higher values of L this bin is sampled more often, and the calculated number N_{eff} begins to depend on the details of an extrapolation of the overflow bin. A power-law fit to Fig. 1.20 shows that N_{eff} varied with L according to

$$N_{eff} \propto (L/1 \text{ cm})^{0.45 \pm 0.1} .$$

Comparing Figs. 1.19 and 1.20, we notice that the exponent is identical, within errors, to that obtained from the theoretical cluster-size distribution. Using the power law, which is only an empirical relation, we may obtain an estimate of the primary ionization density, because it can be expected that the function $N_{eff}(L)$ becomes equal to 1 at $L = \lambda$, the average distance between clusters.

1.3   Gas Ionization by Laser Rays

The light of a narrow pulsed laser beam that traverses a volume of gas, under favourable conditions, is capable of ionizing the gas in the beam so that it imitates a straight particle track. For this to occur, there must be some ionizable molecules in the gas, and the energy density must be sufficiently high, depending on the wavelength. In practice, a small nitrogen laser, emitting pulses of 10^4 W on a few square millimeters at $\lambda = 337$ nm, may be sufficient in an ordinary chamber gas that has not been especially cleansed and therefore contains suitable molecules in some low concentration. This technique has the obvious advantage of producing identical tracks in the same place. Therefore, by taking repeated measurements of the same coordinate, one may form the average, which can be made almost free from statistical variations, if only the number of repeated measurements is made large enough. In practice 100 shots are usually sufficient. The technique has been widely used ever since its first application in 1979 [AND 79].

Chemical compounds used for laser ionization are discussed in Sect. 11.4.

1.3.1 The nth Order Cross-Section Equivalent

The quantum energy of laser light in the visible and the near ultraviolet is much lower than the ionization energies of molecules. It takes two or more such laser photons to ionize the organic molecules present in the chamber gas. Multi-photon ionization processes involving 11 photons have been observed in xenon (see the review by Lambropoulos [LAM 76]). For the ionization of a molecule to occur, the n photons have to be incident on the molecule during the lifetime of the intermediate states. Then the ionization rate varies the nth power of the photon flux because the photons act incoherently in the gas. The probability of n photons arriving in a given time interval is equal to the nth power of the probability of each one of them, and is therefore proportional to the nth power of the photon flux ϕ. Just as the concept of 'cross-section' σ describes one-particle collision rates in units of cm^2, we have an 'nth-order cross-section equivalent' $\sigma^{(n)}$ for n-particle collisions in units of $cm^{2n} s^{n-1}$. In a volume V containing a density N of molecules, the ionization rate R is therefore given by the expression

$$R = \phi^n N V \sigma^{(n)} . \tag{1.75}$$

In fact, n-photon ionization is most easily identified by a measurement of R as a function of ϕ.

A light pulse with cross-sectional area A and duration T contains m photons if the flux ϕ and the energy E are given by

$$\phi = m/(AT) ,$$

$$E = mh\nu = mhc/\lambda .$$

Considering a traversed volume $V = AL$, the specific ionization per unit track length for the n-photon process is

$$\frac{RT}{L} = \frac{m^n}{(AT)^{n-1}} N\sigma^{(n)}$$

$$= \left(\frac{E}{AT}\right)^n ATN \left(\frac{\lambda}{hc}\right)^n \sigma^{(n)} . \tag{1.76}$$

In (1.76), (E/AT) is the power density of the beam.

So far, we have implicitly assumed the duration of the light pulse to be short compared with the inverse transition frequencies between the states involved. The more general situation develops towards a dynamic equilibrium between excitation and de-excitation. The case of two-photon ionization is treated in more detail in Sect. 1.3.2.

The molecular ions created in the ionization do not as a rule resemble the parent molecules very much, because they are cracked in the process. Fragmentation patterns are complicated and usually not understood [REI 85].

1.3.2 Rate Equations for Two-Photon Ionization

For the electrons in the molecules, we distinguish the ground level (0), the intermediate level (1), and the continuum ionization level (2). We denote the population densities in the gas by P_0, P_1, and P_2. The incident radiation stimulates transitions $0 \rightarrow 1$, $1 \rightarrow 2$, and $1 \rightarrow 0$, but there are also spontaneous transitions $1 \rightarrow 0$. Depending on the circumstances, there may be losses from level 1 into other channels. The transition rates, denoted by k_1 to k_4, are determined by the incident flux and by the internal transition mechanism.

The rate equations are written in the following form (the primes denote the time derivatives):

$$P_0'(t) = - k_1 P_0(t) + (k_1 + k_2)P_1(t) \, ,$$
$$P_1'(t) = + k_1 P_0(t) - (k_1 + k_2 + k_3 + k_4)P_1(t) \, , \qquad (1.77)$$
$$P_2'(t) = k_4 P_1 \, .$$

They are symbolized in Fig. 1.24.

The rate per molecule k_1 is taken to be proportional to the incoming flux ϕ of photons; the constant of proportionality is the cross-section σ_{01} for the process $0 \rightarrow 1$:

$$k_1 = \sigma_{01}\phi \, . \qquad (1.78)$$

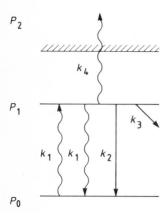

Fig. 1.24. Scheme of the five transition rates used in (1.77a–c)

We want to calculate the transition rate $P_2'(t)$ with which electrons appear in the continuum state. It is proportional to the density $P_1(t)$ of the intermediate state, and the corresponding rate per molecule is

$$k_4 = \sigma_{12}\phi . \tag{1.79}$$

The first two of the rate equations are a system of homogeneous linear differential equations. Before we derive the general mathematical solution, we shall consider some limiting cases in order to better understand the physical content of (1.77).

Before the pulse arrives, $P_0(0)$ is of the order of 10^{13} per cubic centimetre in a gas having a typical concentration of the ionizing molecules of one part per million, and $P_1(0) = 0$. While $P_1(t)$ is slowly built up at the rate $P_1'(t)$, we may consider P_0 as a constant as long as $P_1(t)$ has not developed into a similar order of magnitude. The situation of making only a few ionization electrons with the laser shot is well described by this approximation, which treats the ground state as an infinite reservoir.

(a) *Case of P_0 as an Infinite Reservoir.* We may write the first two of (1.77) in the following form:

$$P_0'(t) = - k_1 P_0 + (k_1 + k_2)P_1(t) , \tag{1.80a}$$

$$P_1'(t) = + k_1 P_0 - (k_1 + k_2 + k_3 + k_4)P_1(t) . \tag{1.80b}$$

The solution of (1.80b) is

$$P_1(t) = \frac{k_1 P_0}{\sum k}(1 - e^{-t\sum k}) , \tag{1.81}$$

where $\sum k = k_1 + k_2 + k_3 + k_4$. Obviously, our approximation is always fulfilled if $k_1 \ll \sum k$ or $\sigma_{01}\phi \ll \sum k$ (using (1.78)). Let us assume that this is the case. Equation (1.81) leads to a production rate of ionization equal to

$$P_2'(t) = k_4 P_1(t) = \frac{k_1 k_4 P_0}{\sum k}(1 - e^{-t\sum k}) . \tag{1.82}$$

Integrated over the duration T of a constant light pulse, the total density reached at the end of the pulse is given by

$$P_2(T) = \frac{k_1 k_4 P_0}{\sum k}[T - (1 - e^{-T\sum k})/\sum k] . \tag{1.83}$$

In the limit of short pulses, i.e. $T \ll 1/\sum k$, the exponential may be expanded to second order and gives us

$$P_2(T) \rightarrow \frac{1}{2}k_1 k_4 P_0 T^2 . \tag{1.84}$$

In the limit of long pulses, i.e. $T \gg 1/\sum k$, the exponential vanishes, so that

$$P_2(T) \rightarrow \frac{k_1 k_4 P_0}{\sum k}T . \tag{1.85}$$

Expression (1.84) is proportional to ϕ^2, the square of the photon flux, because of (1.78) and (1.79). This is also the case for expression (1.85) unless k_1 or k_2 dominates $\sum k$. In our approximation that treats P_0 as an infinite reservoir, we assume that $k_1 \ll \sum k$; then a similar condition holds for k_2 (except for special circumstances in which σ_{12} is orders of magnitude larger than σ_{10}):

$$P_2(T) \to \frac{1}{2}\sigma_{01}\sigma_{12}P_0\phi^2 T^2 \quad \text{(small } T, \text{ small } \phi)\,, \tag{1.86}$$

$$P_2(T) \to \frac{\sigma_{01}\sigma_{12}}{k_2 + k_3}P_0\phi^2 T \quad \text{(large } T, \text{ small } \phi)\,. \tag{1.87}$$

(b) *General Case.* If we want to know what happens when the photon flux is strong, (1.77) have to be solved in a rigorous way. This is achieved by a variable transformation that separates the equations (see any textbook on differential equations, for example [KAM 59].) One finds the eigenvalues s_1, s_2 of the matrix of coefficients from the determinant

$$\begin{vmatrix} -k_1 - s & k_1 + k_2 \\ k_1 & -\sum k - s \end{vmatrix} = 0\,, \tag{1.88}$$

$$s_{1,2} = -\frac{\sum k + k_1}{2} \pm \left[\left(\frac{\sum k + k_1}{2} \right)^2 - k_1(k_3 + k_4) \right]^{1/2}\,. \tag{1.89}$$

These two eigenvalues are real and negative; we take $s_2 < s_1 < 0$.

The solutions to the differential equations (1.77) satisfying the initial conditions $P_1(0) = 0$, $P_2(0)$, are given by

$$P_0(t) = \frac{P_0(0)}{s_1 - s_2}[-(k_1 + s_2)e^{s_1 t} + (k_1 + s_1)e^{s_2 t}]\,, \tag{1.90a}$$

$$P_1(t) = \frac{P_0(0)k_1}{s_1 - s_2}[e^{s_1 t} - e^{s_2 t}]\,, \tag{1.90b}$$

$$P_2(t) = \frac{P_0(0)k_1 k_4}{s_1 s_2}\left[1 + \frac{s_2}{s_1 - s_2}e^{s_1 t} - \frac{s_1}{s_1 - s_2}e^{s_2 t} \right]\,. \tag{1.90c}$$

Equation (1.90c) represents the general solution to the problem. It describes the concentration of ionization electrons as a function of time and of the rate coefficients k_1 to k_4. For small t we recover our expressions (1.84) and (1.86) by developing the exponentials to the second order in $s_1 t$ and $s_2 t$:

$$P_2(t) \to \frac{1}{2}k_1 k_4 P_0(0)t^2\,. \tag{1.91}$$

For large t, using (1.89), we obtain

$$P_2(t) \to \frac{k_4}{k_3 + k_4}P_0(0)\,. \tag{1.92}$$

This enormous concentration of electrons (saturation) would imply that the ground state has been emptied, a situation that should not concern us when we study ionization tracks that are similar to particle tracks.

Equation (1.91), identical to (1.84) and (1.86), gives rise to the definition of the second-order cross-section equivalent described in Sect. 1.3.1. For a laser shot of duration T, identifying R/V with P_2/T, we have

$$\sigma^{(2)} = \frac{1}{2}\sigma_{01}\sigma_{12}T \,. \tag{1.93}$$

Under the conditions that lead to (1.87), we have instead

$$\sigma^{(2)} = \sigma_{01}\sigma_{12}/(k_2 + k_3) \,, \tag{1.94}$$

whereas under the most general conditions, including saturation, $\sigma^{(2)}$ depends not only on T but also on ϕ.

1.3.3 Dependence of Laser Ionization on Wavelength

The ionization yield depends very much on the wavelength. Using tunable dye lasers, Ledingham and co-workers have found an increase of four orders of magnitude, when going from $\lambda = 330$ nm to $\lambda = 260$ nm. In Fig. 1.25 'clean'

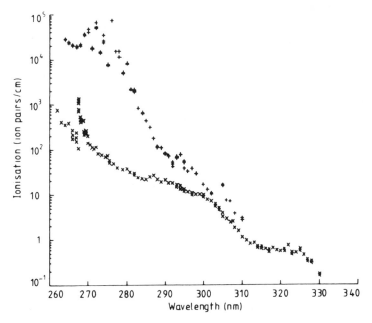

Fig. 1.25. Ionization induced in untreated counter gas [Ar (90%) + CH$_4$ (10%)] by a 1×1 mm^2 pulsed laser beam of 1 μJ (×), compared with the same gas seeded with a small amount of phenol (+) [TOW 86]

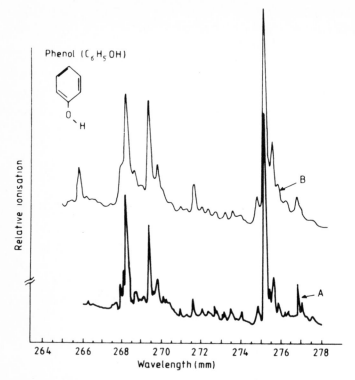

Fig. 1.26. A comparison of the laser-induced R2PI spectrum of counter gas doped with a trace of phenol (B) and the single-photon UV absorption spectrum of phenol [TOW 86]

counter gas is compared with a gas seeded with a small amount of phenol. The fine structure visible around 270 nm was resolved using high-resolution techniques and was compared with the known single-photon UV absorption of phenol (see Fig. 1.26).

The identical wavelength dependence is apparent. We understand this behaviour from (1.93) and (1.94): the wavelength dependence resides almost entirely in the factor σ_{01}, implying that the cross-section for ionization from the intermediate level does not vary so much. The amount of characteristic structure in the spectrum of laser-induced resonant two-photon ionization ('R2PI') makes it a sensitive tool for the identification of molecules [HUR 79].

1.3.4 Laser-Beam Optics

A beam of laser light that is to create ionization as if from a particle must have an approximately constant cross-sectional area along the beam. Relations (1.75) and (1.76) imply that, for example, in two-photon ionization a light pulse with a given energy produces half as many ions per unit track length when it has twice the cross-sectional area. In the interests of a constant high yield and of a narrow deposit of ionization, one would like to have the beam width as small as possible over the largest possible track length.

The photons in a beam occupy a volume in the four-dimensional phase space at each point along the beam. For illustration, we consider the horizontal plane in Fig. 1.27, where an almost parallel beam is shown to be focused into a narrow waist by a lens. After the focus, the beam opens up again. Each photon trajectory is represented by a point in the phase-space diagrams below where the distance from the axis is plotted against the small angle of the trajectory with the axis. Here we deal with geometrical optics only; the wave optical aspects are treated later. In this sense we may regard the optical elements of the beam (lenses, mirrors, light guides, free space, apertures) as determining a trajectory for each photon in a time-independent fashion. Now, Liouville's theorem in statistical mechanics states that in such a passive system the density of photons in phase space is a constant of motion, meaning that it does not change along the beam. As long as there are no photons lost, the occupied volume of cross-sectional area times solid angle is the same everywhere along the beam and the same as just behind the laser. In the phase-space diagrams of Fig. 1.24 the envelopes containing the beam do not change their area. A concentration of the beam in space by a focusing lens is accompanied by a corresponding extension of the solid angle.

There is a principal limit of the size, in phase space, of a photon beam. Heisenberg's uncertainty principle of quantum mechanics states that, in each of the two transverse dimensions separately, the limits are given by

$$\Delta x \Delta p_x > h/4\pi , \qquad (1.95)$$

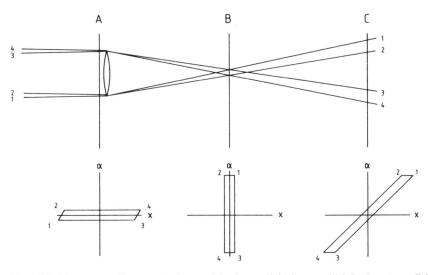

Fig. 1.27. Phase-space diagrams at three points along a light-beam, which is almost parallel at the left and is focussed into B by a lens at A

where Δx and Δp_x are the root mean square widths in space and in momentum [MES 59] and h is Planck's constant. Since the r.m.s. opening angle is

$$\Delta \alpha = \Delta p_x / p = \Delta p_x \lambda / h \ll 1 \,,$$

we have

$$\Delta x \Delta \alpha > \lambda / 4\pi \,. \tag{1.96}$$

An example not very far from the principal limit is provided by the situation where a plane light-wave is delimited by an aperture, say a slit of width $\pm A$. The diffraction caused by the slit will open the range of the angles β of propagation behind the slit so that approximately

$$|\pm \beta| \approx \frac{1}{A} \frac{\lambda}{2\pi} \,. \tag{1.97}$$

Many lasers can be tuned to emit beams near the fundamental limit. If their wavelength is subsequently halved by a frequency doubler, the limit refers to the old wavelength unless the phase space and, inevitably, the power of the beam are reduced afterwards.

For practical purposes we work with approximate full widths W_x and W_α, ignoring the exact distribution of intensity across the beam, and use the order-of-magnitude relation

$$W_x W_\alpha > \lambda \,. \tag{1.98}$$

The highest spatial concentration of light is reached in a beam focus; let it. have a width W_x. Ideally, the arrival angles of the photons do not depend on the point across the focus; let the angular width be W_α. At a distance D behind the focus, every such point has reached a width αD. The statistical ensemble of all the photons has a combined width of W_{tot} given by

$$W_{tot}^2 = W_x^2 + (W_\alpha D)^2 \,. \tag{1.99}$$

The quadratic addition of these two widths is correct in the approximation that the light intensity in the focus is a Gaussian function of both the space and the angular coordinates. For most applications this is not far from reality.

Applying (1.98) and (1.99), the total width of a focused beam, at a distance D away from the focus, is not smaller than

$$W_{tot}^{min} = \left[W_x^2 + \left(\frac{\lambda D}{W_x} \right)^2 \right]^{1/2} \,. \tag{1.100}$$

This is plotted in Fig. 1.28 as a function of D for various values of W_x.

The technique of *light-tracks* simulating particle tracks is particularly well suited for the calibration of coordinate measurements based on the pulse heights of cathode strips, because they find the centre of the track. The one to several mm wide profile of the light-beam is usually not well known, whereas the position of the centre is much better defined. The calibration of coordinates

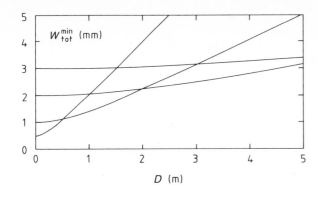

Fig. 1.28. The minimum width of a laser beam at a distance D behind a focus, calculated for various widths at the focus ($\lambda = 1\ \mu m$)

based on drift time measurements requires the profile to be known. First studies of the profile of light-tracks with the aim of improving the drift-coordinate calibration were performed by Fabretti et al. [FAB 90].

References

[ALL 80]   W.W.M. Allison and J.H. Cobb, Relativistic charged particle identification by energy loss, *Ann. Rev. Nucl. Sci.* **30**, 253 (1980)

[AND 79]   M. Anderhub, M.J. Devereux and P.G. Seiler, On a new method for testing and calibrating ionizing-particle detectors, *Nucl. Instrum. Methods* **166**, 581 (1979)

[BIC 75]   H. Bichsel and R.P. Saxon, Comparison of calculational methods for straggling in thin absorbers, *Phys. Rev. A* **11**, 1286 (1975)

[BLU 74]   W. Blum, K. Söchting and U. Stierlin, Gas phenomena in spark chambers, *Phys. Rev. A* **10**, 491 (1974)

[BLU 50]   O. Blunck and S. Leisegang, Zum Energieverlust schneller Elektronen in dünnen Schichten, *Z. Phys.* **128**, 500 (1950)

[BUG 88]   Y.A. Bugadov, G.I. Merson, B. Sitar and V.A. Chechin, Ionization measurements in high energy physics (Energoatomizdat, Moscow 1988). An English translation is in preparation to be published by Springer

[CHE 72]   Chechin, L.P. Kotenko, G.I. Merson and V.C. Ermilova, The relativistic rise of the track density in bubble chambers, *Nucl. Instrum. Methods* **98**, 577 (1972)

[CHE 76]   V.A. Chechin and V.K. Ermilova, The ionization-loss distribution at very small absorber thicknesses, *Nucl. Instrum. Methods* **136**, 551 (1976)

[CHR 71]   L.G. Christophorou, *Atomic and Molecular Radiation Physics* (Wiley, London 1971)

[COM 77]   D. Combecher, W-values of low energy electrons for several gases, in *Proceedings of the 3rd Symposium on Neutron Dosimetry in Biology and Medicine*, ed. by G. Burger and H.G. Ebert, held at Neuherberg, May 1977 (Commission of the European Communities *and* Gesellschaft für Strahlen- und Umweltforschung, Neuherberg, Germany 1978) p. 97

[ERM 69]   V.K. Ermilova, L.P. Kotenko, G.I. Merzon and V.A. Chechin, Primary specific ionization of relativistic particles in gases, *Sov. Phys.-JETP* **29**, 861 (1969)

[ERM 77]   V.K. Ermilova, L.P. Kotenko and G.I. Merzon, Fluctuations and the most probable values of relativistic charged particle energy loss in thin gas layers, *Nucl. Instrum. Methods* **145**, 555 (1977)

[FAB 90]   R. Fabretti, M. Fabre, L. Li, P. Razis, P.G. Seiler and O. Prokofiev, Relation between the lateral distribution of UV laser induced ionization in a drift chamber and the shape of the corresponding signal, *Nucl. Instr. Meth. Phys. Res. A* **287**, 413 (1990)

[FAN 63]   U. Fano, Penetration of protons, alpha particles and mesons, *Ann. Rev. Nucl. Sc.* **13**, 1 (1963)

[FER 40]   E. Fermi, The ionization loss of energy in gases and in condensed materials, *Phys. Rev.* **57**, 485 (1940)

[FIS 91]   H. Fischle, J. Heintze and B. Schmidt, Experimental determination of ionization cluster size distributions in counting gases, *Nucl. Instrum. Methods A* **301**, 202 (1991)

[HAR 73]   F. Harris, T. Katsura, S. Parker, V.Z. Peterson, R.W. Ellsworth, G.B. Yodh, W.W.M. Allison, C.B. Brooks, J.H. Cobb and J.H. Mulvey, The experimental identification of individual particles by the observation of transition radiation in the X-ray region, *Nucl. Instrum. Methods* **107**, 413 (1973)

[HUR 79]   G.S. Hurst, M.G. Payne, S.D. Kramer and J.P. Young, Resonance ionization spectroscopy and one-atom detection, *Rev. Mod. Phys.* **51**, 767 (1979)

[INO 75]   M. Inokuti, Ionization yields in gases under electron irradiation, *Radiat. Res.* **64**, 6 (1975)

[INT 79]   International Commission on Radiation Units and Measurements, Average energy required to produce an ion pair (ICRU Report 31, Washington, DC, 1979)

[JAC 75]   J.D. Jackson, Classical electrodynamics, 2nd edn (Wiley, New York 1975)

[KAM 59]   E. Kamke, Differentialgleichungen, *Lösungsmethoden und Lösungen*, 6th edn (Akademische Verlagsgesellschaft, Leipzig 1959)

[KOB 68]   E.J. Kobetich and R. Katz, Energy deposition by electron beams and delta rays, *Phys. Rev.* **170**, 391 (1968)

[KÖL 84]   K.S. Kölbig and B. Schorr, A program package for the Landau distribution, *Comput. Phys. Commun.* **31**, 97 (1984)

[LAM 76]   P. Lambropoulos, Topics in multiphoton processes in atoms, *Adv. At. and Mol. Phys.* **12**, 87 (1976)

[LAN 44]   L.D. Landau, On the energy loss of fast particles by ionization, *J. Phys. USSR* **8**, 201 (1944). Also in *Collected Papers of L.D. Landau*, ed. by D. Ter Haar (Pergamon, Oxford 1965)

[LAN 60]   L.D. Landau and E.M. Lifshitz, *Electrodynamics of Continuous Media*, Vol. 8 of *Course of Theoretical Physics* (Pergamon, Oxford 1960) p. 84

[LAP 80]   F. Lapique and F. Piuz, Simulation of the measurement by primary cluster counting of the energy lost by a relativistic ionizing particle in argon, *Nucl. Instrum. Methods* **175**, 297 (1980)

[LEH 78]   I. Lehraus, R. Matthewson, W. Tejessy and M. Aderholz, Performance of a large-scale multilayer ionization detector and its use for measurements of the relativistic rise in the momentum range of 20–110 GeV/c, *Nucl. Instrum. Methods* **153**, 347 (1978)

[LEH 82]   I. Lehraus, R. Matthewson and W. Tejessy, Particle identification by dE/dx sampling in high-pressure drift detectors, *Nucl. Instrum. Methods* **196** (1982) 361

[MAR 76]   G.V. Marr and J.B. West, Absolute photoionization cross-section tables for helium, neon, argon and krypton in the VUV spectral regions, *At. Data and Nucl. Data Tables* **18**, 497 (1976)

[MES 59]   A. Messiah, *Mécanique Quantique* (Dunod, Paris 1959)

[PRO 74]   Proc. Symp. on the Jesse Effect and Related Phenomena, Gatlinburg (Tenn.), 1973 [*Radiat. Res.* **59**, 337 (1974)]

[RIE 72]   F.F. Rieke and W. Prepejchal, Ionization cross-sections of gaseous atoms and molecules for high-energy electrons and positrons, *Phys. Rev. A* **6**, 1507 (1972)

[REI 85]   H. Reisler and C. Wittig, Multiphoton ionization of gaseous molecules, in *Photodissociation and Photoionization*, ed. by K.P. Lawley (Wiley, New York 1985)

[ROS 52]   B. Rossi, *High-energy Particles* (Prentice-Hall, Englewood Cliffs, NJ 1952)

[SEL 82]   S.M. Seltzer and M.J. Berger, Evaluation of the stopping power of elements and compounds for electrons and positrons, *Int. J. Appl. Radiation Isotope* **33**, 1189 (1982)

[SÖC 79]   K. Söchting, Measurement of the specific primary excitation of helium and neon by fast charged particles, *Phys. Rev. A* **20**, 1359 (1979)

[STE 84]   R.M. Sternheimer, M.J. Berger and S.M. Seltzer, Density effect for the ionization loss of charged particles in various substances, *At. Data and Nucl. Data Tables* **30**, 261 (1984)

[STE 71]   R.M. Sternheimer and R.F. Peierls, General expression for the density effect for the ionization loss of charged particles, *Phys. Rev. B* **3**, 3681 (1971)

[TOW 86]   M. Towrie, J.W. Cahill, K.W.D. Ledingham, C. Raine, K.M. Smith, M.H.C. Smith, D.T.S. Stewart and C.M. Houston, Detection of phenol in proportional-counter gas by two-photon ionisation spectroscopy, *J. Physics B (At. Mol. Phys.)* **19**, 1989 (1986)

[VAV 57]   P.V. Vavilov, Ionization losses of high-energy heavy particles, *Sov. Phys. JETP* **5**, 749 (1957)

[WAL 79]   A.H. Walenta, J. Fischer, H. Okuno and C.L. Wang, Measurement of the ionization loss in the region of relativistic rise for noble and molecular gases, *Nucl. Instrum. Methods* **161**, 45 (1979)

[WAL 81]   A.H. Walenta, Performance and development of dE/dx counters, *Phys. Scr.* **23**, 354 (1981)

2. The Drift of Electrons and Ions in Gases

The behaviour of the drift chamber depends crucially on the drift of the electrons and ions that are created either by the particles measured or in the avalanches at the electrodes. In addition to the electric drift field, there is often a magnetic field, which is necessary for particle momentum measurement. Obviously we have to understand how the drift velocity vector, in electric and magnetic fields, depends on the properties of the gas molecules, including their density and temperature. We could start by writing down the most general expression and then specialize to the simple practical cases. However, it is more 'anschaulich' to proceed in the opposite way. We will derive the simple cases first and then generalize them to the more rigorous formulation in Sect. 2.3.

2.1 An Equation of Motion with Friction

The motion of charged particles under the influence of electric and magnetic fields, E and B, may be understood in terms of an equation of motion:

$$m\frac{d\boldsymbol{u}}{dt} = e\boldsymbol{E} + e[\boldsymbol{u} \times \boldsymbol{B}] - K\boldsymbol{u} , \qquad (2.1)$$

where m and e are the mass and electric charge of the particle, $\boldsymbol{u}$ is its velocity vector, and K describes a frictional force proportional to $\boldsymbol{u}$, which is caused by the interaction of the particle with the gas. In Sect. 2.2 we will see that, in terms of the more detailed theory involving atomic collisions, (2.1) describes the drift at large t to a very good approximation. Historically, (2.1) was introduced by P. Langevin, who imagined the force $K\boldsymbol{u}$ as a stochastic average over the random collisions of the drifting particle.

The ratio m/K has the dimension of a characteristic time, and we define

$$\tau = \frac{m}{K} . \qquad (2.2)$$

Equation (2.1) is an inhomogeneous system of linear differential equations for the three components of velocity.

The solution for $t \gg \tau$ of (2.1) is a steady state for which $d\boldsymbol{u}/dt = 0$. The drift velocity vector is determined by the linear equation

$$\frac{\boldsymbol{u}}{\tau}\frac{e}{m} - [\boldsymbol{u} \times \boldsymbol{B}] = \frac{e}{m}\boldsymbol{E} \; . \tag{2.3}$$

In order to solve for $\boldsymbol{u}$, we write $(e/m)B_x = \omega_x$ etc., $(e/m)E_x = \varepsilon_x$ etc., and express (2.3) in the form of the matrix equation

$$M\boldsymbol{u} = \varepsilon \; ,$$

$$M = \begin{bmatrix} 1/\tau & -\omega_z & \omega_y \\ \omega_z & 1/\tau & -\omega_x \\ -\omega_y & \omega_x & 1/\tau \end{bmatrix} . \tag{2.4}$$

The solution is obtained by inverting M:

$$\boldsymbol{u} = M^{-1}\varepsilon \; ,$$

$$M^{-1} = \begin{bmatrix} 1 + \omega_x^2\tau^2 & \omega_z\tau + \omega_x\omega_y\tau^2 & -\omega_y\tau + \omega_x\omega_z\tau^2 \\ -\omega_z\tau + \omega_x\omega_y\tau^2 & 1 + \omega_y^2\tau^2 & \omega_x\tau + \omega_y\omega_z\tau^2 \\ \omega_y\tau + \omega_x\omega_z\tau^2 & -\omega_x\tau + \omega_y\omega_z\tau^2 & 1 + \omega_z^2\tau^2 \end{bmatrix}$$

$$\times \frac{\tau}{1 + \omega^2\tau^2} \; , \tag{2.5}$$

where $\omega^2 = \omega_x^2 + \omega_y^2 + \omega_z^2 = (e/m)^2 B^2$ is the square of the cyclotron frequency of the electron. As we will explain in Sect. 2.3, our solution (2.5) is equal to that obtained in the full microscopic theory in the approximation that τ is independent of the collision energy [see the discussion following (2.84)].

A different way of writing (2.5) is the following:

$$\boldsymbol{u} = \frac{e}{m}\tau|\boldsymbol{E}|\frac{1}{1 + \omega^2\tau^2}(\hat{\boldsymbol{E}} + \omega\tau[\hat{\boldsymbol{E}} \times \hat{\boldsymbol{B}}] + \omega^2\tau^2(\hat{\boldsymbol{E}} \cdot \hat{\boldsymbol{B}})\hat{\boldsymbol{B}}) \; . \tag{2.6}$$

Here $\hat{\boldsymbol{E}}$ and $\hat{\boldsymbol{B}}$ denote the unit vectors in the directions of the fields. The drift direction is governed by the dimensionless parameter $\omega\tau$. For $\omega\tau = 0$, $\boldsymbol{u}$ is along $\boldsymbol{E}$; in this case the relation has the simple form

$$\boldsymbol{u} = \frac{e}{m}\tau\boldsymbol{E} = \mu\boldsymbol{E} \; ,$$

$$\mu = \frac{e}{m}\tau \; . \tag{2.7}$$

These equations define the scalar mobility μ as the ratio of drift velocity over electric field in the absence of magnetic field; μ is proportional to the characteristic time τ. For large $\omega\tau$, $\boldsymbol{u}$ tends to be along $\boldsymbol{B}$; if $\hat{\boldsymbol{E}} \cdot \hat{\boldsymbol{B}} = 0$, then large $\omega\tau$ turn $\boldsymbol{u}$ into the direction of $\hat{\boldsymbol{E}} \times \hat{\boldsymbol{B}}$. Reversing the magnetic field direction causes the $\hat{\boldsymbol{E}} \times \hat{\boldsymbol{B}}$ component to change sign because ω always stays positive by definition.

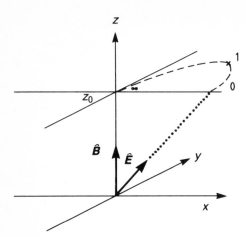

In the presence also of the magnetic field, the mobility is the tensor $(e/m)M^{-1}$ given in (2.5). An illustration of the directional behaviour of u is presented in Fig. 2.1, where the coordinate system is oriented so that the z axis is along $\hat{B}$, and $\hat{E}$ is in the x–z plane. A particle starting at the origin will arrive on the plane z_0 at a point which belongs to a half circle over the diameter connecting the end points of the $\hat{B}$ and $\hat{E}$ vectors in the z_0 plane. This is so because the components of u in (2.6) satisfy the relation

$$\left(\frac{u_x}{u_z} - \frac{1}{2}\frac{E_x}{E_z}\right)^2 + \left(\frac{u_y}{u_z}\right)^2 = \left(\frac{1}{2}\frac{E_x}{E_z}\right)^2$$

for any $\omega\tau$. The component in the direction of $\hat{E}\times\hat{B}$ is largest for $\omega\tau = 1$.

Not only the direction but also the magnitude of u is influenced by the magnetic field. If we distinguish $u(\omega)$, the drift velocity in the presence of B, and $u(0)$, the drift velocity at $B = 0$ and under otherwise identical circumstances, we derive from (2.6)

$$\frac{u^2(\omega)}{u^2(0)} = \frac{1 + \omega^2\tau^2\cos^2\phi}{1 + \omega^2\tau^2} , \tag{2.8}$$

where ϕ is the angle between E and B. This ratio happens to be the same as the one by which the component of u along E, u_E, changes with B:

$$\frac{u_E(\omega)}{u_E(0)} = \frac{1 + \omega^2\tau^2\cos^2\phi}{1 + \omega^2\tau^2} . \tag{2.9}$$

2.1.1 Case of E Nearly Parallel to B

Two cases are of special interest. The first applies to the situation where a magnetic field is almost parallel to the electric field. Using (2.6) with

$E = (0, 0, E_z)$, $B = (B_x, B_y, B_z)$ and $|B_x|, |B_y| \ll |B_z|$, the drift directions are given by the expression

$$\frac{u_x}{u_z} = \frac{-\omega\tau B_y + \omega^2\tau^2 B_x}{(1 + \omega^2\tau^2)B_z} ,$$

$$\frac{u_y}{u_z} = \frac{\omega\tau B_x + \omega^2\tau^2 B_y}{(1 + \omega^2\tau^2)B_z} .$$

(2.10)

Here the terms proportional to $\omega\tau/(1 + \omega^2\tau^2)$ change sign under reversal of B (see (2.6)). Equations (2.10) will be used in Sect. 2.4.4 for the analysis of track distortions.

2.1.2   Case of E Orthogonal to B

The second case of interest applies to the situation where E is at right angles to B. Using (2.6) with $(\hat{E} \cdot \hat{B}) = 0$, $E = (E_x, 0, 0)$, and $B = (0, 0, B_z)$, we have

$$u_x = \frac{(e/m)\tau}{1 + \omega^2\tau^2} |E| ,$$

$$u_y = -\frac{(e/m)\tau}{1 + \omega^2\tau^2} \omega\tau |E| ,$$

$$u_z = 0 ,$$

$$\tan\psi \equiv \frac{u_y}{u_x} = -\omega\tau .$$

(2.11)

From (2.11) we deduce for the magnitude of u

$$|u| = \frac{(e/m)\tau}{\sqrt{1 + \omega^2\tau^2}} |E| = \frac{e}{m}\tau |E| \cos\psi .$$

(2.12)

The product $|E| \cos\psi$ represents the component of the electric field in the drift direction. Compared with the magnetic-field-free case (2.7), we may write the functional dependence on $|E|$ and $|B|$ in the following form:

$$|u| = F(E, B) = F(E \cos\psi, 0) ,$$

(2.13)

which is also known as *Tonks' theorem* [TON 55].

The expression has a simple meaning: whatever the drift direction of the electron, the component of the electric field along this direction already determines the drift velocity for every magnetic field. The derivation of (2.13) is based on a constant value of τ, and it is experimentally verified to good precision (see Fig. 2.26). This will be discussed in more detail in Sects. 2.2.3 and 2.4.6.

2.2   The Microscopic Picture

On the microscopic scale, the electrons or ions that drift through the gas are scattered on the gas molecules so that their direction of motion is randomized in each collision. On the average, they assume a constant drift velocity u in the direction of the electric field E (or, if also a magnetic field is present, in the direction which is given by both fields). The drift velocity u is much smaller than the instantaneous velocity c between collisions.

The gases we deal with are sufficiently rarefied that the distances travelled by electrons between collisions are large in comparison with their Compton wavelengths. So our picture is classical and atomistic. For gas pressures above, say, 100 atm, slow-drifting electrons interact with several atoms simultaneously and require a quantum mechanical treatment [BRA 81, ATR 77].

In order to understand the drift mechanism, we derive some basic relations between the macroscopic quantities of drift velocity u and isotropic diffusion coefficient D on the one hand, and the microscopic quantities of electron velocity c, mean time τ between collisions, and fractional energy loss λ on the other. The microscopic quantities are randomly distributed according to distribution functions treated in Sect. 2.3; in this section we deal with suitable averages. The relevant relations can be expected to be correct within factors of the order of unity.

2.2.1   Drift of Electrons

Let us consider an electron between two collisions. Because of its light mass, the electron scatters isotropically and, immediately after the collision, it has forgotten any preferential direction. Some short time later, in addition to its instantaneous and randomly oriented velocity c, the electron has picked up the extra velocity u equal to its acceleration along the field, multiplied by the average time that has elapsed since the last collision:

$$u = \frac{eE}{m} \tau \ . \tag{2.14}$$

This extra velocity appears macroscopically as the drift velocity. In the next encounter, the extra energy, on the average, is lost in the collision through recoil or excitation. Therefore there is a balance between the energy picked up and the collision losses. On a drift distance x, the number of encounters is $n = (x/u)(1/\tau)$, the time of the drift divided by the average time τ between collisions. If λ denotes the average fractional energy loss per collision, the energy balance is the following:

$$\frac{x}{u\tau} \lambda \varepsilon_E = eEx \ . \tag{2.15}$$

Here the equilibrium energy ε_E carries an index E because it does not contain the part due to the thermal motion of the gas molecules, but only the part taken out of the electric field.

One may wonder about the fact that the average time between collisions is the same as the average time that has elapsed since the last collision. This is because in a completely random series of encounters, characterized only by the average rate $1/\tau$, the differential probability $f(t)\,dt$ that the next encounter is a time between t and $t + dt$ away from any given point $t = 0$ in time is

$$f(t)\,dt = \frac{1}{\tau}\,e^{-t/\tau}\,dt\;,\tag{2.16}$$

independent of the point where the time measurement begins.

Equation (2.14), as compared with (2.7), provides an interpretation and a justification of the constant τ introduced in Sect. 2.1. In the frictional-motion picture, τ was the ratio of the mass the drifting particle over the coefficient of friction. In the microscopic picture, τ is the mean time between the collisions of the drifting particle with the atoms of the gas. For drifting particles with instantaneous velocity c, the mean time τ between collisions may be expressed in terms of the cross-section σ and the number density N:

$$\frac{1}{\tau} = N\sigma c\;.\tag{2.17}$$

Here c is related to the total energy of the drifting electron by

$$\frac{1}{2}mc^2 = \varepsilon = \varepsilon_E + \frac{3}{2}kT\;,\tag{2.18}$$

because the total energy is made up of two parts: the energy received from the electric field and the thermal energy that is appropriate for 3 degrees of freedom (k = Boltzmann's constant, T = gas temperature).

For electron drift in particle detectors, we usually have $\varepsilon_E \gg (3/2)kT$; we can neglect the thermal motion, and (2.14), (2.15), and (2.18) combine to give the two equilibrium velocities as follows:

$$u^2 = \frac{eE}{mN\sigma}\sqrt{\frac{\lambda}{2}}\;,\tag{2.19}$$

$$c^2 = \frac{eE}{mN\sigma}\sqrt{\frac{2}{\lambda}}\;,\tag{2.20}$$

where $\varepsilon = (1/2)mc^2 \simeq \varepsilon_E \gg (3/2)kT$.

Both σ and λ are functions of ε. We notice the important role played by the energy loss in the collisions: with vanishing λ, the drift velocity would become zero. If ε is below the excitation levels of the gas molecules, the scattering is elastic and λ is approximately equal to twice the mass ratio of the collision partners (see Sect. 2.2.2); hence it is of the order of 10^{-4} for electrons.

As an illustration, we show in Fig. 2.2a the behaviour of the effective cross-sections for argon and methane, two typical counter gases used in drift

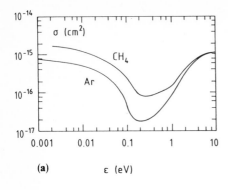

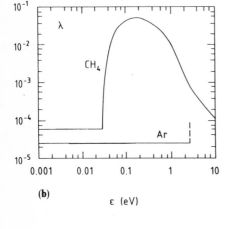

Fig. 2.2. (a) Effective ('momentum transfer') cross-section $\sigma(\varepsilon)$ for argon and methane. (b) Fraction $\lambda(\varepsilon)$ of energy loss per collision for argon and methane. Results of B. Schmidt [SCH 86]; see Sect. 2.4.1

chambers. The method of determining them will be discussed in Sect. 2.4. In both cases we observe a clear dip near $\varepsilon \simeq 0.25$ eV or $\varepsilon \simeq 0.30$ eV. This is the famous Ramsauer minimum [RAM 21] which is due to quantum mechanical processes in the scattering of the electron with the gas molecule [ALL 31]. The noble gases krypton and xenon show the same behaviour but helium and neon do not. For a more recent discussion on the scattering of electrons on molecules the reader is referred to [BRO 66].

The behaviour of $\lambda(\varepsilon)$ for argon and methane is seen in Fig. 2.2b. The threshold for excitation of the argon atom is at 11.5 eV, that for the methane molecule is at 0.03 eV. Under these circumstances it is not surprising that the drift velocity for electrons depends critically on the exact gas composition. Owing to the very different behaviour of the energy loss in molecular gases with respect to noble gases, even small additions (10^{-2}) of molecular gases to a noble gas dramatically change the fractional energy loss λ (cf. Fig. 2.2b), by absorbing collision energy in the rotational states. This change of λ results in an increase of the electron drift velocity by large factors, since the fraction of drifting electrons with energy close to that of the Ramsauer minimum increases and the average σ decreases. Also the functional dependence $u(E)$ is changed.

We note that the dependence of the drift velocity and the electron energy on E and N is always through the ratio E/N. This implies that in two gases operated with different gas pressures and identical temperature, the E fields must be adjusted in proportion to the gas pressures in order to obtain identical u and ε in both gases. Reduced fields are given in units of $V\,cm^{-1}\,Torr^{-1}$. A special unit has also been coined: one Townsend (1 Td) is $10^{-17}\,V\,cm^2$.

In a gas mixture composed of two or more components i with number densities n_i, the effective cross-section $\sigma(\varepsilon)$ and the effective fractional energy loss per collision $\lambda(\varepsilon)$ have to be calculated from the properties of the individual components:

$$\sigma(\varepsilon) = \sum n_i \sigma_i(\varepsilon)/N \ ,$$

$$\lambda(\varepsilon)\,\sigma(\varepsilon) = \sum n_i \lambda_i(\varepsilon)\sigma_i(\varepsilon)/N \ , \qquad (2.21)$$

$$N = \sum n_i \ .$$

The general behaviour of electron drift velocities is that they rise with increasing electric field, then level off or decrease as a result of the combined effects of $\sigma(\varepsilon)$ and $\lambda(\varepsilon)$ as ε increases with increasing E. Examples will be presented in Sect. 2.4 and in Chap. 11.

2.2.2 Drift of Ions

The behaviour of ions differs from that of electrons because of their much larger mass and their chemical reactions. The monograph by McDaniel and Mason [MCD 73] deals with the mobility and diffusion of ions in gases in a comprehensive way. Electrons in an electric field are accelerated more rapidly than ions, and they lose very little energy when colliding elastically with the gas atoms. The electron momentum is randomized in the collisions and is therefore lost, on the average. In electric field strengths that are typical for drift chambers, the electrons reach random energies far in excess of the energy of the thermal motion, and quite often they surpass the threshold of inelastic excitation of molecules in the gas. In this case their mobility becomes a function of the energy loss that is associated with such excitation.

Ions in similar fields acquire, on one mean-free path, an amount of energy that is similar to that acquired by electrons. But a good fraction of this energy is lost in the next collision, and the ion momentum is not randomized as much. Therefore, far less field energy is stored in random motion. As a consequence, the random energy of ions is mostly thermal, and only a small fraction is due to the field. The effect on the diffusion of ions results in this diffusion being orders of magnitude smaller than that of electrons in similar fields. The effect on the mobility is also quite interesting: since the energy scale, over which collision cross-sections vary significantly, is easily covered by the electron random energies reached under various operating conditions, we find rapid and sometimes complicated dependences of electron mobility on such operating conditions

– electric and magnetic field strengths and gas composition being examples. In contrast, the mobility of ions does not vary as much.

In order to understand the randomization in more quantitative terms, we consider the elastic scattering of an ion (or electron) with a molecule of the gas, neglecting the thermal motion.

Elastic Scattering and the Average Losses of Energy and Momentum. We consider the scattering of two particles with masses m and M in their centre-of-mass (c.m.) system in which the incoming momenta are equal and opposite and are denoted by p. In the laboratory frame, M was initially at rest, and the incoming particle had momentum P (see Fig. 2.3). The two frames are transformed into each other by the velocity $v = P/(M + m)$.

For elastic scattering, we have $|p'| = |p| = M|v| = p$. The loss of momentum of particle M along its direction of motion is, in the c.m. system, equal to

$$p(1 - \cos\theta) ,$$

and invariant under the Galilean transformation. At our low momenta the scattering is isotropic, and after averaging over the angles we find the average fractional momentum loss κ:

$$\kappa = \frac{p}{P} = \frac{M}{M + m} . \qquad (2.22)$$

Similarly, the square of the momentum transfer – equally invariant – is

$$(p - p')^2 = 2p^2(1 - \cos\theta) .$$

It is equal to the square of the laboratory momentum of the particle initially at rest; therefore its kinetic energy is $E'_{kin} = p^2(1 - \cos\theta)/M$. After averaging over the angles, the fractional energy loss λ is the ratio of this kinetic energy to that of the other (incoming) particle in the laboratory system:

$$\lambda = \frac{p^2}{M}\frac{2m}{P^2} = \frac{2mM}{(M + m)^2} . \qquad (2.23)$$

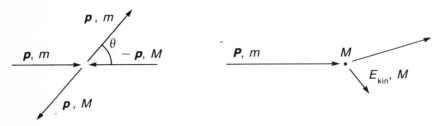

Fig. 2.3. Momentum diagrams for the collision of the drifting particle (mass m) with a gas molecule (mass M) in the centre of mass and in the rest frame of the molecule

We are now in a position to go through the arguments of Sect. 2.2.1 again, applying them to ions.

Expressions for the Mobility of Ions. Let us consider an ion between two collisions. Because of its heavy mass it does not scatter isotropically. Immediately after the collision, superimposed over the random velocity c_r there is a component c_d in the drift direction. Some short time later, it has, in addition, picked up the extra velocity v equal to the acceleration of the ion along the field, multiplied by the average time that has elapsed since the last collision. The drift velocity is the sum of v and c_d:

$$u = v + c_d = \frac{eE}{m}\tau + c_d . \tag{2.24}$$

The value of c_d is given by the average momentum loss in the direction of the drift that takes place in the next encounter; it must be equal to the momentum mv picked up between the encounters:

$$\kappa m \langle c_r + c_d + v \rangle = \kappa m(c_d + v) = mv , \tag{2.25}$$

where the brackets $\langle\ \rangle$ denote the average over the angles. For the magnitude of c_d we calculate $c_d = v(1 - \kappa)/\kappa$, and for the drift velocity (using (2.22)),

$$u = v + c_d = \frac{v}{\kappa} = \frac{e}{m}E\tau\frac{1}{\kappa} = \frac{e}{m}E\tau\left(1 + \frac{m}{M}\right). \tag{2.26}$$

Here τ is given by the relative velocity c_{rel} between the ion and the gas molecule, the cross-section σ, and the number density N as

$$\frac{1}{\tau} = N\sigma c_{rel} . \tag{2.27}$$

We now distinguish two limiting cases of field strength E. If it is low so that the ion random velocity is thermal, we must take for the relative velocity the average magnitude of the difference between the two randomly distributed velocities of the ion (c_{ion}) and the gas atoms (c_{gas}):

$$c_{rel} = \langle|c_{ion} - c_{gas}|\rangle = (c_{ion}^2 + c_{gas}^2)^{1/2} . \tag{2.28}$$

With equipartition of energy, we can express c_{rel} through the temperature T and Boltzmann's constant k:

$$c_{rel}^2 = c_{ion}^2 + c_{gas}^2 = 3kT\left(\frac{1}{m} + \frac{1}{M}\right). \tag{2.29}$$

Using (2.26), (2.27), and (2.29), we obtain

$$u = \left(\frac{1}{m} + \frac{1}{M}\right)^{1/2}\left(\frac{1}{3kT}\right)^{1/2}\frac{eE}{N\sigma} \quad \text{(low E)} . \tag{2.30}$$

It is characteristic of the drift velocity at low fields that it is proportional to E, or that the mobility is independent of E.

For the case of large E, where we can neglect the thermal motion, we proceed as in the case of electrons (where the temperature was always neglected). By combining (2.15), (2.17), (2.24), and (2.26), we solve for u^2 and c^2 and obtain the symmetric forms

$$u^2 = \frac{eE}{m^* N\sigma} \left(\frac{\lambda}{2} \frac{m}{m^*} \right)^{1/2} \tag{2.31}$$

and

$$c^2 = \frac{eE}{m^* N\sigma} \left(\frac{2}{\lambda} \frac{m}{m^*} \right)^{1/2}, \tag{2.32}$$

where the 'reduced mass' m^* between the ion (m) and the gas molecule (M) is defined as

$$\frac{1}{m^*} = \frac{1}{m} + \frac{1}{M}. \tag{2.33}$$

Using (2.23), we may express λ for elastic scattering through the ratio of the masses, with the result that

$$u = \left[\frac{eE}{mN\sigma} \right]^{1/2} \left[\frac{m}{M} \left(1 + \frac{m}{M} \right) \right]^{1/2} \quad \text{(high } E) . \tag{2.34}$$

It is characteristic of the drift velocity at high fields and constant cross-section σ that it is proportional to the square root of E, or the mobility decreases as $1/\sqrt{E}$.

The distinction between 'high' and 'low' E in (2.30) and (2.34) is obviously with respect to the field in which the ion, over one mean free path, picks up an amount of energy equal to the thermal energy. The measured ion drift velocities for noble gases shown in Fig. 2.4 exhibit very well the limit in behaviour of (2.30) and (2.34) with the electric field.

Table 2.1 contains ion mobilities for low fields that were measured in the noble gases. Various approximation methods for the calculation of cross-sections in the scattering of ions on molecules have been treated elsewhere

Table 2.1. Experimental low-field mobilities of various noble gas ions in their parent gas [MCD 73]

Gas	Ion	Mobility $(\text{cm}^2 \, \text{V}^{-1} \text{s}^{-1})$
He	He$^+$	10.40 ± 0.10
Ne	Ne$^+$	4.14 ± 0.2[a]
Ar	Ar$^+$	1.535 ± 0.007
Kr	Kr$^+$	0.96 ± 0.09
Xe	Xe$^+$	0.57 ± 0.05

[a] Average over several measurements

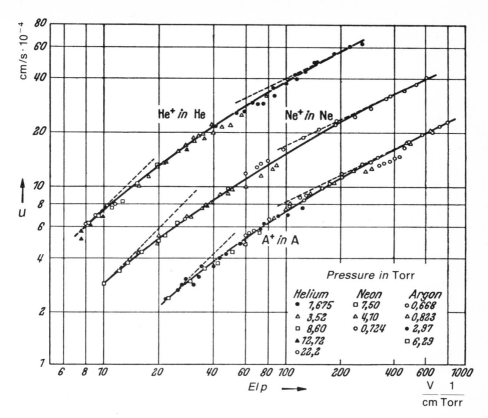

Fig. 2.4. Drift velocity u of positive noble gas ions in their own gas as functions of the reduced electric field. The limiting behaviour at low fields ($u \propto E$) and at high fields ($u \propto \sqrt{E}$) is visible [LAN 57]

[MCD 73]. When ions travel in their parent gas, the original collision cross-section is increased by about a factor of two by the mechanism of resonant charge transfer.

Ion Drift in Gas Mixtures. The gas of a drift chamber is often a composition of two or more constituents. Ion drift in gas mixtures can be understood along the lines of the last subsection. We want to modify (2.26) for the case where one type of ion (mass m) moves between the atoms or molecules with masses M_k ($k = 1$, $2, \ldots$) each present in the gas with its own number density N_k. The gas number density N is equal to the sum over all the N_k. The resulting drift velocity in the field E is called $\bar{u}$; it is equal to

$$\bar{u} = \frac{e}{m} E\bar{\tau} \frac{1}{\bar{\kappa}} . \tag{2.35}$$

Here $\bar{\tau}$ is the mean time between collisions, and $\bar{\kappa}$ is the mean fractional momentum loss in the mixture; they are both quickly calculated from the τ_k and κ_k that belong to each constituent. The overall collision rate is the sum of the individual collision rates, or

$$1/\bar{\tau} = \sum (1/\tau_k) \, . \tag{2.36}$$

The overall momentum loss is the average over all the κ_k, weighted with the relative rates with which each type of molecule is bombarded:

$$\bar{\kappa} = \frac{\sum (\kappa_k/\tau_k)}{\sum (1/\tau_k)} \, . \tag{2.37}$$

Combining (2.35) to (2.37), we obtain

$$\bar{u} = \frac{e}{m} E \frac{1}{\sum (\kappa_k/\tau_k)} \, . \tag{2.38}$$

This equation can be cast into a form that is known as *Blanc's law* if we denote by u_k the drift velocity of the ion in the pure component k at the full density N:

$$u_k = \frac{e}{m} E \frac{\tau_k}{\kappa_k} \frac{N_k}{N} \, , \tag{2.39}$$

where we have used (2.17). Expressions (2.38) and (2.39) give us Blanc's law:

$$\frac{1}{\bar{u}} = \sum \frac{N_k}{N} \frac{1}{u_k} \, ; \tag{2.40}$$

it states that the inverse drift velocities in the individual gas components add in proportion to the relative component densities to yield the inverse drift velocity of the mixture.

Blanc's law holds for the case of low electric fields. The case of high fields is slightly more complicated because the energy partition of the drifting ion into random and directed motion must be taken into account. The reader is referred to the monograph of McDaniel and Mason [MCD 73].

In practice, (2.40) is quite well fulfilled. In Fig. 2.5 we show mobilities, measured by the Charpak group [SCH 77], as functions of the relative gas concentration for several binary mixtures of argon with the common quench gases: isobutane, methane, and carbon dioxide.

The production of positive ions in molecular gases is a fairly complex phenomenon, and without a clear identification – perhaps using a mass spectrometer – one is usually ignorant of the identity of the migrating charged bodies. For the purposes of drift chambers it often suffices to know their drift velocities. However, the chemical deposits on the wire electrodes, as a result of chamber operation, can be evaluated with more complete knowledge about which ions really migrate toward the cathodes.

One must read Table 2.2, which contains measured ion mobilities, with these remarks in mind. For example, the notation '$[(OCH_3)_2CH_2]^+$' means the

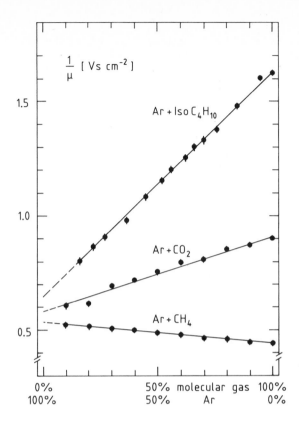

Fig. 2.5. The inverse of the mobility of the positive molecular ions as a function of the relative volume concentration of the two components in three gas mixture. Measurements by [SCH 77]

Table 2.2. Measured mobilities of various ions in some gases used in drift chambers [SCH 77]

Ion notation	Gas	Mobility $(cm^2\ V^{-1}\ s^{-1})$
$[CH_4]^+$	Ar	1.87
$[CO_2]^+$	Ar	1.72
$[IsoC_4H_{10}]^+$	Ar	1.56
$[(OCH_3)_2CH_2]^+$	Ar	1.51
$[CH_4]^+$	CH_4	2.26
$[CO_2]^+$	CO_2	1.09
$[IsoC_4H_{10}]^+$	$IsoC_4H_{10}$	0.61
$[(OCH_3)_2CH_2]^+$	$IsoC_4H_{10}$	0.55
$[(OCH_3)_2CH_2]^+$	$(OCH_3)_2CH_2$	0.26

ion(s) travelling towards the cathode in a mixture of argon + methylal. It is assumed here that the argon ions that are created simultaneously transfer their charge to the methylal molecules, thus creating more '$[(OCH_3)_2CH_2]^+$' ions. This is elaborated below.

Charge Transfer. When the migrating ions in a drift chamber collide with molecules that have an ionization potential smaller than the energy available in the ion, there is the possibility that a charge-exchange process takes place, which neutralizes the ion and creates a new ion. This process happens with cross-sections that are of a similar order of magnitude to the other ion–molecule scattering cross-sections. Therefore the rate of ion transformation through charge transfer is correspondingly high and proportional to the concentration of the molecules to be ionized.

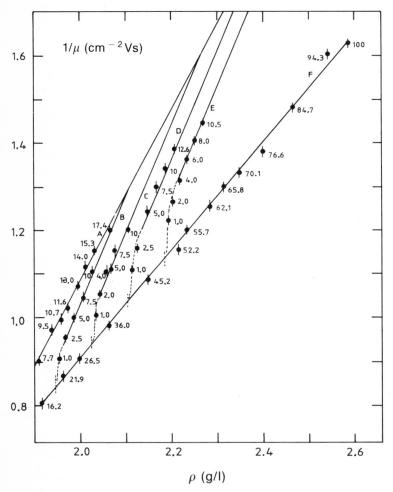

Fig. 2.6. Inverse mobility of ions in binary and ternary gas mixture composed of argon–isobutane–methylal, measured by [SCH 77] and plotted as a function of the gas density. A: argon–methylal; B to E: argon–isobutane–methylal, with argon in different proportions (B: 80%, C: 70%, D: 60%, and E: 50%) F: argon–isobutane. The numbers close to the experimental points represent the percentage concentration of methylal (A to E) or isobutane (F), respectively

The Charpak group have found that 1% of methylal added to binary argon + isobutane mixtures made a significant change in the ion mobility over 2 cm of drift in their field (a few hundred to a thousand V/cm). This has to be interpreted as a charge transfer process from the ions in the original mixture to the methylal molecules, thus creating '$[(OCH_3)_2 CH_2] +$' ions with lesser mobility. Their result [SCH 77] is presented in Fig. 2.6, where the inverse measured mobility is plotted against the density of the various binary and ternary mixtures. Straight lines correspond to Blanc's law; they connect mixtures in which the migrating ion(s) is (are) supposedly the same. We must therefore imagine the motion of ions in drift chambers as a dynamic process in which the various ionic species, which were originally produced in the avalanches or any other collision process, disappear quickly and leave only the one type that has the lowest ionization potential.

Practical drift chamber gases – apart from the principal components that were intentionally mixed together by the experimenter – always contain some impurities; they are difficult to control below the level of 10^{-4}. The mere fact that pulsed lasers of modest energy are capable of producing ionization in 'clean' gas (see Sect. 1.3 for details) is proof of the presence of impurities with low ionization potentials. It would obviously be interesting to know the identity of migrating ions at the end of the 2-m-long drift space of a large time projection chamber (TPC).

2.2.3 Inclusion of Magnetic Field

When we consider the influence of a magnetic field on drifting electrons and ions, the first indication may be provided by the value of the mobility μ of these charges. In particle chamber conditions, this is of the order of magnitude of $\mu \simeq 10^4 \, cm^2 \, V^{-1} \, s^{-1}$ for electrons (see Fig. 2.17 and use (2.7)), whereas for ions the order of magnitude is $\mu = 1 \, cm^2 \, V^{-1} \, s^{-1}$ (see Tables 2.1 and 2.2). Typical magnetic fields B available to particle experimenters are limited, so far, by the magnetic susceptibility of iron, and the order of magnitude is $1 \, T$ or $10^{-4} \, V \, s \, cm^{-2}$. We know from Sect. 2.1 that it is the numerical value of $\omega\tau = (e/m)B\tau$ that governs the effects of the magnetic field on the drift velocity vector. Using (2.7), which stated that $\mu = (e/m)\tau$, we find

$$\omega\tau = B\mu \simeq \begin{cases} 10^{-4} & \text{for ions} \\ 1 & \text{for electrons} \end{cases}$$

in order of magnitude. Therefore, the effect of such magnetic fields on ion drift is negligible, and we concentrate on electrons. This has the advantage that we may assume that the colliding body scatters isotropically in all directions, owing to its small mass.

When the magnetic field is added to the considerations of Sect. 2.2.1, we can describe the most general case in a coordinate system in which B is along z, and E has components E_z and E_x. An electron between collisions moves according to the equation of motion,

$$m \frac{d\boldsymbol{v}}{dt} = e\boldsymbol{E} + e[\boldsymbol{v} \times \boldsymbol{B}] \, , \tag{2.41}$$

which in our case is written as

$$\dot{v}_x = \varepsilon_x + \omega v_y \, ,$$
$$\dot{v}_y = - \omega v_x \, , \tag{2.42}$$
$$\dot{v}_z = \varepsilon_z \, ,$$

where $\omega \equiv (e/m)B$ and $\varepsilon \equiv (e/m)E$. Electrons have their direction of motion randomized in each collision, and we are interested in the extra velocity picked up by the electron since the last collision. Hence we look for the solution of (2.42) with $\boldsymbol{v} = 0$ at $t = 0$. It is given by

$$v_x(t) = (\varepsilon_x/\omega) \sin \omega t \, ,$$
$$v_y(t) = (\varepsilon_x/\omega) (\cos \omega t - 1) \, , \tag{2.43}$$
$$v_x(t) = \varepsilon_z t \, .$$

Before we can identify $\boldsymbol{v}$ with the drift velocity, we must average over t, using (2.16), the probability distribution of t. This was also done when deriving (2.14), which, being a linear function of time, required t to be replaced by τ, the mean time since the last collision. The drift velocity for the present case is given by

$$u_x = \langle v_x(t) \rangle = \frac{\varepsilon_x}{\omega} \int_0^\infty \frac{1}{\tau} e^{-t/\tau} \sin \omega t \, dt = \frac{\varepsilon_x \tau}{1 + \omega^2 \tau^2} \, ,$$
$$u_y = \langle v_y(t) \rangle = \frac{\varepsilon_x}{\omega} \int_0^\infty \frac{1}{\tau} e^{-t/\tau} (\cos \omega t - 1) dt = \frac{- \varepsilon_x \omega \tau^2}{1 + \omega^2 \tau^2} \, , \tag{2.44}$$
$$u_z = \langle v_z(t) \rangle = \varepsilon_z \int_0^\infty \frac{1}{\tau} e^{-t/\tau} t \, dt = \varepsilon_z \tau \, .$$

We immediately recognize that the result is the same as for (2.5) and (2.6), derived from the macroscopic equation of motion using the concept of friction. The mobility tensor (2.5) is therefore the same in the macroscopic and the microscopic picture. It will only have to be modified in Sect. 2.3, where the electron velocity has a probability distribution over a range of values, rather than a single value.

We can now state the condition of energy balance in the presence of a magnetic field. If the angle between the two fields is ϕ, the energy transferred to the electron per unit distance along $\boldsymbol{E}$, equal to the corresponding collision loss, is

$$eE = \frac{1}{2} \frac{\lambda mc^2}{u_E(\omega)\tau} = \frac{1}{2} \frac{\lambda m^2 c^2 (1 + \omega^2 \tau^2)}{eE\tau^2 (1 + \omega^2 \tau^2 \cos^2 \phi)} \, . \tag{2.45}$$

Here we have used (2.9). On the other hand, the square of the drift velocity was given by (2.7) and (2.8),

$$u^2 = \left(\frac{e}{m} E\tau\right)^2 \frac{1 + \omega^2\tau^2 \cos^2\phi}{1 + \omega^2\tau^2} . \tag{2.46}$$

The two expressions (2.45) and (2.46) determine the two equilibrium velocities u and c. First we notice that the ratio

$$\frac{u^2}{c^2} = \frac{\lambda}{2} \tag{2.47}$$

is given by the fractional energy loss per collision alone; in particular, it is independent of both fields. In the situation without magnetic field, the two electron velocities were determined by the cross-section and the energy loss in (2.19) and (2.20). With magnetic field, the corresponding relations are more complicated owing to the presence of the terms proportional to $\omega^2\tau^2 = \omega^2/(N^2\sigma^2 c^2)$. Whereas in the most general case (2.45) and (2.46) could be solved for c^2 by a graphical or numerical method, we present the closed solution for two important practical cases.

Case of E Parallel to B. The first case concerns the drift volume of the TPC; here we have $\cos^2\phi = 1$ and therefore

$$c^2 = \frac{eE}{mN\sigma} \sqrt{\frac{2}{\lambda}} , \tag{2.48}$$

exactly as in the absence of magnetic field.

Case of E Orthogonal to B. The second case concerns the drift volume of the axial wire drift chamber, where $\cos^2\phi = 0$; here

$$c^2 = \left[\left(\frac{eE}{mN\sigma}\right)^2 \frac{2}{\lambda} + \left(\frac{\omega^2}{2N^2\sigma^2}\right)^2\right]^{1/2} - \left[\frac{\omega^2}{2N^2\sigma^2}\right] . \tag{2.49}$$

We note that the random velocity c is reduced in the presence of a magnetic field orthogonal to E. The same is true for the drift velocity because of (2.47). Any two values of E and B that produce the same solution for c in (2.49) lead to the same drift velocity. In particular, if E_1 is the electric field that produces certain values of c and u in the absence of magnetic field ($B_1 = 0$), then there is a corresponding E_2 for the same values of c and u at some non-zero B_2. In order to find an expression for E_2 in terms of E_1 and B_2, we rewrite (2.49) for the two pairs of fields in the following form:

$$-\left(\frac{eE_1}{mN\sigma}\right)^2 \frac{2}{\lambda} + c^4 = 0 ,$$

$$\left(\frac{ecB_2}{mN\sigma}\right)^2 - \left(\frac{eE_2}{mN\sigma}\right)^2 \frac{2}{\lambda} + c^4 = 0 . \tag{2.50}$$

Note that λ and σ are the same in both cases because they are functions of c alone. From (2.50) we deduce that

$$E_2^2 = E_1^2 \left[1 + \left(\frac{eB_2}{c\,mN\sigma} \right)^2 \right] = E_1^2(1 + \omega_2^2\tau^2)\,, \tag{2.51}$$

which is the microscopic justification for (2.12) and (2.13). One cannot understand (2.12) from a derivation using a constant friction term τ because τ does not remain constant when E is varied (see (2.7) and Fig. 2.17). Rather, as we have seen, τ also depends on B in such a way that it comes back to its old value for the appropriate combination of B and E.

Next, we wish to evaluate (2.49) for large values of the B field. In the limit $\omega\tau \gg 1$ the second term in the square root of (2.49) becomes much larger compared to the first term (using (2.46) and (2.47) it can be shown that the two terms are equal when $\omega\tau \simeq 2.2$), and we have to first order in $1/\omega\tau$

$$c^2 = \left(\frac{eE}{m\omega} \right)^2 \frac{2}{\lambda}\,, \tag{2.52}$$

$$u^2 = \left(\frac{eE}{m\omega} \right)^2, \quad u = \frac{E}{B}\,, \tag{2.53}$$

which is remarkable because c no longer depends on σ.

Equation (2.53) has the following significance. In the presence of a magnetic field B, orthogonal to E and strong enough for $\omega\tau$ to be large, the drift velocity u approaches a universal value E/B, which is the same for all gases.

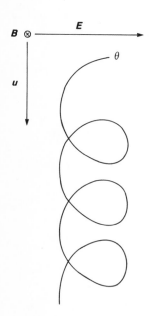

Fig. 2.7. Motion of an electron *in vacuo* under the influence of orthogonal E and B fields: B goes into the paper, the electron stays in the plane of the paper

In order to study the deviations from the universal expression (2.53), we may carry the expansion of the square root in (2.49) to the second order. Using (2.47) and (2.53) we find

$$u = \frac{E}{B}\left(1 - \frac{1}{\omega^2 \tau^2}\right) \quad (\omega\tau \gg 1) .$$ (2.54)

This means that the universal limit (2.53) is approached rapidly from below as B is increased.

This physical content of (2.53) can best be understood by comparison with the motion of an electron *in vacuo*: under the influence of crossed electric and magnetic fields, the electron is accelerated by the E field and bent by the B field; it performs a cycloidal motion in the plane orthogonal to B (see (2.43) and Fig. 2.7). When averaging over the periodic part, one finds that the electron drifts with the constant speed E/B in the direction of $-\hat{E} \times \hat{B}$. This motion *in vacuo* is the limit of the gas drift for large values of $\omega\tau$, the mean number of turns (expressed in radians) between collisions. Experimentally it can be reached either by increasing the B field or by reducing the gas density, which is roughly proportional to $1/\tau$.

2.2.4 Diffusion

As the drifting electrons or ions are scattered on the gas molecules, their drift velocity deviates from the average owing to the random nature of the collisions. In the simplest case the deviation is the same in all directions, and a point-like cloud of electrons which then begins to drift at time $t = 0$ from the origin in the z direction will, after some time t, assume the following Gaussian density distribution:

$$n = \left(\frac{1}{\sqrt{4\pi Dt}}\right)^3 \exp\left(\frac{-r^2}{4Dt}\right) ,$$ (2.55)

where $r^2 = x^2 + y^2 + (z - ut)^2$; D is the diffusion constant because n satisfies the continuity equation for the conserved electron current Γ:

$$\Gamma = nu - D\nabla n ,$$ (2.56)

$$\frac{\partial n}{\partial t} + \nabla \cdot \Gamma = 0 ,$$ (2.57)

$$\frac{\partial n}{\partial t} + n\nabla \cdot u - D\nabla^2 n = 0 .$$ (2.58)

The diffusion constant D, defined by (2.56), makes the mean squared deviation of the electrons equal to $2Dt$ in any one direction from their centre. (This is a special case of (2.75), which deals with the case of anisotropic diffusion.)

In order to express the diffusion constant in terms of the microscopic picture, we suppose that one electron or ion starts at time $t = 0$ and has a velocity c;

hence, according to (2.16), there is probability distribution of free path l equal to

$$g(l)\,dl = \frac{1}{l_0}\,e^{-l/l_0}\,dl\,, \tag{2.59}$$

where $l_0 = c\tau$ is the mean free path. The simplest case is the one where scattering is isotropic with respect to the drift direction. We consider this case first. It applies to ions at low electric fields E $(E/(N\sigma) \ll kT)$ and, to a good first approximation, to electrons.

Consider a time t at which a large number, n, of encounters have already occurred; then $n = t/\tau$. The mean square displacement in one direction, say x, is

$$\int \left(\sum_1^n l_i \cos\theta_i\right)^2 \prod_1^n g(l_k)\,dl_k\,\frac{d\cos\theta_k}{2} = n\frac{2}{3}l_0^2 = \frac{2}{3}\frac{l_0^2}{\tau}t\,. \tag{2.60}$$

Hence $l_i \cos\theta_i$ is the displacement along x between the $(i-1)$th and the ith collision (Fig. 2.8); the $\cos\theta_i$ are uniformly distributed between -1 and $+1$ according to our assumption. The l_i are distributed with the probability density (2.59). Note that the mixed terms in the sum vanish. The part proportional to $2t$ is the diffusion coefficient D,

$$D = \frac{l_0^2}{3\tau} = \frac{cl_0}{3} = \frac{c^2\tau}{3} = \frac{2}{3}\frac{\varepsilon}{m}\tau\,. \tag{2.61}$$

Recalling the expression for the electron mobility μ,

$$\mu = \frac{e}{m}\tau\,,$$

we notice that the electron energy can be determined by a measurement of the

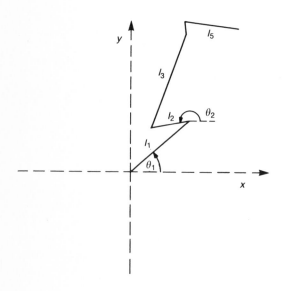

Fig. 2.8. Electron paths for the derivation of (2.60)

ratio D/μ:

$$\varepsilon = \frac{3}{2} \frac{De}{\mu} .$$ (2.62)

When the diffusing body has thermal energy, $\varepsilon = (3/2)kT$, (2.62) takes the form

$$\frac{D}{\mu} = \frac{kT}{e} ,$$

which is known as the *Nernst–Townsend formula*, or the *Einstein formula*. (For historical references, see [HUX 74].)

The energy determines the diffusion width σ_x of an electron cloud which, after starting point-like, has travelled over a distance L:

$$\sigma_x^2 = 2Dt = \frac{2DL}{\mu E} = \frac{4\varepsilon L}{3eE} .$$ (2.63)

In drift chambers we therefore require small electron energies at high drift fields in order to have σ_x^2 as small as possible. In the literature one finds the concept of characteristic energy, ε_k, which is related to our ε by the relation $\varepsilon_k = (2/3)\varepsilon$.

In Fig. 2.9 we show the variation of ε_k with the electric field strength, measured with electrons drifting in the two gases, argon and carbon dioxide, which represent somewhat extreme cases concerning the change-over from

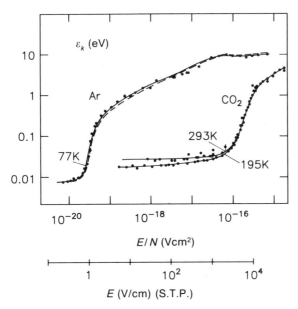

Fig. 2.9. Characteristic energy of electrons in Ar and CO_2 as a function of the reduced electric field. The electric field under normal conditions is also indicated. The parameters refer to the different temperatures at which the measurements were made [SCH 76]

thermal behaviour to field-dominated behaviour. In argon a field strength as low as 1 V/cm produces electron energies distinctly larger than thermal ('hot gas'). In carbon dioxide the same behaviour occurs only at field strengths above 2 kV/cm ('cold gas'). The reason is a large value of the relative energy loss λ in CO_2, due to the internal degrees of freedom of the CO_2 molecule, which are accessible at low collision energies.

For ions and electrons with thermal energy $\varepsilon = (3/2)kT$, (2.63) shows that the diffusion width of a cloud is independent of the nature of the gas and is proportional to the square root of the absolute temperature: the 'thermal limit' is given by

$$\sigma_x = \left(\frac{2kTL}{eE}\right)^{1/2}.$$

(2.64)

2.2.5 Electric Anisotropy

Until 1967 it had always been assumed that the diffusion of drifting electrons in gases has the isotropic form implied by (2.55). But then Wagner et al. [WAG 67] discovered experimentally that the value of electron diffusion along the electric field can be quite different from that in the perpendicular direction. The diffusion of drifting ions is also often found to be non-isotropic.

When ions collide with the gas molecules, they retain their direction of motion to some extent because the masses of the two collision partners are similar, and therefore the instantaneous velocity has a preferential direction along the electric field (see Sect. 2.2.2). This causes the diffusion to be larger in the drift direction; the mechanism is at work for ions travelling in high electric fields E ($eE/(N\sigma) \geq kT$), and we do not treat it here because it plays no role in the detection of particles.

In the case of electrons, there is almost no preferential direction for the instantaneous velocity. We will describe the electron diffusion anisotropy following the semiquantitative treatment of Partker and Lowke [PAR 69], which is restricted to energy loss by elastic collisions. The essential point of the argument is that the mobility of the electrons assumes different values in the leading edge and in the centre of the travelling cloud if the collision rate is a function of electron energy. This change of mobility inside the cloud is equivalent to a change of diffusion in the longitudinal direction.

We express the energy balance (2.15) in the drifting cloud more precisely by making use of the electron current introduced in (2.51); it now contains the diffusion term $D\nabla n$. Let the drift be in the z direction ($\boldsymbol{E} = E\hat{z}$). Using the collision frequency $v \equiv 1/\tau$ instead of τ and dropping the index from ε, the energy balance takes the form

$$nv\lambda\varepsilon = e\boldsymbol{E}\cdot\boldsymbol{\Gamma} = e\mu E^2 n - eED\frac{\partial n}{\partial z},$$

(2.65)

where v, μ, λ and D are functions of ε. In principle, we can solve this equation for ε in terms of $(1/n)\,(\partial n/\partial z)$. Here we will put λ constant, as it is in the case of elastic scattering, and develop $v(\varepsilon)$ around the equilibrium point, which is given when $\partial n/\partial z = 0$:

$$\varepsilon_0 = \frac{1}{m\lambda}\left(\frac{eE}{v_0}\right)^2 .$$

The functions D and m are approximated by their expressions (2.61) and (2.62). Putting $\varepsilon = \varepsilon_0 + \Delta\varepsilon$ and $v = v_0 + (\partial v/\partial\varepsilon)\Delta\varepsilon$ in (2.65), the variation of energy inside the cloud can be expressed to first order as

$$\Delta\varepsilon = -\frac{2\varepsilon_0^2}{3eE[1 + 2(\partial v/\partial\varepsilon)_0\,(\varepsilon_0/v_0)]}\frac{1}{n}\frac{\partial n}{\partial z} , \qquad (2.66)$$

$(\partial v/\partial\varepsilon)_0$ being the derivative of the collision frequency with respect to the energy, evaluated at ε_0. Equation (2.66) shows that ε is larger in the leading edge of the cloud and smaller in the trailing edge, and from (2.62) it follows that – unless $(\partial v/\partial\varepsilon) = 0$ – the mobility μ also is a function of position in the cloud (see Fig. 2.10):

$$\mu = \mu_0\left(\frac{1}{v_0}\frac{\partial v}{\partial\varepsilon}\,\Delta\varepsilon\right) .$$

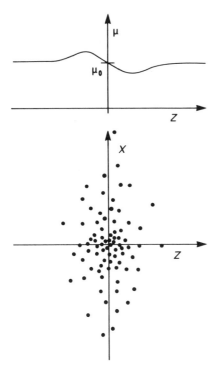

Fig. 2.10. Mobility variation inside an electron cloud travelling in the z direction

We now rearrange the terms in the expression for the electron current and find

$$\boldsymbol{\Gamma} = \mu_0 \, En\hat{z} - D \left(\frac{\partial n}{\partial x} \hat{x} + \frac{\partial n}{\partial y} \hat{y} \right) - D \left(1 - \frac{\gamma}{1 + 2\gamma} \right) \frac{\partial n}{\partial z} \hat{z} \, ,$$

where $\gamma \equiv (\varepsilon_0/v_0)(\partial v/\partial \varepsilon)$. Obviously the diffusion in the drift direction has changed and is no longer equal to the diffusion in the perpendicular direction. We distinguish the two diffusion coefficients by the indices L and T. The ratio is

$$\frac{D_L}{D_T} = \frac{1 + \gamma}{1 + 2\gamma} \, . \tag{2.67}$$

Instead of the density distribution (2.55) of the diffusing cloud of electrons, we have in the anisotropic case

$$n = \frac{1}{\sqrt{4\pi D_L t}} \left(\frac{1}{\sqrt{4\pi D_T t}} \right)^2 \exp \left[- \frac{x^2 + y^2}{4D_T t} - \frac{(z - ut)^2}{4D_L t} \right] . \tag{2.68}$$

2.2.6 Magnetic Anisotropy

Let us now consider the effect of a magnetic field B along z. The electric field is in the x–z plane and we assume there is no *electric anisotropy*. The magnetic field causes the electrons to move in helices rather than in straight lines between collisions. Projection onto the x–y plane yields circles with radii

$$\rho = \frac{c}{\omega} \sin \theta \, , \tag{2.69}$$

where $\omega = (e/m)B$ is the cyclotron frequency of the electron, c its velocity and θ the angle with respect to the z axis. Projection onto the x–z plane yields sinusoidal curves. Figure 2.11a, in contrast to Fig. 2.8, shows how the random propagation of the electron is diminished by the magnetic field. We must repeat our calculation (2.60) with the new orbits.

Looking at Fig. 2.11b we describe the motion of the electron which has suffered a collision at the origin by the orbit

$$x(l) = \rho \left[\sin \left(\frac{\omega l}{c} - \phi \right) + \sin \phi \right] ,$$

$$y(l) = \rho \left[\cos \left(\frac{\omega l}{c} - \phi \right) - \cos \phi \right] , \tag{2.70}$$

$$z(l) = l \cos \theta \, ,$$

where l is the length of the trajectory and ϕ is the starting direction in the x–y plane. More precisely, the derivatives with respect to l at the origin, giving the

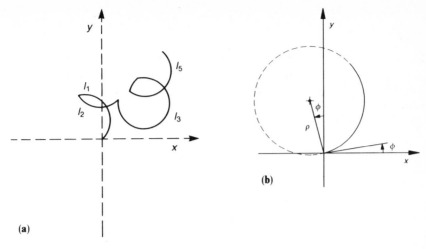

Fig. 2.11. (a) Electron paths causing the magnetic anisotropy of diffusion. **(b)** First part of the trajectory showing the direction projected into the x–y plane

initial direction of the electron, are the following:

$$x'(0) = \sin\theta\cos\phi \;,$$

$$y'(0) = \sin\theta\sin\phi \;,$$

$$z'(0) = \cos\theta \;.$$

The mean square displacement of the electron after the first collision is given by an integration over the solid angle and over the distribution (2.59) of path lengths:

$$\langle x^2 \rangle = \frac{1}{4\pi l_0} \int e^{-l/l_0} x^2(l)\, d\phi\, d\cos\theta\, dl \qquad (2.71)$$

and similarly for $\langle y^2 \rangle$ and $\langle z^2 \rangle$. It is not difficult to integrate (2.71), using (2.70) and (2.69). The result is

$$\langle x^2 \rangle = \langle y^2 \rangle = \frac{2}{3}\frac{l_0^2}{1+\omega^2 l_0^2/c^2} = \frac{2}{3}\frac{l_0^2}{1+\omega^2\tau^2}\;,$$

$$\langle z^2 \rangle = \frac{2}{3}l_0^2 \;.$$

In comparison to (2.60), the magnetic field has caused the diffusion along x and y (then perpendicular to the magnetic field) to be reduced by the factor

$$\frac{D_T(\omega)}{D_T(0)} = \frac{1}{1+\omega^2\tau^2}\;, \qquad (2.72)$$

whereas the longitudinal diffusion is the same as before:

$$D_L(\omega) = D_L(0) \ . \tag{2.73}$$

If there is an E field as well as a B field in the gas, the electric and magnetic anisotropies combine. In the most general case of arbitrary field directions, the diffusion is described by a 3×3 tensor: let the B field be along the z axis of a right-handed coordinate system S, and let the drift direction $\hat{u}$, which is at an angle β with respect to B, have components along $\hat{z}$ and $\hat{x}$. The electric anisotropy is along $\hat{u}$, and the diffusion tensor is diagonal in the system S' which is rotated around $\hat{y}$ by the angle β.

In order to describe the two anisotropies in S, we must transform the diagonal tensor S' to S before we multiply by the diagonal tensor that represents the magnetic anisotropy. If the electric anisotropy is equal to D_L/D_T and the magnetic one is equal to $D(0)/D(\omega) = 1/\eta$, we get for the combined tensor S

$$D_{ik} = \begin{pmatrix} \cos\beta & 0 & \sin\beta \\ 0 & 1 & 0 \\ -\sin\beta & 0 & \cos\beta \end{pmatrix} \begin{pmatrix} D_T & 0 & 0 \\ 0 & D_T & 0 \\ 0 & 0 & D_L \end{pmatrix} \begin{pmatrix} \cos\beta & 0 & -\sin\beta \\ 0 & 1 & 0 \\ \sin\beta & 0 & \cos\beta \end{pmatrix}$$

$$\times \begin{pmatrix} \eta & 0 & 0 \\ 0 & \eta & 0 \\ 0 & 0 & 1 \end{pmatrix},$$

$$D_{ik} = \begin{pmatrix} \eta(D_T\cos^2\beta + D_L\sin^2\beta) & 0 & (D_L - D_T)\sin\beta\cos\beta \\ 0 & \eta\, D_T & 0 \\ \eta(D_L - D_T)\sin\beta\cos\beta & 0 & D_T\sin^2\beta + D_L\cos^2\beta \end{pmatrix}. \tag{2.74}$$

The importance of diffusion for drift chambers is in the limitation for the coordinate measurement. Hence, we are interested in the deviation along a given direction of an electron that has been diffusing for a time t. We treat the general case of a diffusion tensor D_{ij} and a direction $\hat{a}$ given by the three cosine α_k ($\alpha_1^2 + \alpha_2^2 + \alpha_3^2 = 1$), both expressed in the same coordinate system. We make use of the continuity equation (2.58) for the density $n(x_1, x_2, x_3)$, normalized so that $\int n\, dx = 1$. For a time-independent and homogeneous field the drift velocity is constant, and we have

$$\frac{dn}{dt} = D_{ik} \frac{\partial^2 n}{\partial x_i \partial x_k}$$

(summation over identical indices is always understood). The rate of the mean square deviation along $x' = \alpha_i x_i$ is given by

$$\frac{d\langle x'^2 \rangle}{dt} = \int_{-\infty}^{+\infty} (\alpha_i x_i)^2 \frac{dn}{dt}\, dx = \int_{-\infty}^{+\infty} \alpha_i \alpha_j x_i x_j D_{nm} \frac{\partial^2 n}{\partial x_n \partial x_m}\, dx = 2\alpha_i \alpha_j D_{ij}.$$

In the last integral all the 81 terms vanish except when the powers of the x_i match the powers of the derivatives: one shows by two partial integrations that

$$\int_{-\infty}^{+\infty} x_i x_j \frac{\partial^2 n}{\partial x_n \partial x_m}\, \mathrm{d}x = 2\delta_{im}\delta_{jn}\,,$$

where $\delta_{ik} = 1$ if $i = k$, and zero otherwise. We have used the fact that electron density and its derivatives vanish at infinity.

If a point-like ensemble of electrons begins to diffuse at time zero, then after a time t it has grown so that the mean square width of the cloud in the direction $\hat{a}$ has the value

$$\langle x'^2 \rangle = 2\alpha_i \alpha_j D_{ij} t \,. \tag{2.75}$$

The isotropic case in which x'^2 is independent of $\hat{\alpha}$ is obviously given by a D_{ij} which is the unit matrix multiplied with the isotropic diffusion constant. The factor 2 in (2.75) is also present in (2.55) and in the comparison between (2.60) and (2.61). Equation (2.75) implies that the width of the cloud is calculated only from the symmetric part of D_{ij}; furthermore, the diffusion tensor must be positive definite, otherwise our cloud would shrink in some direction – a thermodynamic impossibility.

2.2.7 Electron Attachment

During their drift, electrons may be absorbed in the gas by the formation of negative ions. Whereas the noble gases and most organic molecules can form only stable negative ions at collision energies of several electronvolts (which is higher than the energies reached during the drift in gas chambers), there are some molecules that are capable of attaching electrons at much lower collision energies. Such molecules are sometimes present in the chamber gas as impurities. Among all the elements, the largest electron affinities, i.e. binding energies of an electron to the atom in question, are found with the halogenides (3.1–3.7 eV) and with oxygen (~ 0.5 eV). Therefore we have in mind contaminations due to air, water, and halogen-containing chemicals.

Our account must necessarily be brief; for a thorough discussion of the atomic physics of electron attachment, the reader is referred to Massey et al. [MAS 69]. Modern developments may be followed in the Proceedings of the International Conference on the Physics of Electronic and Atomic Collisions, see especially [CHR 81]. One distinguishes two-body and three-body processes.

The two-body process involves a molecule M that may or may not disintegrate into various constituents A, B, . . . , one of which forms a negative ion:

$$e^- + M \to M^-$$

or

$$e^- + M \to A^- + B + \ldots$$

The break-up of the molecule owing to the attaching electron is quite common and is called *dissociative attachment*. The probability of the molecule staying intact is generally higher at lower electron energies. The rate R of attachment is given by the cross-section σ, the electron velocity c, and the density N of the attaching molecule:

$$R = c\sigma N = kN .\tag{2.76}$$

The rate is proportional to the density. The constant of proportionality is called the two-body attachment rate-constant k.

As examples of strong attachment of slow electrons, we depict in Fig. 2.12 the measured cross-sections of some freons. The rate constants of many other halogen-containing compounds are known [MCC 81]; some of them reach several $10^{-8}\,\mathrm{s}^{-1}\,\mathrm{cm}^3$ at low (thermal) electron energies.

Among the three-body processes, the *Bloch–Bradbury* process is the best known [BLO 35, HER 69]. It plays an important role in the absorption of electrons by oxygen molecules. At energies below 1 eV, the electron is attached to the oxygen molecule that forms the excited and unstable state O_2^{*-}. Unless

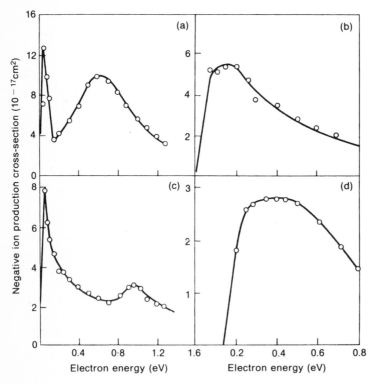

Fig. 2.12a–d. Cross-sections for the production of negative ions by slow electrons in (**a**) CCl_4, (**b**) CCl_2F_2, (**c**) CF_3I and (**d**) BCl_3, observed by Buchel'nikova [BUC 58]

the energy of excitation is carried away by a third body X during the lifetime of the excitation, the O_2^{*-} ion will lose its electron, which is then no longer attached. The lifetime τ is of the order of 10^{-10} s; experimental determinations span the range between 0.02×10^{-10} and 1.0×10^{-10} s, and theoretical calculations come out between 0.88×10^{-10} and 3×10^{-10} s [HAT 81]. The rate R of effective attachment depends on τ, as well as on the two collision rates and the cross-sections for the two processes,

$$\sigma_1: \quad e^- + O_2 \rightarrow O_2^{*-} \,,$$

$$\sigma_2: \quad O_2^{*-} + X \rightarrow O_2^- + X \,.$$

Let us look at a swarm of electrons in a gas that contains O_2 molecules with density $N(O_2)$ and third bodies X with density $N(X)$. Per electron, let the rate dn/dt of formation of the O_2^{*-} state be $1/T_1$; if the rate of spontaneous decay is n/τ and the rate of de-excitation through collision with third bodies n/T_2, then the equilibrium number n per electron of O_2^{*-} states is given by

$$\frac{1}{T_1} = \frac{n}{\tau} + \frac{n}{T_2} \,,$$

$$n = \frac{1}{T_1} \frac{\tau T_2}{\tau + T_2} \,.$$

Therefore, the rate R of effective attachment of one electron is

$$R = \frac{n}{T_2} = \frac{1}{T_1} \frac{\tau}{\tau + T_2} \,.$$

The rates $1/T_1$ and $1/T_2$ may be expressed by the cross-sections a and relative velocities c:

$$\frac{1}{T_1} = c_1 \sigma_1 N(O_2) \,,$$

$$\frac{1}{T_2} = c_2 \sigma_2 N(X) \,.$$

In our application, c_1 is the instantaneous electron velocity and c_2 the velocity of the thermal motion between the molecules O_2 and X.

For ordinary pressure and temperature we have $T_2 \gg \tau$, and therefore the rate is given by

$$R = \tau c_1 c_2 \sigma_1 \sigma_2 N(O_2) N(X) = k_1 N(O_2) N(X) \,. \tag{2.77}$$

It is proportional to the product of the two densities. The three-body attachment coefficient k depends on the electron energy through c_1 and σ_1, on the temperature through c_2, and on the nature of the third body through σ_2. The rate varies linearly with each concentration and quadratically with the total gas density.

In recent years another three-body process has been identified as making an important contribution to the absorption of electrons on O_2 molecules. The O_2 molecules and some other molecules X form *unstable van der Waals molecules*:

$$O_2 + X \leftrightarrow (O_2X) \,.$$

These disintegrate if hit by an electron, which finally remains attached to the O_2 molecule:

$$e^- + (O_2X) \rightarrow O_2^- + X \,.$$

Since the concentration of van der Waals molecules in the gas is proportional to the product $N(O_2)N(X)$, the dependence on pressure and concentration ratios in a first approximation is the same as in the Bloch–Bradbury process.

The relative importance of these two three-body attachment processes has been measured with the help of an oxygen-isotope effect [SHI 84]. The authors find examples where the van der Waal mechanism dominates, whilst the Bloch–Bradbury process is weak; also at higher gas pressures, the van der Waals mechanism makes a strong contribution.

For the operation of drift chambers, the absorption of electrons is a disturbing side effect that can be avoided by using clean gases. Therefore we leave out further refinements that are discussed in the literature.

Some measured three-body attachment coefficients for O_2 are shown in Fig. 2.13 as functions of the electron energy. It can be seen that the H_2O

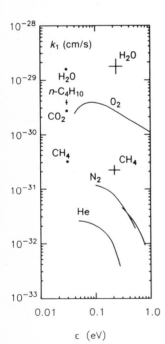

Fig. 2.13. Some three-body attachment coefficients for O_2 quoted from Massey et al. [MAS 69]. The two crosses are measurements by Huk et al. [HUK 88]; for the hydrocarbons at thermal energy, see [SHI 84]

molecule is much more effective as the third body than the more inert atoms or the nitrogen molecule.

The three-body O_2 attachment coefficient involving the methane molecule is large enough to cause electron losses in chambers where methane is a quenching gas. For example (compare Table 10.4), with 20% CH_4 at 8.5 bar, an oxygen contamination of 1 ppm will cause an absorption rate of 2000/s, or roughly 3%/m of drift at a velocity of 6 cm/µs. More information on electron attachment is collected in Chap. 11.

2.3 Results from the Complete Microscopic Theory

In this section we quote without proof results for mobility and diffusion in the framework of the complete microscopic theory. The reader is referred to Huxley and Crompton [HUX 74] and to Allis [ALL 56] for a more detailed discussion.

2.3.1 Distribution Function of Velocities

The principal approximation of Sect. 2.2 was to take a single velocity c to represent the motion between collisions of the drifting electrons. In reality, these velocities are distributed around a mean value according to a distribution function

$$f_0(c)\,dc\ ,$$

which represents the isotropic probability density of finding the electron in the three-dimensional velocity interval $dc_x\,dc_y\,dc_z$ at c. Therefore the distribution function is normalized so that

$$4\pi \int_0^\infty c^2 f_0(c)\,dc = 1\ .$$

The term 'isotropic' actually refers to the first, isotropic, term of an expansion in Legendre functions of the probability distribution for the vector c.

The shape of the distribution depends on the way in which the energy loss and the cross-section vary with the collision velocity. The regime of elastic scattering as well as a good part of the regime of inelastic scattering can be described by two functions: an effective cross-section $\sigma(c)$, and the fractional energy loss $\lambda(c)$ ($\sigma(c)$ is sometimes called the momentum transfer cross-section). Electron drift velocities that are appropriate for drift chambers are achieved by electric fields that are high enough to make the random energy of the electron much higher than the thermal energy of the gas molecules. Therefore we may neglect thermal motion. Absorption or production of electrons (ionization) is equally excluded.

With these assumptions the distribution of the random velocities can be derived in the framework of the Maxwell–Boltzmann transport equations [HUX 74, SCH 86, ALL 56]:

$$f_0(c) = \text{const} \exp\left[-3\int_0^\varepsilon \lambda \frac{\varepsilon \, d\varepsilon}{\varepsilon_l^2} \right] = \text{const} \exp\left[\frac{-3m}{2e^2 E^2} \int_0^\varepsilon \lambda v^2 \, d\varepsilon \right], \qquad (2.78)$$

where $\varepsilon = mc^2/2$, and $\varepsilon_l = eEl_0 = eE/(N\sigma)$ is the energy gained through a mean free path l_0 in the direction of the field. The distribution function can also be written as an integral over the collision frequency $v = cN\sigma$. The second form of (2.78) lends itself to the generalization owing to the presence of a magnetic field. For a discussion of the range of validity of our assumptions, the reader is also referred to the quoted literature.

In order to make a picture of such velocity distributions, we consider two limiting cases.

● If both the collision time τ and the fractional energy loss are independent of c, then $f_0(c)$ is the Maxwell distribution. Using (2.77) and (2.78),

$$f_0(c) \rightarrow \frac{1}{(\alpha\sqrt{\pi})^3} \exp\left[-\left(\frac{c}{\alpha}\right)^2 \right], \qquad (2.79)$$

with

$$\alpha^2 = \frac{4}{3\lambda}\left(\frac{eE}{m}\tau\right)^2 .$$

● If both the mean free path $l_0 = c\tau$ and the fractional energy loss are independent of c – an approximation that is sometimes useful in a limited range of energies – then $f_0(c)$ assumes the form of a *Druyvesteyn distribution* (Fig. 2.14):

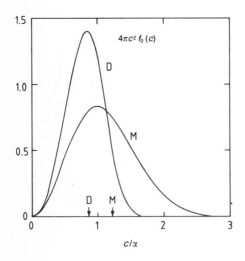

Fig. 2.14. Normalized Maxwell and Dryvesteyn distributions according to (2.79) and (2.80). The probability density is given in units of c/α. The r.m.s. values are indicated

$$f_0(c) \rightarrow \frac{1}{\alpha^3 \pi \Gamma(3/4)} \exp\left[-\left(\frac{c}{\alpha}\right)^4 \right], \tag{2.80}$$

with

$$\alpha^4 = \frac{8}{3\lambda}\left(\frac{eE}{m} l_0\right)^2.$$

Once $f_0(c)$ is known, the mobility and diffusion are calculated from appropriate averages over all velocities c.

2.3.2 Drift

The drift velocity is given by

$$u = \frac{4\pi}{3} \frac{e}{m} \frac{E}{N} \int_0^\infty f_0(c) \frac{d}{dc}\left[\frac{c^2}{\sigma(c)}\right] dc = -\frac{4\pi}{3} \frac{e}{m} E \int_0^\infty \frac{c^3}{v(c)} \frac{d}{dc} f_0(c) dc . \tag{2.81}$$

The two expressions are related to each other by a partial integration and by the fact that the mean collision frequency $v(c)$ is $N\sigma(c)c$. The drift velocity depends on E and N only through the ratio E/N.

Another way of writing (2.81) is to use brackets for denoting the average over the velocity distribution. With a little algebra, we obtain

$$u = \frac{e}{m}\left(\langle \tau \rangle + \left\langle \frac{c}{3} \frac{d\tau}{dc} \right\rangle\right) E . \tag{2.82}$$

This shows that expression (2.14) is recovered not only in the obvious case that τ is independent of c, but already when only the average $\langle c \, d\tau/dc \rangle$ vanishes.

In the case of a constant σ and λ, using the Druyvesteyn distribution (2.80), we may derive the mean square random velocity $\langle c^2 \rangle$ and the square of the drift velocity from (2.81). We find that

$$\langle c^2 \rangle = \frac{eE}{mN\sigma} \sqrt{\frac{2}{\lambda}} \, 0.854 ,$$

$$u^2 = \frac{eE}{mN\sigma} \sqrt{\frac{\lambda}{2}} \, 0.805 .$$

A comparison with (2.19) and (2.20) shows that there are extra factors of 0.854 and 0.805 resulting from the finite width of the velocity distribution.

2.3.3 Inclusion of Magnetic Field

In Sect. 2.2.3 we have seen that the addition of a magnetic field to electrons drifting in an electric field will change their effective random velocity c unless the two fields are parallel, so there has to be a change in the distribution function. Following Allis and Allen [ALL 37] (but see also [MAS 69, HUX 74]), we quote the result for the case where the two fields are orthogonal to each other. The distribution takes the form

$$f_0(c) = \text{const} \exp\left[-\frac{3m}{2e^2 E^2} \int_0^\varepsilon \lambda(\varepsilon) [v^2(\varepsilon) + \omega^2] \, d\varepsilon \right],$$ (2.83)

where the cyclotron frequency $\omega = (e/m)B$ is, of course, independent of ε.

This function depends on two constants of nature e and m, on two functions of the electron energy $\sigma(\varepsilon)$ and $\lambda(\varepsilon)$, and on three numbers (E, B, N) in the hands of the experimenter, of which two (say E/N and B/N) are independent. For arbitrary orientation of the fields, f_0 would depend on the angle between the two fields.

In the presence of a magnetic field, the drift-velocity vector depends on both the electric and magnetic fields. The corresponding mobility tensor is μ_{ik}, so that $u_i = \mu_{ik} E_k$ (summation over common indices is understood). The three diagonal components and the six off-diagonal terms are given by the following expressions:

$$\mu_{ii} = -\frac{4\pi}{3} \frac{e}{m} \int_0^\infty \frac{c^3(v^2 + \omega_i^2)}{v(v^2 + \omega^2)} \frac{d}{dc} f_0(c) \, dc,$$

$$\mu_{ik} = -\frac{4\pi}{3} \frac{e}{m} \int_0^\infty \frac{c^3(\omega_i \omega_k + v\varepsilon_{ikj}\omega_j)}{v(v^2 + \omega^2)} \frac{d}{dc} f_0(c) \, dc.$$ (2.84)

where the fully antisymmetric symbol ε_{ikj} represents a sign-factor that is equal to $+1$ for all even permutations of 123, to -1 for all odd ones, and to zero if any two elements are equal; $\omega_1, \omega_2, \omega_3$ are the components of the magnetic field, multiplied by e/m; $\omega^2 = \omega_1^2 + \omega_2^2 + \omega_3^2$; $f_0(c)$ is the velocity distribution function appropriate for the two fields. As in the case of the scalar mobility, if v is independent of c then one recovers the expression (2.5) derived from the equation of motion with friction. Deviations from this relation can therefore be expected to the extent that v varies with c. See also the discussion in Sect. 2.4.5.

We notice that the number density N of gas molecules is a scaling factor, not only for the electric field but also for the magnetic field. If in (2.84) we divide numerator and denominator by $\omega^2 = (e/m)^2 B^2$, then all terms containing ωs appear either as ω_i/ω or as $v/\omega = N\sigma c/\omega$.

The dependence of the drift-velocity vector on the three scalar quantities E, B, and N may therefore be written in the form

$$u_i = \mu_{ik}\left(\frac{E}{N}, \frac{B}{N}, \phi\right) \frac{E_k}{N},$$ (2.85)

where ϕ is the angle between the fields.

This relation is important when it comes to a discussion of the gas pressure in drift chambers (Chap. 11).

2.3.4 Diffusion

As given in the quoted literature, the isotropic diffusion coefficient is derived from the velocity distribution function through the integral

$$D = 4\pi \int_0^\infty \frac{c^2}{3v(c)} f_0(c) c^2 \, dc \,. \tag{2.86}$$

We note that expression (2.61) represented the case of one unique velocity,

$$4\pi c^2 f_0(c) = \delta(c - \langle c \rangle) \,.$$

The magnetic anisotropy caused by a B field may be described by the 3×3 tensor D_{ik}. Its elements are given by

$$\begin{aligned} D_{ii} &= \frac{4\pi}{3} \int_0^\infty \frac{c^2(v^2 + \omega_i^2)}{v(v^2 + \omega^2)} f_0(c) c^2 \, dc \,, \\ D_{ik} &= \frac{4\pi}{3} \int_0^\infty \frac{c^2(\omega_i \omega_k + v\varepsilon_{ikj}\omega_j)}{v(v^2 + \omega^2)} f_0(c) c^2 \, dc \,. \end{aligned} \tag{2.87}$$

For the electric anisotropy of diffusion, we must retain in the velocity distribution its dependence on position; the reader is referred to the literature [PAR 69].

2.4 Applications

In this section we discuss few practical applications of the formulae derived in the previous sections.

2.4.1 Determination of $\sigma(\varepsilon)$ and $\lambda(\varepsilon)$ from Drift Measurement

In the drift of ions and electrons in gases, the quantities that are important for track localization are the drift velocity and the diffusion tensor. In this capter we have described how they are derived as functions of E and B from the two functions of the collision velocity , $\sigma(c)$ and $\lambda(c)$, the effective scattering cross-section, and the fraction of energy loss per collision. However, these two functions are generally not known from independent measurements, and one goes in the opposite direction: starting from measurements of $u(E)$ and $D(E)/\mu(E)$, one tries to find, with some intuition, $\sigma(c)$ and $\lambda(c)$, in such a way that they reproduce the measurements using the formulae described in Sect. 2.3. The first example using this procedure was applied to electrons in argon by Frost and Phelps in 1964 [FRO 64].

In the recent work of B. Schmidt [SCH 86], considerable progress is visible not only in the accuracy of his measurements – which is near 1% in the drift velocity and 5% in D/μ – but also in the inclusion of the electric anisotropy of diffusion. He studied the noble gases and some molecular gases – chiefly methane – which are useful in drift chambers. In Fig. 2.15 we see the CH_4 drift-velocity measurements, and in Fig. 2.16 the measurements of the transverse and longitudinal diffusion. Superimposed we find curves that represent the recalculated values. The effective (momentum transfer) cross-section of methane and

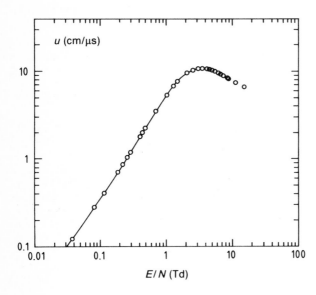

Fig. 2.15. Measured and recalculated electron drift velocity in CH_4 [SCH 86]

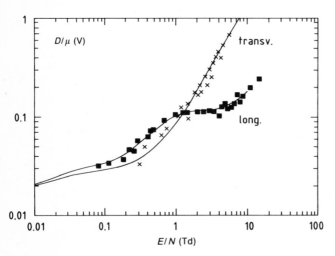

Fig. 2.16. Measured and recalculated transverse and longitudinal electron diffusion in CH_4 [SCH 86]. The ratio of the diffusion coefficients over the mobility is shown as a function of the reduced electric field

the energy loss factor of methane used in their calculation have already been shown in Figs. 2.2a and 2.2b as functions of the electron energy. The consistency of these results leaves no doubt that the method is correct, although the assessment of the errors to $\sigma(\varepsilon)$ and $\lambda(\varepsilon)$ remains difficult. It is possible to give $\lambda(\varepsilon)$ a form that is consistent with plausible assumptions about the quadrupole moments of the methane molecule. Recently, similar computations were done by Biagi [BIA 89]. A special case concerning water contamination is treated in Sect. 11.3.3.

There are microwave relaxation measurements that allow an alternative derivation of the cross-sections of some gases; also, there are quantum-mechanical calculations for the noble gases. Where a comparison of cross-sections can be made, they are identical within factors of 2 to 5, and often much better. According to the calculations, the minimum is sharper than that derived from the drift measurement because the wide energy distribution smears it out. For more details, the reader is referred to the literature.

There is a very large number of measurements of electron drift velocities and diffusion coefficients. Many of them are collected in two compilations [FEH 83] and [PEI 84], which also contain other material that is of interest with respect to drift chambers.

Measurements in magnetic fields are rare; we present some examples in Sect. 2.4.7. Precision reproductions of magnetic data, and determinations of $\sigma(\varepsilon)$ and $\lambda(\varepsilon)$ from them, have not yet been done.

2.4.2 Example: Argon–Methane Mixture

In this subsection we want to collect and interpret some drift and diffusion measurements for argon + methane mixtures without magnetic field. This gas has been investigated more systematically than most others, and is being used in several large TPCs.

The measurements, done by Jean-Marie et al. [JEA 77], of the drift velocity as a function of the electric field E at atmospheric pressure are presented in Fig. 2.17. For all mixing ratios R there is a rising part at low E and a falling part at high E. The maxima occur at values E_p depend almost linearly on the CH_4 concentration, up to $E_p = 900$ V/cm for pure CH_4. For every concentration, the maximum occurs at the electric-field value where the average electron energy is at the Ramsauer minimum of the momentum transfer cross-section ($e = 0.3$ eV for CH_4 and 0.25 eV for Ar). According to (2.19), the peak behaviour of u^2 is determined mostly by the $\sigma(\varepsilon)$ in the denominator because the factor $\sqrt{\lambda(\varepsilon)}$ varies much less.

We can actually measure the values of ε for every E and R, and check that the Ramsauer minimum coincides with the velocity peak. For this we make use of the systematic measurements of the width σ_x of the single-electron diffusion conducted by the Berkeley group [PEP 76]. Their results have been rescaled to 1 cm drift length at 760 Torr, and are plotted in Fig. 2.18. They exhibit a fairly

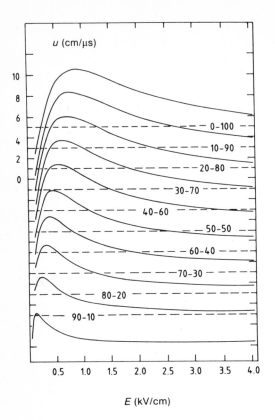

Fig. 2.17. Drift velocity u in an argon–methane mixture as a function of the electric field E, measured by Jean-Marie et al. [JEA 77]. The lowest curve is for 90% Ar + 10% CH_4 etc.

constant behaviour of the diffusion when E is changed, and an increase for smaller CH_4 concentrations.

Referring to our equation (2.63), we may calculate the electron energy for every E and every mixing ratio R:

$$\varepsilon(E, R) = \frac{3}{4} \frac{\sigma_x^2(E, R)}{1 \text{ cm}} eE , \tag{2.88}$$

This has been graphed in Fig. 2.19, where we have also marked on the horizontal scale the velocity peaks for the various values of R. We observe that the measured electron energies are in the Ramsauer minimum at the velocity peaks (see Fig. 2.2a).

2.4.3 Experimental Check of the Universal Drift Velocity for Large $\omega\tau$

In a drift experiment with crossed electric and magnetic fields, Merck [MER 89] measured drift angles and drift velocities at atmospheric pressure in Ar + CH_4

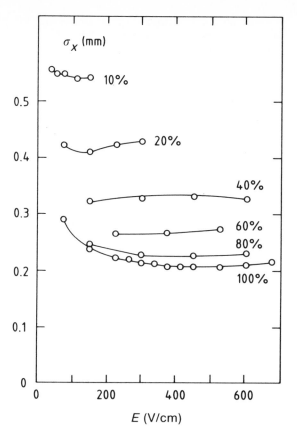

Fig. 2.18. Width of the diffusion cloud after a drift of 1 cm for various mixtures of CH_4 and Ar at 760 Torr. The upper group of points refer to 10% CH_4 + 90% Ar, etc. Measurements by the Berkeley TPC group [PEP 76] were done at 600 Torr over a drift length of 15 cm

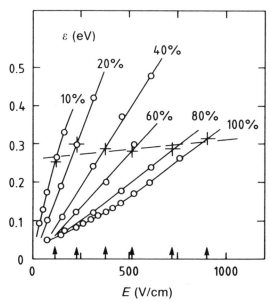

Fig. 2.19. Electron energy defined in (2.63) as a function of the electric field for various mixtures of CH_4 + Ar at 760 Torr. The data of Fig. 2.18 were used. The arrows on the horizontal scale indicate the position of the maxima of the drift velocities according to Fig. 2.17

mixtures that are known to have large values of τ under typical drift-chamber conditions. Using the tangent of the drift angle as measure for $\omega\tau$ (2.11), he found the drift velocities that are plotted as a function of $\omega\tau$ in Fig. 2.20. We observe that they do approach the limit

$$u \to \frac{E}{B}$$

quickly from below. However, the approximation (2.54) is not very good, as

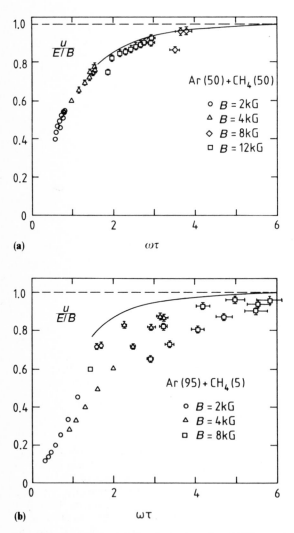

(a)

(b)

Fig. 2.20a, b. Measured drift velocity in units of E/B as a function of the tangent of the observed drift angle, for various magnetic fields. (a) Ar (50%) + CH₄ (50%), (b) Ar (95%) + CH₄ (5%) [MER 89]

different combinations of E and B that lead to the same $\omega\tau$ do not yield the same u, especially at low concentrations of CH_4. We would expect that the more detailed theory involving the electron energy distribution is required in order to explain these variations.

2.4.4 A Measurement of Track Displacement as a Function of Magnetic Field

In a TPC with almost perfect parallelism between B and E, the small angle a of misalignment causes a displacement of every track element that is proportional to the drift length L. If we imagine the coordinate system aligned with the E field along z and with small extra B components along x and y, there will be displacements at the wire plane, equal to $\delta_x = Lu_x/u_z$ and $\delta_y = Lu_y/u_z$; according to (2.6), these are

$$\delta_x = L\left(\alpha_x \frac{\omega^2\tau^2}{1+\omega^2\tau^2} - \alpha_y \frac{\omega\tau}{1+\omega^2\tau^2} \right),$$

$$\delta_y = L\left(\alpha_x \frac{\omega\tau}{1+\omega^2\tau^2} + \alpha_y \frac{\omega^2\tau^2}{1+\omega^2\tau^2} \right),$$

(2.89)

with $\alpha_x = B_x/B_z$ and $\alpha_y = B_y/B_z$. It is by reversing the magnetic field that the different contributions can be separated. The symmetric part of δ_x is

$$\Delta_x^S = \frac{1}{2}[\delta_x(B) + \delta_x(-B)] = L\alpha_x \frac{\omega^2\tau^2}{1+\omega^2\tau^2},$$

(2.90)

and its antisymmetric part is

$$\Delta_x^A = \frac{1}{2}[\delta_x(B) - \delta_x(-B)] = L\alpha_y \frac{-\omega\tau}{1+\omega^2\tau^2}.$$

(2.91)

In Fig. 2.21 we see measured displacements [AME 85] for two different values of L and a range of values of B. Fitting the symmetric part in Fig. 2.21 with the functional form of (2.90) yielded $\tau = 0.29 \times 10^{-10}$ s, and the antisymmetric part fitted with (2.91) yielded $\tau = 0.35 \times 10^{-10}$ s, for the particular conditions of the experiment (argon (91%) + methane (9%), 1 bar, $E = 110$ V/cm, $u = 5.05$ cm/μs). These values of τ are not the same because they represent averages over the distribution functions, which are written explicitly in (2.84).

2.4.5 A Measurement of the Magnetic Anisotropy of Diffusion

In a TPC with parallel E and B fields, the transverse width σ_{diff} of the electron cloud originating from a laser track was measured for various drift distances L and magnetic fields B [AME 86]. Since this width is related to the diffusion

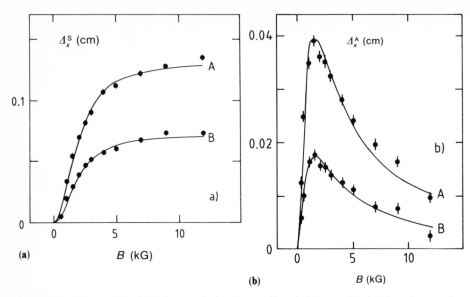

Fig. 2.21a, b. Measured track displacements due to a small angle between the electric and magnetic fields; (a) symmetric part, (b) antisymmetric part. The curves are fits to (2.90) and (2.91). Curves A: L = 126 cm; curves B: $L = 65$ cm [AME 85]

constant D and the known drift velocity u by

$$\sigma_{\text{diff}}^2(L, B) = 2D(B)L/u ,$$

we determined $D(B)$. The quantity actually observed, σ_1^2, was the quadratic sum of σ_{diff} and of a constant, σ_0, given by the width of the laser beam and the electrode configuration (see Sect. 5.2.1 for signal formation at the pads). Figure 2.22 depicts the observations.

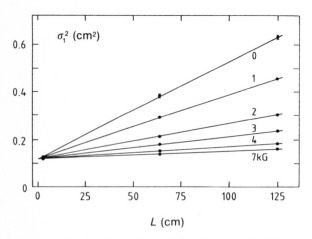

Fig. 2.22. Measured dependence of σ_1^2 on the magnetic field and the drift length. The slope decreases with increasing B as the diffusion is reduced [AME 86]

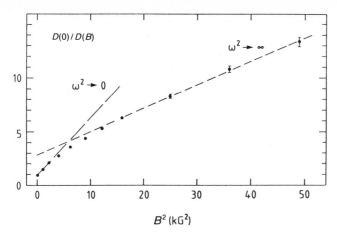

Fig. 2.23. The factor by which the magnetic field reduces the transverse diffusion in the example elaborated in the text. It is a linear function of B^2 for both small and large B [AME 86]

If we express the factor $D(0)/D(B)$ by which the magnetic field reduces the diffusion constant as a function of B^2, we see a linear behaviour for small B^2 as well as for large B^2. The transition region between the two straight lines is around $6\,\mathrm{kG}^2$ (see Fig. 2.23). The interpretation is the following.

Referring to (2.84) and (2.87) we note that the off-diagonal elements of the diffusion tensor do not contribute to σ_{diff}. We write

$$D_{11}(\omega) = \frac{4\pi}{3} \int_0^\infty \frac{c^2 v}{v^2 + \omega^2} f_0(c) c^2 \, dc = \left\langle \frac{c^2 v}{3(v^2 + \omega^2)} \right\rangle .$$

This expression can be evaluated in the two limiting cases of ω^2 small and ω^2 large compared with v_0^2, the square of a typical collision frequency v_0. For this typical frequency we may take the value derived from the electron mobility μ in the limit in which v is independent of c: $v_0 = (e/m)/\mu$. We find in lowest order

$$\frac{D_{11}(0)}{D_{11}(B)} = 1 + \omega^2 \tau_1^2, \quad \omega^2 \ll v_0^2,$$

$$\frac{D_{11}(0)}{D_{11}(B)} = C + \omega^2 \tau_2^2, \quad \omega^2 \gg v_0^2.$$

(2.92)

The values of

$$\tau_1^2 = \frac{\langle c^2/v^3 \rangle}{\langle c^2/v \rangle},$$

$$\tau_2^2 = \frac{\langle c^2/v \rangle}{\langle c^2 v \rangle},$$

$$C = \frac{\langle c^2/v \rangle \langle c^2 v^3 \rangle}{\langle c^2 v \rangle^2} \tag{2.93}$$

are independent of ω, because the two fields are parallel to each other.

Since v is a function of c and is therefore distributed over a wide range of values, the averages in (2.93) are all different. In fact the straight lines in Fig. 2.23 are described by $\tau_1 = (0.4 \pm 0.02) \times 10^{-10}$ s, $\tau_2 = (0.266 \pm 0.006) \times 10^{-10}$ s, and $C = 2.8 \pm 0.2$ for the particular conditions investigated, which were argon (91%) + methane (9%) at 1 bar, $E = 110$ V/cm, and $u = 5.05$ cm/µs, implying $v_0 = 4.2 \times 10^{10}$ s^{-1}.

The two straight lines cross each other at $\omega = (3.6 \pm 0.3) \times 10^{10}$ s^{-1}, where ω/v_0 is nearly 1. If v were independent of c, or if the distribution of c were extremely narrow, we would have had $\tau_1 = \tau_2$ and $C = 1$.

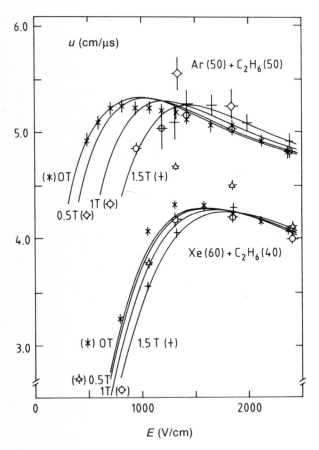

Fig. 2.24. Magnitude of the drift velocity in crossed electric and magnetic fields for two different gas mixtures, plotted as a function of the electric field (at atmospheric pressure) for four different values of B. The curves represent calculations by Ramanantsizhena [RAM 79], the points measurements by Daum et al. [DAU 78]

2.4.6   Calculated and Measured Electron Drift Velocities in Crossed Electric and Magnetic Fields

Drift chambers that have to operate in a magnetic field B often have their proportional wires parallel to B, hence their electrons drifting in crossed electric and magnetic fields.

Daum et al. [DAU 78] have measured the drift-velocity vector for a variety of suitable gas mixtures as functions of both fields. Typical velocities $|u|$ are shown in Fig. 2.24, and typical drift angles ψ, measured against the direction of E towards $E \times B$, are shown in Fig. 2.25. We have selected two examples of gas mixtures, argon + ethane, and xenon + ethane, which exhibit different mobilities and, therefore, different values of $\omega\tau$ at a given magnetic field.

These measurements have been compared with calculations based on the formalism explained in Sect. 2.3, by Ramanantsizhena [RAM 79]. He used elastic cross-sections for the e–C_2H_6 collision, which are based on the work of

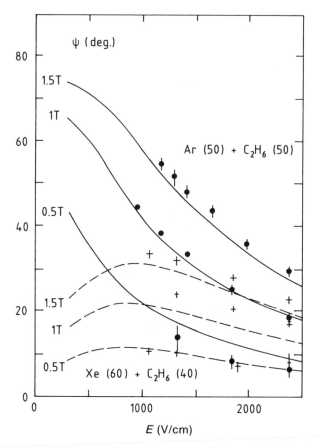

Fig. 2.25. Drift angle in crossed electric and magnetic fields for the same conditions as in Fig. 2.24

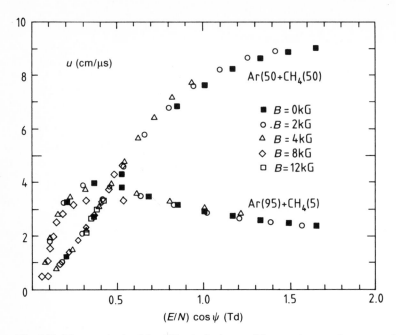

Fig. 2.26. The magnitude of the drift velocity in two different mixtures of argon–methane, measured at various magnetic fields by Merck [MER 89], and plotted as a function of the electric field component along the drift direction. The measured points fall essentially on universal curves that are independent of the magnetic field

Duncan and Walker [DUN 74], and inelastic cross-sections derived from Palladino and Sadoulet [PAL 75]. Comparing the measurements with the calculations, we notice that agreement exists, mostly at the level of 5 to 10%, occasionally worse. In addition we observe the small variation of the calculated maximum of the drift velocity as the magnetic field is changed: according to Tonks' theorem it should stay the same, but with a realistic distribution function it decreases by 1 or 2% for the gas mixtures of Fig. 2.24 as B goes up from 0 to 1.5 T.

In Fig. 2.26 we see drift velocities measured by Merck [MER 89] in crossed electric and magnetic fields as functions of the electric field component along the drift direction. We note that the points belonging to different B fall on the same line characteristic of the gas, within the measurement error, which is 2 or 3% (only for the small CH_4 concentration is there a 12% deviation near the peak). Obviously Tonks' theorem, (2.13) and (2.51), is quite well confirmed.

References

[ALL 31] W.P. Allis and P.M. Morse, Theorie der Streung langsamer Elektronen an Atomen, *Z. Phys.* **70**, 567 (1931)

[ALL 37] W.P. Allis and H.W. Allen, Theory of the Townsend method of measuring electron diffusion and mobility, *Phys. Rev.* **52**, 703 (1937)

[ALL 56] W.P. Allis, Motions of ions and electrons, in *Handbuch der Physik*, vol. **XXI**, ed. by S. Flügge, (Springer, Berlin Heidelberg, 1956), p. 383

[AME 85] S.R. Amendolia et al., Calibration of field inhomogeneities in a time projection chamber with laser rays, *Nucl. Instrum. Methods Phys. Res.* **A235**, 296 (1985)

[AME 86] S.R. Amendolia et al., Dependence of the transverse diffusion of drifting electrons on magnetic field, *Nucl. Instrum. Methods Phys. Res.* **A244**, 516 (1986)

[ATR 77] V.M. Atrazhev and I.T. Iakubov, The electron drift velocity in dense gases, *J. Phys.* **D10**, 2155 (1977)

[BIA 89] S.F. Biagi, A multiform Boltzmann analysis of drift velocity, diffusion, gain and magnetic effects in argon–methane–water-vapour mixtures, *Nucl. Instr. Meth.* **A283**, 716 (1989)

[BLO 35] F. Bloch and N.E. Bradbury, On the mechanism of unimolecular electron capture, *Phys. Rev.* **48**, 689 (1935)

[BRA 81] G.L. Braglia and V. Dallacasa, Multiple scattering theory of electron mobility in dense gases, in [CHR 81], p. 83

[BRO 66] S.C. Brown, *Introduction to Electrical Discharges in Gases* (Wiley, New York 1966)

[BUC 58] I.S. Buchel'nikova, Cross sections for the capture of slow electrons by O_2 and H_2O molecules and molecules of halogen compounds, *Zh. Eksper. Teor. Fiz.* **35**, 1119 (1958); *Sov. Phys.-JETP* **35** (8), 783 (1959)

[CHR 71] L.G. Christophorou, *Atomic and Molecular Radiation Physics* (Wiley, London 1971)

[CHR 81] L.G. Christophorou (ed.), *Proc. Second International Swarm Seminar, Oak Ridge, USA, 1981* (Pergamon, New York 1981)

[DAU 78] H. Daum et al., Measurements of electron drift velocities as a function of electric and magnetic fields in several gas mixtures, *Nucl. Instrum. Methods* **152**, 541 (1978)

[DUN 74] C.W. Duncan and I.C. Walker, Collision cross-sections for low-energy electrons in some simple hydro-carbons, *J. Chem. Soc. Faraday Trans.* **II.70**, 577 (1974)

[FEH 83] J. Fehlmann and G. Viertel, Compilation of data for drift chamber operation, IHP Detector Group Note, ETH Zurich, 1983, unpublished

[FRO 64] L.S. Frost and A.V. Phelps, Momentum transfer cross-sections for slow electrons in He, Ar, Kr and Xe from transport coefficients, *Phys. Rev. A* **136**, 1538 (1964)

[HAT 81] Y. Hatano and H. Shimamori, Electron attachment in dense gases, in [CHR 81], p. 103

[HER 69] A. Herzenberg, Attachment of flow electrons to oxygen molecules, *J. Chem. Phys.* **51**, 4942 (1969)

[HUK 88] M. Huk, P. Igo-Kemenes and A. Wagner, Electron attachment to oxygen, water, and methanol, in various drift chamber gas mixtures, *Nucl. Instrum. Methods Phys. Res.* **A 267**, 107 (1988)

[HUX 74] L.G.H. Huxley and R.W. Crompton, The diffusion and drift of electrons in gases (Wiley, New York 1974)

[JEA 77] B. Jean-Marie, V. Lepeltier and D. L'Hote, Systematic measurement of electron drift velocity and study of some properties of four gas mixtures: Ar–CH_4, Ar–C_2H_4, Ar–C_2H_6, Ar–C_3H_8, *Nucl. Instrum. Methods* **159**, 213 (1979)

[LAN 57]   Landolt-Börnstein, vol. IV/3, *Eigenschaften des Plasmas*, no. 44315, 6th edn. (Springer, Berlin Heidelberg 1957)

[MAS 69]   H.S.W. Massey, E.H.S. Burhop and H.B. Gilbody, *Electronic and Ionic Impact Phenomena*, 2nd edn. (Clarendon, Oxford 1969), 4 vols.

[MCC 81]   D.L. McCorkle, L.G. Christophorou and S.R. Hunter, Electron attachment rate constants and cross-sections for halocarbons, in [CHR 81], p. 21

[MCD 73]   E.W. McDaniel and E.A. Mason, *The Mobility and Diffusion of Ions in Gases* (Wiley, New York 1973)

[MER 89]   M. Merck, Messung des Driftgeschwindigkeitsvectors von Elektronen in Gasen im gekreuzten elektrischen und magnetischen Feld, Diploma Thesis, Munich University, 1989, unpublished

[PAL 75]   V. Palladino and B. Sadoulet, Application of classical theory of electrons in gases to drift proportional chambers, *Nucl. Instrum. Methods* **128**, 323 (1975)

[PAR 69]   J.H. Parker, Jr. and J.J. Lowke, Theory of electron diffusion parallel to electric fields, *Phys. Rev.* **181** (1969): I—Theory, p. 290; II—Application to real gases, p. 302

[PEI 84]   A. Peisert and F. Sauli, Drift and diffusion of electrons in gases: a compilation, CERN report 84-08 (1984)

[PEP 76]   Johns Hopkins University, LBL, UC Riverside, and Yale University (PEP Expt. No. 4), Proposal for a PEP facility based on the time projection chamber (TPC), Stanford report SLAC-Pub-5012 (1976)

[RAM 21]   C. Ramsauer, Über den Wirkungsquerschnitt der Gasmoleküle gegenüber langsamen Elektronen, *Ann. Phys.* **66**, 546 (1921)

[RAM 79]   P. Ramanantsizhena, Vitesse de migration des électrons et autres coefficients de transport dans les mélanges argon-éthane et xenon-éthane. Influence d'une induction magnétique, Thèse, Univ. Louis-Pasteur, Strasbourg, 1979, unpublished

[SCH 76]   G. Schultz, Etude d'un detecteur de particules a très haute précision spatiale (chambre a drift), Thèse, Univ. Louis-Pasteur, Strasbourg, 1976, unpublished

[SCH 77]   G. Schultz, G. Charpak and F. Sauli, Mobilities of positive ions in some gas mixtures used in proportional and drift chambers, *Rev. Phys. Appl.* (*France*) **12**, 67 (1977)

[SCH 86]   B. Schmidt, Drift and Diffusion von Elektronen in Methan and Methan-Edelgas-Mischungen, Dissertation, Univ. Heidelberg, 1986, unpublished

[SHI 84]   H. Shimamori and H. Hotta, Mechanism of thermal electron attachment to O_2 isotope effect studied with $^{18}O_2$ in rare gases and some hydrocarbons, *J. Chem. Phys.* **81**, 1271 (1984)

[TON 55]   L. Tonks, Particle transport, electric currents and pressure balance in a magnetically immobilized plasma, *Phys. Rev.* **97**, 1443 (1955)

[WAG 67]   E.B. Wagner, F.J. Davies and G.S. Hurst, Time-of-flight investigations of electron transport in some atomic and molecular gases. *J. Chem. Phys.* **47**, 3138 (1967)

3. Electrostatics of Wire Grids and Field Cages

The electric field in a drift chamber must provide two functions: drift and amplification. Whereas in the immediate vicinity of the thin proportional or 'sense' wire the cylindrical electric field provides directly the large field strengths required for charge amplification, the drift field must be created by a suitable arrangement of electrodes that are set at potentials supplied by external voltage sources. It is true that drift fields have also been created by depositing electric charges on insulators – such chambers are described in Sect. 10.4 – but we do not treat them here. Also, charge amplification is not necessarily confined to proportional wires – the parallel-plate chamber is discussed in Chap. 13.

There is a large variety of drift chambers, and they have all different electrode arrangements. An overview of existing chambers is given in Chap. 10, where we distinguish three basic types. In the volume-sensitive chambers (types 2 and 3) the functions of drift and amplification are often more or less well separated, either by special wire grids that separate the drift space from the sense wire or at least by the introduction of 'field' or 'potential' wires between the sense wires. The drift field, which fills a space large compared to the amplification space, then has to be defined at its boundaries; these make up the 'field cage'. For a uniform field, the electrodes at the boundaries are at graded potentials in the field direction and at constant potentials orthogonal to it.

Also the area-sensitive chambers (type 1) have often been built with 'potential' or 'field-shaping' wires to provide a better definition and a separate adjustment of the drift field. In this chapter we want to discuss some elements that are typical for the volume-sensitive chambers with separated drift and amplification spaces: one or several grids with regularly spaced wires in conjunction with a field cage. Although directly applicable to a time projection chamber (TPC), the following considerations will also apply to many type 2 chambers.

Electrostatic problems of the most general electrode configuration are usually solved by numerical methods, for example using relaxation techniques [WEN 58]. Computer codes exist for wire chamber applications [VEN 89]. In this chapter we develop analytical methods taking advantage of simple geometry, in order to establish the main concepts. A classic treatment is found in Morse and Feshbach [MOR 53].

3.1   Wire Grids

3.1.1   The Electric Field of an Ideal Grid of Wires Parallel to a Conducting Plane

We assume a reference system with the x–y plane coincident with the conducting plane (pad plane) and the y axis along the direction of the wires of the grid. The z axis is perpendicular to the plane (see Fig. 3.1).

The potential is a function of the coordinates x and z only, because the problem has a translational symmetry along the y direction. We assume the zero of the potential on the conducting plane ($z = 0$). The complex potential of a single line of charge λ per unit length placed at $U' = x' + iz'$ is

$$\phi(U) = -\frac{\lambda}{2\pi\varepsilon_0}\ln\frac{(U - U')}{(U - \bar{U}')}, \tag{3.1}$$

where $U = x + iz$ is the coordinate of a general point and $\bar{U}' = x' - iz'$ is the complex conjugate of U'. MKS units are used throughout.

The potential of the whole grid is obtained by adding up the contributions of each wire:

$$\phi(U) = -\frac{\lambda}{2\pi\varepsilon_0}\sum_{k=-\infty}^{k=+\infty}\ln\frac{(U - U'_k)}{(U - \bar{U}'_k)},$$

where U'_k is the coordinate of the kth wire.

All the wires of the grid are equispaced with a pitch s: therefore

$$U'_k = x_0 + ks + iz_0 \quad (k = \ldots -2, -1, 0, 1, 2, \ldots),$$

where x_0 and z_0 are the coordinates of the 0th wire of the grid. The potential of the grid can be written as

$$\phi(U) = -\frac{\lambda}{2\pi\varepsilon_0}\sum_{k=-\infty}^{k=+\infty}\ln\frac{(U - U'_0 - ks)}{(U - \bar{U}'_0 - ks)}.$$

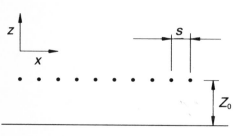

Fig. 3.1. An ideal grid of wires parallel to a conducting plane

This summation can be computed [ABR 65]:

$$\phi(U) = -\frac{\lambda}{2\pi\varepsilon_0}\ln\frac{\sin[(\pi/s)(U - U_0')]}{\sin[(\pi/s)(U - \bar{U}_0')]}$$

and the corresponding real potential is

$$V(x,z) = \mathrm{Re}\,\phi(U)$$

$$= -\frac{\lambda}{4\pi\varepsilon_0}\ln\frac{\sin^2[(\pi/s)(x - x_0)] + \sinh^2[(\pi/s)(z - z_0)]}{\sin^2[(\pi/s)(x - x_0)] + \sinh^2[(\pi/s)(z + z_0)]}. \qquad (3.2)$$

Figure 3.2a shows $V(x, z)$ as function of z at two different values of x. In the chosen example the grid was placed at $z_0 = s$. We notice that such a grid behaves almost everywhere like a simple layer of charge with surface density λ/s. Only in the immediate vicinity of the wires (distances much smaller than the pitch) can the structure of the grid be seen (compare with Fig. 3.2b, which shows the potential of a simple layer of charge). Therefore it is possible to superimpose different grids or plane electrodes at distances larger than the pitch without changing the boundary conditions of the electrostatic problem. We have an example of the general two-dimensional problem where a potential ϕ that varies periodically in one direction x has every Fourier component $\approx \cos(2\pi n x/s)$ damped along the transverse direction z according to the factor $\mathrm{e}^{-2\pi n z/s}$. This

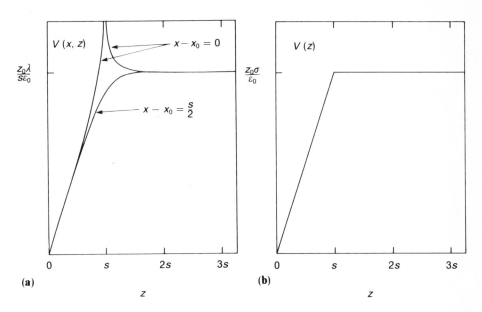

Fig. 3.2. a The potential $V(x, z)$ of a grid of wires situated a distance z_0 from a conducting plane at $x_0, x_0 + s, \ldots$ with pitch $s = z_0$ and a linear charge density λ per wire. **b** The potential $V(z)$ of a plane of charge situated a distance z_0 from a conducting plane, with a surface charge density σ

holds because every Fourier component must satisfy Laplace's equation

$$\frac{\partial^2 \phi}{\partial x^2} + \frac{\partial^2 \phi}{\partial z^2} = 0$$

outside the wires. The solution is

$$\phi = \sum_n C_n e^{-2\pi nz/s} \cos(2\pi nx/s),$$

the C_n being constants.

At a distance from the grid comparable or larger than the pitch the potential assumes the value

$$V(x, z) = \frac{z\lambda}{\varepsilon_0 s} \quad \text{for } z < z_0, z_0 - z \gg \frac{s}{2\pi},$$

$$V(x, z) = \frac{z_0 \lambda}{\varepsilon_0 s} \quad \text{for } z_0 < z, z - z_0 \gg \frac{s}{2\pi}.$$

(3.3)

On the surface of the wires the potential assumes the value evaluated at $(x - x_0)^2 + (z - z_0)^2 = r^2$. We replace in first order of r/s the hyperbolic functions by their arguments and take for $\sinh(2\pi z_0/s)$ the positive exponential. The potential of the wire is then

$$V(\text{wire}) = \frac{\lambda z_0}{\varepsilon_0 s}\left(1 - \frac{s}{2\pi z_0}\ln\frac{2\pi r}{s}\right).$$

(3.4)

This means that the wire grid, although it behaves like a simple sheet with area charge density λ/s has some 'transparency'. We could have given the grid a zero potential and the conducting plane a potential V. Then, beyond the grid the potential would have been

$$\frac{\lambda}{2\pi\varepsilon_0}\ln\frac{2\pi r}{s}$$

and not zero as on the grid – some of the potential of the plane behind can be 'seen through the grid'. In other words, the potential beyond the grid is different from the potential on the grid by the difference between the electric field on the two sides of the grid, multiplied by the length $(s/2\pi)\ln(2\pi r/s)$.

The electric field that can be computed from (3.2) is given here for convenience:

$$E_x(x, z) = \frac{\lambda}{2s\varepsilon_0}\sin\left[\frac{2\pi}{s}(x - x_0)\right]\left[\frac{1}{A_1} - \frac{1}{A_2}\right],$$

$$E_z(x, z) = \frac{\lambda}{2s\varepsilon_0}\left\{\frac{\sinh[(2\pi/s)(z - z_0)]}{A_1} - \frac{\sinh[(2\pi/s)(z + z_0)]}{A_2}\right\},$$

(3.5)

where

$$A_1 = \cosh\left[\frac{2\pi}{s}(z - z_0)\right] - \cos\left[\frac{2\pi}{s}(x - x_0)\right],$$

$$A_2 = \cosh\left[\frac{2\pi}{s}(z + z_0)\right] - \cos\left[\frac{2\pi}{s}(x - x_0)\right].$$

3.1.2 Superposition of the Electric Fields of Several Grids and of a High-Voltage Plane

Knowing the potential created by one plane grid (3.2) and its very simple form (3.3) valid at a distance d ($d \gg s/2\pi$) out of this plane, we want to be able to superimpose several such grids, in order to accommodate not only the sense wires but also all the other electrodes: the near-by field and shielding wires and the distant high-voltage electrode.

In order to be as explicit as possible we will present the specific case of a TPC. These chambers have more grids than most other volume-sensitive drift chambers, and the simpler cases may be derived by removing individual grids from the following computations.

We want to calculate the electric field in a TPC with the geometry sketched in Fig. 3.3. The pad plane is grounded, and we have four independent potentials: the high-voltage plane (V_p), the zero-grid wire (V_z), the field wires (V_f) and the sense wires (V_s). We have to find the relations between those potentials and the charge induced on the different electrodes.

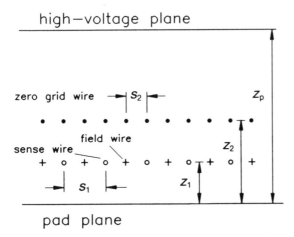

Fig. 3.3. Basic grid geometry of a TPC: The sense-wire/field-wire plane is sandwiched between two grounded planes – the zero-grid-wire plane and the pad plane. The high-voltage plane is a large distance away from them

Since the distance d between the grid planes satisfies the condition $d \gg s/2\pi$, the total electric field is generally obtained by superimposing the solution of each grid. This assumes that all the wires inside a single grid are at the same potential, but it is not the case for the sense-wire and field-wire grids where the sense wires and the field wires are at different potentials and the two grids are at the same z. In this case the superposition is still possible owing to the symmetry of the geometry: the potential induced by the sense-wire grid on the field wires is the same for all the field wires and vice-versa.

The potential induced by the sense-wire grid on the field wires can easily be computed because the field-wire grid and the sense-wire grid have the same pitch. Evaluating formula (3.2) at the position of any field wire we find that

$$V\left(x_0 + \frac{s_1}{2} + ks_1, z_1\right) = \frac{\lambda_s}{2\pi\varepsilon_0} \ln\left[\cosh\frac{2\pi z_1}{s_1}\right] \approx \frac{\lambda_s z_1}{\varepsilon_0 s_1} - \frac{\lambda_s}{2\pi\varepsilon_0}\ln 2 . \qquad (3.6)$$

We can now use (3.3), (3.4) and (3.6) to calculate the potential of each electrode as the sum of the contributions of the charges induced on each grid and on the high-voltage plane:

$$V_s = \frac{\lambda_s z_1}{\varepsilon_0 s_1}\left(1 - \frac{s_1}{2\pi z_1}\ln\frac{2\pi r_s}{s_1}\right) + \frac{\lambda_f z_1}{\varepsilon_0 s_1}\left(1 - \frac{s_1}{2\pi z_1}\ln 2\right) + \frac{\lambda_z z_1}{s_2 \varepsilon_0} + \frac{\sigma_p z_1}{\varepsilon_0} ,$$

$$V_f = \frac{\lambda_s z_1}{\varepsilon_0 s_1}\left(1 - \frac{s_1}{2\pi z_1}\ln 2\right) + \frac{\lambda_f z_1}{\varepsilon_0 s_1}\left(1 - \frac{s_1}{2\pi z_1}\ln\frac{2\pi r_f}{s_1}\right) + \frac{\lambda_z z_1}{s_2 \varepsilon_0} + \frac{\sigma_p z_1}{\varepsilon_0} ,$$

$$V_z = \frac{\lambda_s z_1}{s_1 \varepsilon_0} + \frac{\lambda_f z_1}{s_1 \varepsilon_0} + \frac{\lambda_z z_2}{\varepsilon_0 s_2}\left(1 - \frac{s_2}{2\pi z_2}\ln\frac{2\pi r_z}{s_2}\right) + \frac{\sigma_p z_2}{\varepsilon_0} ,$$

$$V_p = \frac{\lambda_s z_1}{s_1 \varepsilon_0} + \frac{\lambda_f z_1}{s_1 \varepsilon_0} + \frac{\lambda_z z_2}{s_2 \varepsilon_0} + \frac{\sigma_p z_p}{\varepsilon_0} ,$$

$$(3.7)$$

where σ_p is the surface charge density on the high-voltage plane, $\lambda_s, \lambda_f, \lambda_z$ are the charges per unit length on the wires, and r_s, r_f, r_z are the radii of the wires.

We can define a surface charge density σ, for each grid, as the charge per unit length, divided by the pitch (λ_i/s_i), and (3.7) can be written as

$$\begin{pmatrix} V_s \\ V_f \\ V_z \\ V_p \end{pmatrix} = A \begin{pmatrix} \sigma_s \\ \sigma_f \\ \sigma_z \\ \sigma_p \end{pmatrix} , \qquad (3.8)$$

where A is the matrix of the potential coefficients. The matrix A can be inverted to give the capacitance matrix (this is the solution of the electrostatic problem):

$$\begin{pmatrix} \sigma_s \\ \sigma_f \\ \sigma_z \\ \sigma_p \end{pmatrix} = A^{-1} \begin{pmatrix} V_s \\ V_f \\ V_z \\ V_p \end{pmatrix} . \qquad (3.9)$$

Once the capacitance matrix A^{-1} is known we can calculate the charges induced on each electrode for any configuration of the potentials. The electric field is given by the superposition of the drift field with the fields of all the grids, given by (3.5).

In normal operating conditions the electric field in the amplification region is much higher than the drift field. In this case $\lambda_s/s_1 \gg |\sigma_p|$ and the charge induced on the high-voltage plane can be approximated by

$$\sigma_p = \varepsilon_0 \frac{V_p - V_z}{z_p - z_2} .$$

Then the matrix A of (3.8) can be reduced to a 3×3 matrix, neglecting the last row and the last column,

$$A = \frac{1}{\varepsilon_0} \begin{pmatrix} z_1 - \dfrac{s_1}{2\pi} \ln \dfrac{2\pi r_s}{s_1} & z_1 - \dfrac{s_1}{2\pi} \ln 2 & z_1 \\ z_1 - \dfrac{s_1}{2\pi} \ln 2 & z_1 - \dfrac{s_1}{2\pi} \ln \dfrac{2\pi r_f}{s_1} & z_1 \\ z_1 & z_1 & z_2 - \dfrac{s_2}{2\pi} \ln \dfrac{2\pi r_z}{s_2} \end{pmatrix} . \tag{3.10}$$

Tables 3.1 and 3.2 give the coefficients of the matrix A of (3.10) and of its inverse in a standard case:

$$z_1 = 4 \text{ mm}, \; z_2 = 8 \text{ mm}, \; s_1 = 4 \text{ mm}, \; s_2 = 1 \text{ mm} ,$$

$$r_s = 0.01 \text{ mm}, \; r_f = r_z = 0.05 \text{ mm} .$$

Using (3.7) we can now compute the potential of the high-voltage plane when it is uncharged ($\sigma_p = 0$):

$$V_p = V(\infty) = \sigma_s \frac{z_1}{\varepsilon_0} + \sigma_f \frac{z_1}{\varepsilon_0} + \sigma_z \frac{z_2}{\varepsilon_0} . \tag{3.11}$$

The electric field in the drift region is zero.

Table 3.1. Matrix of the potential coefficients (m^2/farad) referring to the grids s, f and z in the standard case

6.64	3.56	4.00
3.56	5.62	4.00
4.00	4.00	8.28

$A = 1.13 \times 10^8$

Table 3.2. Capacitance Matrix (farad/m^2) referring to the grids s, f and z in the standard case

0.25	-0.11	-0.07
-0.11	0.32	-0.10
0.07	-0.10	0.21

$A^{-1} = 8.85 \times 10^{-9}$

In order to produce a drift field E we have to set the high-voltage plane at a potential

$$V_p = -E(z_p - z_2) + V(\infty) .$$ (3.12)

(E is defined positive and is the modulus of the drift field.)

3.1.3 Matching the Potential of the Zero Grid and of the Electrodes of the Field Cage

When we set the potential of the high-voltage plane V_p to the value defined by (3.12), the potential in the drift region ($z - z_2 \gg s_2/2\pi$) is given by

$$V(z) = -E(z - z_2) + V(\infty) .$$ (3.13)

Using equations (3.10) and (3.11) one can show that

$$V_z = V(\infty) - \frac{s_2}{s\pi\varepsilon_0} \sigma_z \ln \frac{\pi r_z}{s_2} ,$$

and (3.13) can be written as

$$V(z) = -E(z - z_2) + V_z + \frac{s_2}{s\pi\varepsilon_0} \sigma_z \ln \frac{2\pi r_z}{s_2} .$$ (3.14)

Figure 3.4 shows $V(z) - V_z$ as function of $z - z_2$ in the geometry discussed in the previous section with $V_s = 1300$ V, $V_g = V_z = 0$ and with a superimposed

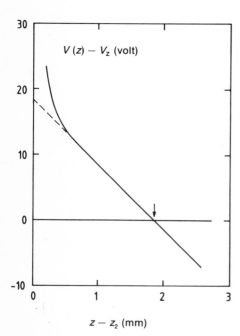

Fig. 3.4. Potential in the region of the zero grid as a function of z in presence of the drift field; example as discussed in the text

drift field of 100 V/cm. We notice that in this configuration the equipotential surface of potential V_z is shifted into the drift region by about 2 mm. This effect has to be taken into account when the potential of the electrodes of the field cage has to be matched with the potential of the zero grid and when one has to set the potential of the gating grid (see Sect. 3.2.2).

After the adjustment of all the potentials on the grids and the high-voltage plane the field configuration is established. We show in Fig. 3.5 the field lines for the typical case of a TPC corresponding to Fig. 3.3 with $z_1 = 4$ mm, $z_2 = 8$ mm, $s_1 = 4$ mm, $s_2 = 1$ mm, $r_s = 0.01$ mm, $r_f = r_z = 0.05$ mm, $V_s = 1300$ V, $V_f = V_z = 0$, $E_p = 100$ V/cm. One observes that the drift region is filled with a very uniform field, but also in the amplification region we find homogeneous domains as expected from (3.3). The electric field lines reach the sense wires from the drift region along narrow paths.

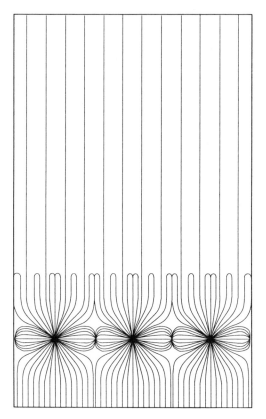

Fig. 3.5. Field lines for the typical case of a TPC with electrodes as in Fig. 3.3 ($z_1 = 4$ mm, $z_2 = 8$ mm, $s_1 = 4$ mm, $s_2 = 1$ mm, $r_s = 0.01$ mm, $r_f = r_z = 0.05$ mm, $V_s = 1300$ V, $V_f = V_z = 0$, $E_p = 100$ V/cm)

3.2 An Ion Gate in the Drift Space

It is possible to control the passage of the electrons from the drift region into the amplification region with a gate, which is an additional grid ('gating grid'), located inside the drift volume in front of the zero grid and close to it. This is important when the drift chamber has to run under conditions of heavy background.

In this section we deal with the principal (electrostatic) function of the ion gate and how it is integrated into the system of all the other electrodes. Chapter 8 is devoted to a discussion of the behaviour of ion gates and their transparency under various operating conditions, including alternating wire potentials, and magnetic fields.

The potential of the wires of the gating grid can be regulated to make the grid opaque or transparent to the drifting electrons. In this section we neglect the effect of the magnetic field and assume that the electrons follow the electric field lines.

In the approximation that the electric field in the amplification region is much higher than the drift field, the variation of the potentials of the gating grid will not change appreciably the charge distribution on the electrodes in the amplification region, and we can schematize the gating grid as a grid of wires placed in between two infinite plane conductors: the high-voltage plane and the zero grid. The geometry is sketched in Fig. 3.6.

3.2.1 Calculation of Transparency

If the electrons follow the electric field lines we can compute the transparency T, i.e. the fraction of field lines that cross the gating grid, from the surface charge on the high-voltage plane and on the gating grid:

$$T = 1 - \frac{\sigma_g^+}{|\sigma_p|}, \tag{3.15}$$

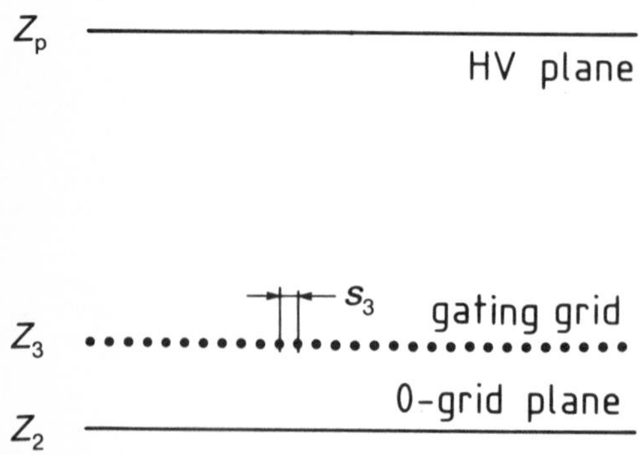

Z_p ———————————————————————
HV plane

Z_3 •••••••••••••••••••• s_3 gating grid
0-grid plane

Z_2 ———————————————————————

Fig. 3.6. Scheme for the inclusion of the gating grid

where σ_p is the surface charge density on the high-voltage plane (negative) and σ_g^+ is the surface charge density of positive charges on the gating grid.

If we approximate the wires of the grid by lines of charge λ per unit length, the surface charge density on the grid is simply λ/s, and the condition of full transparency is

$$\lambda \leq 0 .$$

This approximation is no longer correct when the absolute value of λ is so small that we have to consider the variation of charge density over the surface of the wire. A wire 'floating' in an external electric field E is polarized by the field, which produces a surface charge density σ_D on the wire. It depends on the azimuth (see Fig. 3.7):

$$\sigma_D = 2E\varepsilon_0 \cos \theta , \tag{3.16}$$

where θ is the angle between E and the radius vector from the center to the surface of the wire [PUR 63].

In the general case the wire has, in addition to this polarization charge, a linear charge λ. The total surface charge density on the wire is

$$\sigma_w = \frac{\lambda}{2\pi r} + 2E\varepsilon_0 \cos \theta , \tag{3.17}$$

r being the radius of the wire. An illustration is Fig. 3.8. When

$$\frac{|\lambda|}{4\pi\varepsilon_0} \gg Er$$

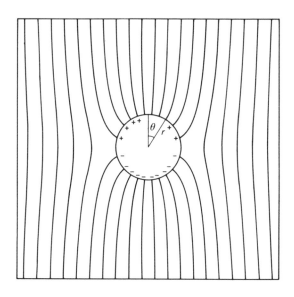

Fig. 3.7. Electric field lines near an uncharged wire floating in a homogeneous field

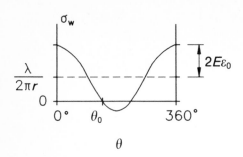

Fig. 3.8. Dependence on azimuth of the surface charge density σ_w located on the gating grid wire

the polarization effect can be neglected. This is the case of the zero-grid, sense and field wires, as discussed in the previous section.

There may be both positive and negative charges on the wires of the gating grid. In order to calculate T we must now count the positive charges only. This can be done using (3.17):

$$\lambda^+ = 0 \qquad\qquad \text{when} \quad \frac{\lambda}{2\pi r} < -2E\varepsilon_0 \;,$$

$$\lambda^+ = \lambda \qquad\qquad \text{when} \quad \frac{\lambda}{2\pi r} > 2E\varepsilon_0 \;,$$

$$\lambda^+ = \frac{\lambda\theta_0}{\pi} + 4E\varepsilon_0 r \sin\theta_0$$
$$\qquad\qquad\qquad\qquad \text{when} \quad -2E\varepsilon_0 < \frac{\lambda}{2\pi r} < 2E\varepsilon_0 \;, \qquad\qquad (3.18)$$
$$\theta_0 = \arccos\frac{-\lambda}{4\pi\varepsilon_0 Er}$$

and the surface charge density of positive charge on the gating grid is

$$\sigma_g^+ = \frac{\lambda^+}{s_3} \;. \qquad\qquad\qquad (3.19)$$

Using (3.15) and (3.19) we obtain the condition for full transparency in the general case:

$$\frac{\lambda_g}{2\pi r_g} \leq -2E\varepsilon_0 \quad \text{or} \quad \frac{\sigma_g}{2\pi r_g} s_3 \leq -2E\varepsilon_0 \;. \qquad\qquad (3.20)$$

In the limiting condition of full transparency, the electric field between the gating grid and the zero-grid increases by a factor

$$1 + 4\pi\frac{r_g}{s_3}$$

with respect to the drift field.

Using (3.18) and (3.19) we can calculate the transparency in the special case of $\sigma_g = 0$:

$$T = 1 - \frac{4r_g}{s_3} \;.$$

The opacity of the gating grid in this configuration is twice the geometrical opacity. To illustrate the situation of limiting transparency the field lines above and below the fully open gate are displayed in Fig. 3.9.

The drift field lines for our standard case are shown in Fig. 3.10. We observe that the drifting electrons arrive on a sense wire on one of four roads through the zero grid – a consequence of the ratio of $s_1/s_3 = 4$.

In order to compute the transparency as a function of the gating-grid potential we have to calculate the capacity matrix. Following the scheme of Sect. 3.1 we obtain

$$\begin{pmatrix} V_p - V_z \\ V_g - V_z \end{pmatrix} = \frac{1}{\varepsilon_0} \begin{pmatrix} z_p - z_2 & z_3 - z_2 \\ z_3 - z_2 & (z_3 - z_2) - \frac{s_3}{2\pi} \ln \frac{2\pi r_g}{s_3} \end{pmatrix} \begin{pmatrix} \sigma_p \\ \sigma_g \end{pmatrix} \tag{3.21}$$

and

$$\begin{pmatrix} \sigma_p \\ \sigma_g \end{pmatrix} = K \begin{pmatrix} z_3 - z_2 - \frac{s_3}{2\pi} \ln \frac{2\pi r_g}{s_3} & -(z_3 - z_2) \\ -(z_3 - z_2) & z_p - z_2 \end{pmatrix} \begin{pmatrix} V_p - V_z \\ V_g - V_z \end{pmatrix}, \tag{3.22}$$

where

$$K = \frac{\varepsilon_0}{(z_p - z_2)\left(z_3 - z_2 - \frac{s_3}{2\pi} \ln \frac{2\pi r_g}{s_3}\right) - (z_3 - z_2)^2}$$

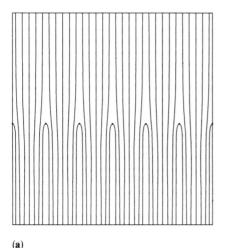
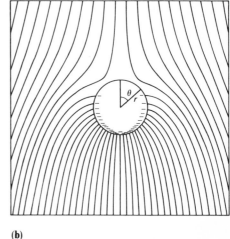

(a)   (b)

Fig. 3.9a, b. Field lines in the limiting case of full transparency. (a) Neighbourhood above and below the gate (the electrons drift from above). (b) Enlargement around the region around one wire

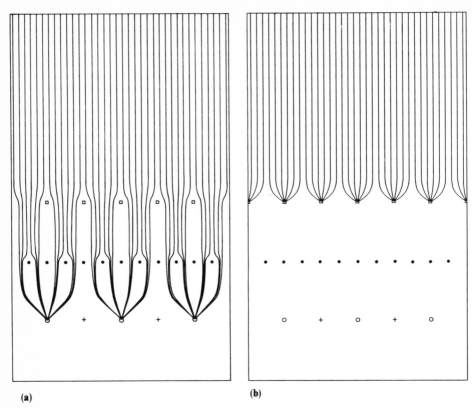

(a)   (b)

Fig. 3.10a, b. Drift field lines in a standard case ($z_1 = 4$ mm, $z_2 = 8$ mm, $z_3 = 12$ mm, $s_1 = 4$ mm, $s_2 = 1$ mm, $s_3 = 2$ mm). The field lines between the sense wires and the other electrodes have been omitted for clariy. *Squares*: gating grid; *black circles*: zero-grid; *open circles*: sense wires; *crosses*: field wires. **(a)** gating grid open (maximal transparency), **(b)** gating grid closed

and V_g is the potential of the gating grid. In the following we neglect the second term in the denominator of the factor K, since $z_p - z_2 \gg z_3 - z_2$ in the standard case.

Using (3.20) and (3.22) we can calculate the minimum value of V_g needed for the full transparency of the gating grid:

$$V_g - V_z \leq \frac{4\pi r_g}{s_3} \frac{V_p - V_z}{z_p - z_2}\left(z_3 - z_2 - \frac{s_3}{2\pi}\ln\frac{2\pi r_g}{s_3}\right) + \frac{z_3 - z_2}{z_p - z_2}(V_p - V_z) . \quad (3.23)$$

The second term of (3.23) is the potential difference that makes $\sigma_g = 0$. The first term is the additional difference needed to eliminate the positive charges from the gating grid. In a standard configuration the two terms are comparable.

In Fig. 3.11 we compare the transparency calculated according to (3.15–3.22) for our standard case with measurements performed on a model of the ALEPH TPC [AME 85-1]. There is good agreement between theory and experiment.

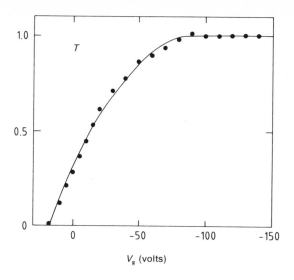

Fig. 3.11. Electron transparency of a grid with pitch 2 mm, as a function of the common potential V_g applied to the wires. The electrode configuration corresponds to Fig. 3.3. The points represent measurements by [AME 85-1]. The line is a function calculated using (3.15–3.23)

3.2.2   Setting of the Gating Grid Potential with Respect to the Zero-Grid Potential

In Sect. 3.3 it was shown that because of the high field in the amplification region the average potential at the position of the zero-grid does not coincide with V_z. The potential in the drift region, where the gating grid has to be placed, is given by (3.14).

This formula can be extrapolated at $z = z_2$:

$$V(z_2) = V_z + \frac{s_2}{2\pi\varepsilon_0} \sigma_z \ln \frac{2\pi r_z}{s_2} , \qquad (3.24)$$

giving the effective potential of the zero-grid seen from the region where the gating grid has to be placed.

In Sect. 3.2.1 we calculated the condition of full transparency, approximating the zero-grid as a solid plane at a potential V_z and referring to it the potential of the gating grid. In the real case V_g has to be referred to the effective potential $V(z_2)$ given by (3.24).

From what has been shown so far one can deduce that by making V_g sufficiently positive all drift field lines terminate on the gating grid and the transparency T is 0. Figure 3.10b shows how the drift field lines terminate on the gating grid.

3.3   Field Cages

The electric field in the drift region has to be as uniform as possible and ideally similar to that of an infinitely large parallel-plate capacitor. The ideal boundary

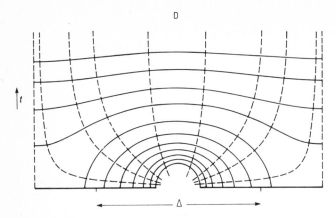

Fig. 3.12. Electric field configuration near the strips of a field cage. Equipotential lines (*broken*) and electric field lines (*full*) are shown. D: drift space; Δ: electrode pitch; t: coordinate in the drift space

condition on the field cage is then a linear potential varying from the potential of the high-voltage membrane to the effective potential of the zero-grid.

This boundary condition can be constructed, in principle, by covering the field cage with a high-resistivity uniform material. A very good approximation can be obtained covering the inner surface of the field cage with a regular set of conducting strips perpendicular to the electric field, with a constant potential difference ΔV between two adjacent strips:

$$\Delta V = E\Delta \ ,$$

where Δ is the pitch of the electrode system.

The exact form of the electric field produced by this system of electrodes can be calculated with conformal mapping taking advantage of the symmetry of the boundary conditions [DUR 64]. Figure 3.12 shows the electric field lines and the equipotentials near the strips in a particular case when the distance between two strips is 1/10 of the strip width. The electric field very near to the strips is not uniform and there are also field lines that go from one strip to the adjacent one. The transverse component essentially decays as $\exp(-2\pi t/\Delta)$ where t is the distance from the field cage (see Sect. 3.1), and when $t = \Delta$ the ratio between the transverse and the main component of the electric field is about 10^{-3}. At larger distances it becomes completely negligible, showing that this geometry of the electrodes is indeed a good approximation of the ideal case. However, the insulators present some problems.

3.3.1 The Difficulty of Free Dielectric Surfaces

The Dirichlet boundary conditions of the electrostatic problem to be solved in some volume require the specification of a potential at every point of a closed boundary surface. When conducting strips are involved that are at increasing potentials, insulators between them are unavoidable, and the potential on these cannot be specified. If the amount of free charge deposited on these surfaces were known everywhere, one would have mixed boundary conditions, partly Dirichlet, partly Neumann, and a solution could be found. In high-voltage field cages there is always some gas ionization, and the amount of free charges ready to deposit on some insulator is infinitely large. How can this uncertainty in the definition of the drift field be limited? There are several solutions to this difficulty:

Controlled (small, surface or volume) conductivity of the insulator: For every rate of deposit of ions there is some value of conductivity allowing the transport of these ions sufficiently fast to the next electrode, so that their disturbing effect is limited.

Retracted insulator surfaces: If the electrodes have the form indicated in Fig. 3.13a or b, any field E that may develop at the bottom of the ditch between conductors owing to charge deposit will be damped by a factor of the order of $\exp(-\pi d/s)$ at the edge of the drift space. (For details of this electrostatic problem, see e.g. [JAC 75, p. 72].) A maximum field E is given by the breakdown strength of the particular gas.

Thin insulator with shielding electrodes covering the gap from behind: Apart from the system of main electrodes, there is another set separated by a thin layer of insulator, and staggered by a half-step according to Fig. 3.14. The purpose of these shielding electrodes at intermediate potentials is to regularize the field in

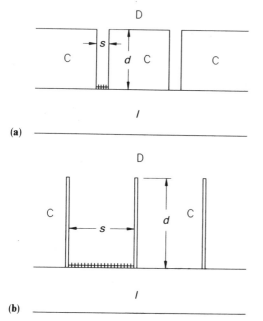

(a)

(b)

Fig. 3.13a, b. Field-cage electrode configuration with retracted insulator surfaces. I: insulator, C: conductor, D: drift space. **(a)** small gap, **(b)** large gap

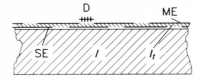

Fig. 3.14. Field-cage electrode configuration with secondary electrode strips (SE) covering the gaps between the main electrode strips (ME) behind a thin insulator foil (It). D: drift space; I: insulator

the drift region, but also to produce virtual mirror charges of any free charges that may have deposited in the gap, thus reducing their effect in the drift space.

If we assume that by one of the measures described above the adverse effects of any charge deposit are sufficiently reduced, we may imagine a smooth (linear) transition of potential between neighbouring electrodes, and the exact form of the electric field produced by the system of electrodes at increasing potentials can be calculated as discussed at the beginning of this section.

Electrostatic distortions were studied by Iwasaki et al. [IWA 83].

3.3.2 Irregularities in the Field Cage

In this subsection we present some studies that have arisen in practice when estimating the tolerances that had to be respected in the ALEPH TPC for the conductivity of the insulator, for the match of the gating grid and for the resistors in the potential divider. We quote the three relevant electrostatic solutions as examples of similar problems in chambers with different geometry.

If all the electrodes of the field cage are set at the correct potential, the drift field is uniform and parallel to the axis of the TPC in the whole drift volume. A wrong setting of the potentials produces a transverse component of the electric field and causes a distortion of the trajectories of the drifting electrons. In order to study this effect in detail the drift volume of the TPC is schematized as a cylinder of length L and radius A (see Fig. 3.15). The potentials defined on the surface of the cylinder define the boundary conditions of the electrostatic problem and the drift field inside the volume. If the potentials on the boundary do not vary with ϕ, the electric field can only have a transverse component in the radial direction.

In the absence of magnetic field the electrons drift along the electric field line; an electron placed at the position (r_0, z_0) reaches the end of the drift volume at a radial position

$$r = r_0 + \int_{z_0}^{0} \frac{E_r(r, z)}{E_z} \, dz \ .$$

The radial component of the field can be calculated solving the Poisson equation in cylindrical coordinates [JAC 75, p. 108]. Since the ideal setting of the potential does not produce radial electric field components it is convenient to use as a boundary condition the difference between the actual and the ideal

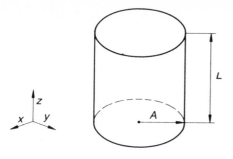

Fig. 3.15. Scheme of a cylindrical TPC drift region

potentials. This is possible because the Poisson equation is linear in the potential. Although we discuss the examples in a cylindrical geometry, they apply to other geometries in a similar way.

Non-Linearity in the Resistor Chain. The potential of the strips of the field cage are defined connecting them to a linear resistor chain. The strips are insulated from the external ground by an insulator that has a finite resistivity and that is in parallel with the resistors of the chain. It can be shown that the resistance to ground of a strip placed at a position z is

$$R(z) = R_{tot} \left[\frac{z}{L} - \frac{1}{6} \left(\frac{z}{L} \right)^3 \frac{R_{tot}}{R_{man}} \right],$$

where R_{tot} is the resistance of the whole resistor chain and R_{man} is the resistance of the insulator mantel across the wall of the insulator. This non-linear resistance produces a potential distribution that differs from the linear one by

$$\Delta V(z) = \left[\frac{z}{L} - \left(\frac{z}{L} \right)^3 \right] V_p \frac{R_{tot}}{R_{man}} \qquad \frac{R_{tot}}{R_{man}} \ll 1,$$

where V_p is the potential of the central electrode. This error potential can be approximated by

$$\Delta V(z) \approx \frac{0.38}{6} V_p \frac{R_{tot}}{R_{man}} \sin \left(\pi \frac{z}{L} \right).$$

Following Jackson's formalism [JAC 75] it can be shown that the radial displacement of an electron drifting from the point (r, z) is

$$\Delta r \approx \frac{0.38}{6} \frac{R_{tot}}{R_{man}} L \frac{i J_1 \left(\dfrac{i\pi r}{L} \right)}{J_0 \left(\dfrac{i\pi A}{L} \right)} \cos \left(\frac{\pi z}{L} - 1 \right),$$

where J_0 are the Bessel functions of order 0 and 1. Figure 3.16 shows a plot of Δr as function of z for different values of r assuming $L = A = 2$ m and $R_{tot}/R_{man} = 10^{-3}$.

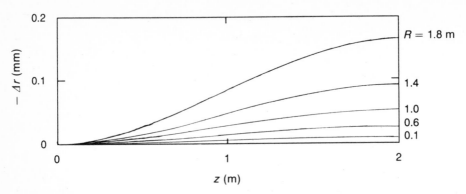

Fig. 3.16. Radial displacement r of the arrival point of an electron, caused by a resistance R_{man} of the insulator mantle a thousand times the value of the resistor chain, Δr is shown as a function of the starting point z, R in the example $A = L = 2\,\text{m}$

Mismatch of the Gating Grid. The potentials of the field cage have to match that of the last wire plane as discussed in Sect. 3.1. An error ΔV in the potential of the last electrode of the field cage produces a radial displacement

$$\Delta r = 2L\frac{\Delta V}{V_p}\sum_n \frac{J_1(x_n r/A)}{J_1(x_n)x_n}\left(\frac{\cosh[\pi(L-z)(x_n/A)]-\cosh(Lx_n/A)}{\sinh(Lx_n/A)}\right),$$

where $x_n = k_n A$ and $J_0(k_n) = 0$.

Figure 3.17 shows a plot of Δr as a function of z for different values of r assuming $L = A = 2\,\text{m}$ and $\Delta V/V_p = 10^{-4}$.

Resistor Chain Containing One Wrong Resistor. If in the voltage-divider chain one of the resistors (placed at $z = \bar{z}$) has a wrong value, it induces an error in the potential of the field cage:

$$\Delta V(z) = \frac{-\Delta V}{L}z\,,\quad z < \bar{z}\,,$$

$$\Delta V\left(1-\frac{z}{L}\right),\qquad z > \bar{z}\,,$$

$$\Delta V = V_p\frac{\Delta R}{R_{tot}}\,,$$

where ΔR is the error in the resistance of that particular resistor and R_{tot} is the total resistance of the chain. The induced radial displacement is

$$\Delta r = L\frac{\Delta R}{r_{tot}}\left[\frac{1}{\pi}\sum\frac{iJ_1(in\pi r)/L}{J_0(in\pi A)/L}\frac{1}{n}\cos\left(\frac{n\pi\bar{z}}{L}\right)\left(\cos\frac{n\pi z}{L}-1\right)\right].$$

Figure 3.18 shows a plot of Δr as a function of z for different values of r for a particular case $L = A = 2\,\text{m}$, $z = 1.4\,\text{m}$, $\Delta R/R_{tot} = 1/400$.

We have computed in this subsection three specific cases of field-cage problems in the spirit of showing the order of magnitude of the relevant electron

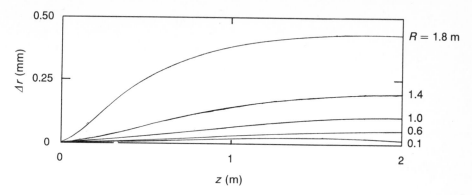

Fig. 3.17. Radial displacement Δr of the arrival point of an electron, caused by a voltage mismatch ΔV between the last electrode of the field cage and the last wire plane, amounting to $\Delta V/V_\mathrm{p} = 10^{-4}$ of the drift voltage. Δr is shown as a function of the starting point (z, R) in the example $A = L = 2$ m

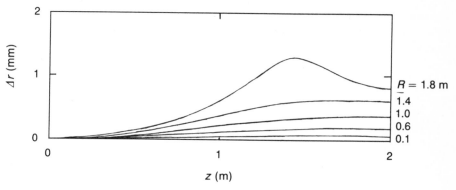

Fig. 3.18. Radial displacement Δr of the arrival point of an electron, caused by one wrong resistor value $R^* = R + \Delta R$, where $\Delta R/R_\mathrm{tot} = 1/400$ of the value of the total resistor chain. Δr is shown as a function of the starting point (z, R) in the example $A = L = 2$ m

displacements. The design of the drift chamber must be such that the displacement Δr induced by such irregularities remains small in comparison to the required accuracy.

3.4 Wire Displacements Due to Gravitational and Electrostatic Forces

Throughout this chapter we have assumed that the wires in the grids are placed in their ideal position and that they are perfectly strung.

In reality the wires are subject to gravitational and electrostatic forces that tend to displace them. Since they are glued on the frame, the combined effect of these forces and of the restoring mechanical tension results in an elastic deformation of the wire.

If T is the mechanical tension of the wire, the restoring force per unit length R in the direction perpendicular to the wire is

$$R = T\frac{d^2x}{dy^2}, \tag{3.25}$$

where $x(y)$ is the displacement perpendicular to the wire and y is the coordinate along the wire direction.

The gravitational force per unit length acting on the wire is $F = g\rho\sigma$, where ρ is the density of the wire and σ its cross section.

The electrostatic force acting on the wire is $F = \lambda E$, where λ is the charge per unit length and E is the electric field acting on the wire which is generated by the other electrodes. If this field is constant it creates a constant force per unit length, similar to the gravitational one. This is the case of the wires of the zero-grid of the TPC discussed in Sect. 3.1. If the wire is placed in a position of electrostatic equilibrium (this is the case of the sense wires of the TPC), a small deviation from this position generates a force acting on the wire. This force in general tends to displace the wire from its equilibrium position and has the form

$$F = \frac{V^2}{2}\frac{dC}{dx},$$

where V is the potential of the wire and dC/dx is the variation of its capacitance per unit length due to the displacement. It can be shown that

$$F \simeq \frac{V^2}{2}4\pi\varepsilon_0\frac{1}{[a\ln(a/r)]^2}x = kx, \tag{3.26}$$

where r is the radius of the wire and a the typical distance of the wire from the other electrodes.

Using (3.25) we obtain the equation for the equilibrium position of the wire:

$$T\frac{d^2x}{dy^2} + kx + g\rho\sigma = 0. \tag{3.27}$$

If we assume that the wire has been glued in its ideal position, the boundary conditions of this equation are $x(0) = x(L) = 0$, where L is the length of the wire.

3.4.1 Gravitational Force

In order to study the effect of gravity alone we solve (3.27) in the special case $k = 0$:

$$x(y) = \frac{1}{2}\frac{g\rho\sigma}{T}(yL - y^2).$$

Table 3.3. Maximum stresses before deformation of typical wire materials and corresponding sagittas for 1-m-long wires

Material	T_c/σ (kg/mm^2)	Sagitta(μm) of a 100 cm long wire strung at T_c
Al	4... 16	21...84
Cu	21... 37	30...53
Fe	18... 25	39...54
W	180...410	6...13

The sagitta of the wire is

$$x(L/2) = s_g = \frac{L^2 g\rho\sigma}{8T} \tag{3.28}$$

and is proportional to the inverse of the mechanical tension. If the tension is increased the effect is reduced, but the tension cannot be arbitrarily increased since non-elastic deformations take place. The maximum tension T_c that can be applied to a wire is proportional to its cross section, and the ratio T_c/σ is constant. The minimum sagitta of a wire of given length is independent of the wire cross section. The values of the critical tension and typical sagittas for 100-cm-long wires of different materials are shown in Table 3.3. Usually the wires are strung to a tension close to the critical one in order to reduce the sagging. Inspecting Table 3.3 we notice that among the various materials, tungsten is the one that allows the smallest sagittas but at the expense of quite a large tension, for a given diameter of the wire. Since the tension of the wires is held by the endplates of the chamber, a large tension requires stiff endplates. In the design of a chamber one usually compromises between these two parameters.

3.4.2 Electrostatic Forces

The last term of (3.27) can be modified using (3.28):

$$T\frac{d^2x}{dy^2} + kx + \frac{8s_g T}{L^2} = 0 \, .$$

The solution of this equation is

$$x(y) = \frac{8s_g T}{L^2 k}\left(\frac{\cos\sqrt{k/T}(y - L/2)}{\cos\sqrt{k/T}(L/2)} - 1\right) .$$

The sagitta of the wire under the combined effect of electrostatic and gravitational forces is

$$x(L/2) = \frac{8s_g T}{L^2 k}\left(\frac{1}{\cos\sqrt{k/T}(L/2)} - 1\right) = s_g\frac{2}{q^2}\left(\frac{1}{\cos q} - 1\right). \tag{3.29}$$

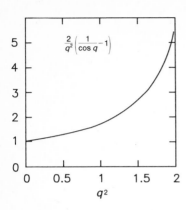

Fig. 3.19. Amplification factor of the gravitational sag, owing to electrostatic forces, according to (3.29)

The electrostatic forces amplify the sagitta produced by the gravitational forces: Fig. 3.19 shows a plot of this amplification factor as a function of q^2. The amplification factor diverges when q approaches $\pi/2$ and the position of wire is no longer stable. Using (3.26) we obtain the stability condition

$$q^2 = 4\pi\varepsilon_0 \frac{1}{[a\ln(a/r)]^2} \frac{V^2}{2T} \frac{L^2}{4} \le 1 \; .$$

Since T cannot exceed the critical tension in order to obtain stable conditions, one has to compromize between the length of the wire and its gain.

It is interesting to solve (3.27) neglecting the gravitational term and using as boundary conditions $x(0) = 0$ and $x(L) = \delta$. This represents the case of an error in the positioning of the wire. The solution is

$$x(y) = \delta \frac{\sin\sqrt{k/T}\,y}{\sin\sqrt{k/T}\,L} \; .$$

Also in this case the positioning error is amplified by the electrostatic forces and we get an unstable condition when $q = \pi/2$.

References

[ABR 65]   M. Abramowitz and I.A. Segun, *Handbook of Mathematical Functions*, (Dover, New York 1965) p. 75

[AME 85-1] S.R. Amendolia et al., Influence of the magnet field on the gating of a time projection chamber, *Nucl. Instrum. Methods A* **234**, 47 (1985)

[DUR 64]   Durand, *Electrostatique* vol. II (Masson et C. edts., Paris 1964)

[IWA 83]   H. Iwasaki, R.J. Maderas, D.R. Nygren, G.T. Przybylski, and R.R. Sauerwein, Studies of electrostatic distortions with a small time projection chamber, in *The Time Projection Chamber* (A.I.P. Conference Proceedings No. 10 TRIUMF, Vancouver, Canada, 1983)

[JAC 75]   J.D. Jackson, *Classical Electrodynamics* (Wiley, New York 1975)

[MOR 53]   P.M. Morse and H. Feshbach, *Methods of Theoretical Physics* (McGraw Hill, New York 1953).

[PUR 63]   E.M. Purcell, *Electricity and Magnetism* (Berkeley Physics Course Vol. II, McGraw Hill, New York 1963) Problem 9.11

[VEN 89]   R. Venhof, *Drift Chamber Simulation Program Garfield*, CERN/DD Garfield Manual (1984).

[WEN 58]   G. Wendt, Statische Felder und stationäre Ströme in *Handbuch der Physik*, Vol. XVI (Springer, Berlin Heidelberg 1958) p. 148

4. Amplification of Ionization

4.1 The Proportional Wire

Among all the amplifiers of the feeble energy deposited by a particle on its passage through the matter, the proportional wire is a particularly simple and well-known example. When it is coupled with a sensitive electronic amplifier, a few – even single – electrons from the ionization of a particle can be observed, which create ionization avalanches in the high electric field near the surface of the thin wire. The development of the proportional counter started as long ago as 1948; for an early review, see [CUR 58]. The proportional counter is also treated in text books on counters in general, for example those of Korff [KOR 55] and Knoll [KNO 79]. The latter contains a basic list of modern reference texts. The classical monograph by Raether [RAE 64] deals with electron avalanches in a broader context.

As an electron drifts towards the wire it travels in an increasing electric field E, which, in the vicinity of the wire at radius r, is given by the linear charge density λ on the wire:

$$E = \frac{\lambda}{2\pi\varepsilon_0} \frac{1}{r} .$$

In the absence of a magnetic field, the path of the electron will be radial. The presence of a magnetic field modifies the path (as discussed in Chaps. 2 and 6) but, if the electric field is strong enough, the trajectory terminates in any case on the wire.

Once the electric field near the electron is strong enough that between collisions with the gas molecules the electron can pick up sufficient energy for ionization, another electron is created and the avalanche starts. At normal gas density the mean free path between two collisions is of the order of microns; hence the field that starts the avalanche is of the order of several 10^4 V/cm, and the wire has to be thin, say a few 10^{-3} cm, for 1 or 2 kV.

As the number of electrons multiplies in successive generations (Fig. 4.1), the avalanche continues to grow until all the electrons are collected on the wire. The avalanche does not, in general, surround the wire but develops preferentially on the approach side of the initiating electrons (cf. Sect. 4.3). The whole process develops in the longitudinal direction over as many free paths as there are generations, say typically over 50 to 100 μm. This part ends after a fraction of

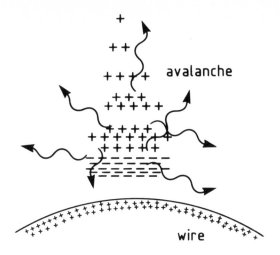

+

+ +

+ + + + +   avalanche

+ + +
+ + + + +

+ + + +
+ + + + + +
+ + + + + +
⁻ ⁻ ⁻ ⁻ ⁻
⁼ ⁼ ⁼ ⁼ ⁼

wire

Fig. 4.1. Schematic development of the proportional wire avalanche

a nanosecond, under normal gas conditions. The physical processes inside the avalanche are quite complicated, as they involve single and multiple ionization, optical and metastable excitations, perhaps recombinations, and energy transfer by collisions between atoms, much the same as discussed in Sect. 1.1. The de-excitation of metastable states may in principle increase the duration of the avalanche up to the collision time between the molecules, but not much is known about this.

The proportional wire owes its name to the fact that the signal is proportional to the number of electrons collected. This proportionality is possible to the extent that the avalanche-induced changes of the electric field remain negligible compared to the field of the wire. Referring to Sect. 4.5.1, we may characterize the regime of proportionality as that in which the charge density in the avalanche is negligible compared with the linear charge density of the wire. The charge density of the avalanche is given by the product of the gain factor and the number of electrons that start the avalanche, divided by its width.

A very interesting role is played by the photons, which are as abundant as electrons because the relevant cross-sections are of the same order of magnitude. A small fraction of them will be energetic enough to ionize the gas, in all probability ionizing a gas component with a low ionization potential. If it now happens that these ionizing photons travel further, on the average, than the longitudinal size of the avalanche, then the electrons they produce will each give rise to another full avalanche and the counter may break down. In order to formulate a stability criterion, we refer to the avalanche of Fig. 4.2, created by one electron e_1. Let us call n_e the number of electrons in the avalanche and n_{ph} the number of energetic photons, and use q to denote the average probability that one of them ionizes the gas outside the radius r_{min}. Breakdown occurs if

$$n_{ph}q > 1 \, , \qquad xn_eq > 1 \, ,$$

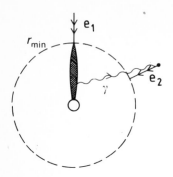

Fig. 4.2. Mechanism of breakdown by photoionization of the gas outside the cylinder that contains the full avalanche

where x is the ratio n_{ph}/n_e. This ratio may be expected to stay roughly constant as n_e is changed by a variation of the wire voltage.

A similar consideration applies also to photons from the avalanche that reach the conducting surfaces of the cathodes, where they may create free electrons by the photoelectric effect.

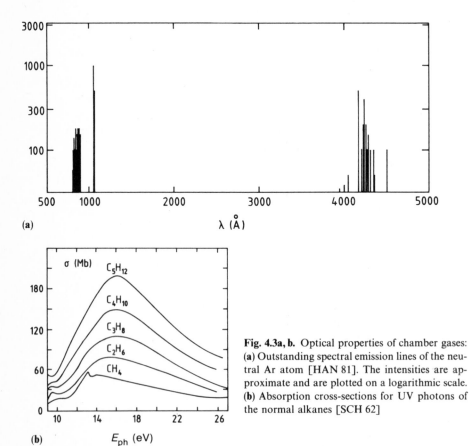

Fig. 4.3a, b. Optical properties of chamber gases: (a) Outstanding spectral emission lines of the neutral Ar atom [HAN 81]. The intensities are approximate and are plotted on a logarithmic scale. (b) Absorption cross-sections for UV photons of the normal alkanes [SCH 62]

Table 4.1. Common quench gases

Methane	CH_4
Ethane	C_2H_6
Propane	C_3H_8
Butane	C_4H_{10}
Pentane	C_5H_{12}
Isobutane	$(CH_3)_2CHCH_3$
Carbon dioxide	CO_2
Ethylene	$(C_2H_2)_2$
Methylal	$CH_2(CH_2OH)_2$

It is because of these far-travelling photons that an organic 'quench gas' needs to be present. Its effect is to reduce q allowing larger values of n_e, and hence larger gain. Organic molecules, with their many degrees of freedom, have large photoabsorption coefficients over a range of wavelengths that is wider than that for noble gas atoms. Some examples are given in Fig. 4.3, which also contains a picture of the most prominent emission lines of the argon. The photoabsorption spectrum of argon was shown as a graph in Fig. 1.3 in the context of a calculation of particle ionization. Table 4.1 contains a list of common quench gases. Inorganic quench gases have been studied by Dwurazny, Jelen, and Rulikowska–Zabrebska [DWU 83].

4.2 Beyond the Proportional Mode

If the avalanche amplification is increased beyond the region of proportionality, the space charge of the positive ions *reduces* appreciably the field near the head of the avalanche, that is the field experienced by the electrons between the wire and the positive cloud. The amplification is smaller for any subsequent increment and we are in the regime of 'limited proportionality'.

Near the tail of the avalanche, however, we have an *increase* of the electric field owing to the positive ions, especially once the fast-moving electrons have disappeared into the wire. This situation leads to two different kinds of multiplicative processes, depending on the behaviour of the photons:

- If the UV absorption of the quench gas is very strong, the photons in the avalanche produce ionization near their creation point, including the region near the tail of the avalanche where the electric field is particularly large. This starts the phenomenon of the 'limited streamer', which is a backward-moving multiplication process in the sense of a series of avalanches whose starting points move further and further away from the wire. In a paper by Atac, Tollestrup, and Potter [ATA 82], it is suggested that energetic photons are created by the recombination process with cool electrons when the field at the avalanche head is sufficiently reduced. Under the influence of the growing charge, the starting points of successive new avalanches move further outwards. Once the streamer has reached a certain length that depends on the wire voltage, typically 1 to 3 mm, it comes to an end. The total charge is almost independent of the amount of charge that originally started the process. Figure 4.4 shows a typical pulse height distribution of the limited streamer for individual electrons (b) and for ∼100 electrons (a) [BAT 85]: they give approximately the same total charge. The self-quench occurs because the electric field becomes weaker as the avalanches are produced further away from the wire. But the exact mechanism is not well understood. More details can be found in the articles by Alekseev et al. [ALE 80], Iarocci [IAR 83] and Atac et al. [ATA 82] and in the references

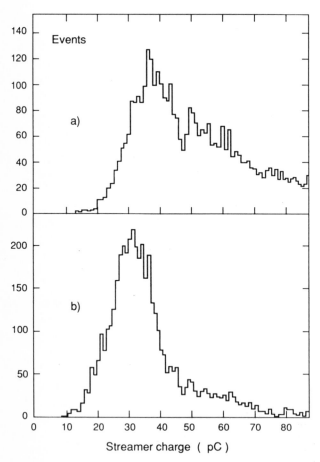

Fig. 4.4a, b. Streamer charge distribution for **(a)** β-rays and **(b)** photoelectrons from [BAT 85]. Aluminium tube $1 \times 1 \, \mathrm{cm}^2$ cell size, 50 μm wire, gas mixture Ar(65%) + Isobutane(65%), wire potential 3.7 kV

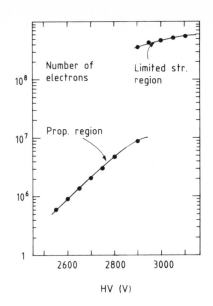

Fig. 4.5. Collected charge as a function of the high voltage, measured by ATAC et al. [ATA 82] on a $100\,\mu m$ diameter wire in a tube $12 \times 12\,mm^2$, filled with $Ar(49.3\%) + C_2H_6(49.3\%) + CH_3CH_2OH(1.4\%)$

quoted therein. A summary and overview has been given by Alexeev, Kruglov, and Khazins [ALE 82].

As the high voltage of a wire tube with the correct quenching gas is increased (Fig. 4.5), the collected charge at first follows a nearly exponential mode (proportional mode), then flattens off as the avalanche charge density approaches the wire charge density (limited proportional mode), and then suddenly jumps to the limited streamer mode where the charge multiplication is one and a half orders of magnitude larger. This behaviour is reported in Fig. 4.5. The collected charge continues to rise more slowly up to the general breakdown of the counter. F.E. Taylor argues that the size of the limited streamer is essentially given by electrostatic considerations [TAY 90].

• If the UV absorption of the gas is so low that ionizing photons can travel distances that are comparable to the counter dimension, then multiplicative processes may develop along the full length of the wire. The wire is said to be in Geiger mode if the external electric circuit terminates the discharge with the help of a large HV feed resistor. The size of the signal is therefore also independent of the original charge. The Geiger counter is described, for example, in [KNO 79] and [KOR 55].

4.3 Lateral Extent of the Avalanche

The electrons from an ionization track, which are collected on a wire and start an avalanche there, do not usually arrive at the same point in space, but there is some spread between them. This happens not only because the track was originally extended in space, and every electron is guided on its drift path to a different place near the wire, but also because of the diffusion in the gas during the drift. The proportional avalanche has a lateral extent that is at least as large as this spread.

In the avalanche process itself there is a lateral development associated with the multiplication of charges, and this is mainly due to diffusion of electrons, the electrostatic repulsion of charges, and the propagation of ionizing photons. The intrinsic lateral size of an avalanche therefore depends on the gas (collision cross-section and UV absorption), on the number of charges in the avalanche and their density and on the electron energy that is obtained in the various parts of the multiplication process. If one wants to know whether the charges go fully around the wire or stay on one side, then the wire diameter also plays an important role.

The spread of avalanches on a wire has been studied experimentally by Okuno et al. [OKU 79], by observing the signals from the positive ions in a segmented cathode tube surrounding the wire. In a mixture of $Ar(90\%) + CH_4(10\%)$, the avalanches, created by a ^{55}Fe source on a wire of 25 μm diameter, occupied only 100° in azimuth (FWHM) in the proportional region (total charge below 10^6 electrons). When the voltage was raised, the avalanche started to surround the anode wire. Figure 4.6 shows how the

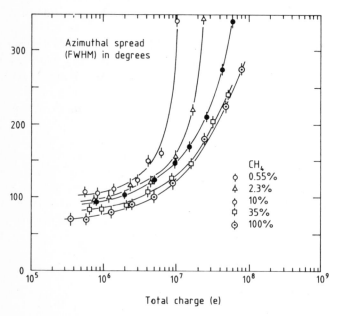

Fig. 4.6. Angular spread (FWHM) of the avalanche from ^{55}Fe X-rays in various Ar + CH_4 mixtures as a function of the total charge in the avalanche. The anode wire had a diameter of 25 μm [OKU 79]

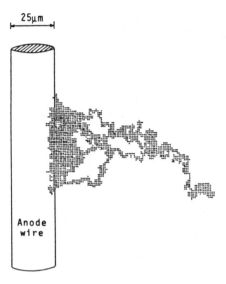

Fig. 4.7. Two-dimensional displays of the electron density in a small avalanche created by a Monte Carlo simulation from a single electron. Photon ionization was neglected [MAT 85]

azimuthal width increased with the total charge of the avalanche. The increase came at smaller total charges when the concentration of the quenching gas was smaller. The influence of the UV photons was more pronounced at higher charge multiplication. The X-ray photon of the ^{55}Fe source has an energy of 5.9 keV and creates approximately 227 electrons. The finite size of the electron cloud produced by the X-ray photon can be neglected.

In a detailed Monte Carlo simulation of the scattering processes involved in the multiplication, Matoba et al. have shown how a small avalanche develops in three dimensions [MAT 85]. In Fig. 4.7 we reproduce a picture of their electron density. In recent years the computational techniques for the simulation of such processes have been considerably advanced; see for example [GRO 89]. We may expect more detailed insight into the dynamics of the avalanche as computer codes are developed that describe not only the various collision phenomena between the electrons, ions and gas molecules but also the important effect of the photons.

4.4 Amplification Factor (Gain) of the Proportional Wire

The multiplication of ionization is described by the first Townsend coefficient α. If multiplication occurs, the increase of the number of electrons per path ds is given by

$$dN = N\alpha\, ds \, . \tag{4.1}$$

The coefficient α is determined by the excitation and ionization cross sections of the electrons that have acquired sufficient energy in the field. It also depends on the various transfer mechanism discussed in Chap. 1. Therefore, no fundamental expression exists for α and it must be measured for every gas mixture. It also depends on the electric field E and increases with the field because the ionization cross-section goes up from threshold as the collision energy ε increases. If the gas density ρ is changed while keeping the distribution of ε fixed (that is, at fixed E/ρ), then α changes proportionally with the density because all the linear dimensions in the gas scale with the mean free collision length (cf. Chap. 2). Therefore we may write the functional dependence of α as

$$\alpha\left(\frac{E}{\rho}, \rho\right) = \frac{\alpha_0}{\rho_0}\rho = f\left(\frac{E}{\rho}\right)\rho \, . \tag{4.2}$$

In Fig. 4.8 we show measurements with pure noble gases in the form of α over the gas pressure as a function of the electric field over the gas pressure, at normal temperature, as obtained by von Engel [ENG 56].

The amplification factor on a wire is given by integrating (4.1) between the point s_{min} where the field is just sufficient to start the avalanche and the wire radius a:

$$N/N_0 = \exp \int_{s_{min}}^{a} \alpha(s)\, ds = \exp \int_{E_{min}}^{E(a)} \frac{\alpha(E)}{dE/ds}\, dE \, . \tag{4.3}$$

Here N and N_0 are the final and initial number of electrons in the avalanche;

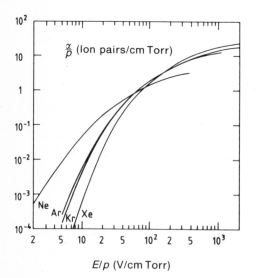

Fig. 4.8. First Townsend ionization coefficient α measured for some noble gases as a function of the electric field E, collected by von Engel [ENG 56]. The plots show α/p versus E/p, where p is the gas pressure

dE/ds is the electric field gradient. The minimal field E_{min} needed for ionization to multiply is given by the energy required to ionize the molecules in question, divided by the mean free path between collisions; therefore E_{min} must be proportional to the gas density.

The electric field near a wire whose radius is small compared with the distance to other electrodes is given by the charge per unit length, λ, as a function of the radius:

$$E(r) = \frac{\lambda}{2\pi\varepsilon_0 r}, \tag{4.4}$$

where $\varepsilon_0 = 8.85 \times 10^{-12}$ A s/(V m), and λ can be derived from the wire voltages and the capacitance matrix (cf. Chap. 3). Inserting (4.4) into (4.3) we have

$$N/N_0 = \exp \int_{E_{min}}^{E(a)} \frac{\lambda\alpha(E)}{2\pi\varepsilon_0 E^2} \, dE . \tag{4.5}$$

The ratio N/N_0 is usually called *gain*. In the following paragraphs we use $G = N/N_0$ to indicate it.

4.4.1 The Diethorn Formula

Diethorn [DIE 56] derived a useful formula for G by assuming α to be proportional to E. Looking at Fig. 4.8, this is not unreasonable for heavy noble gases between 10^2 and 10^3 [V/cm Torr], a typical range of fields near thin wires. Inserting $\alpha = \beta E$ into (4.5) gives

$$\ln G = \frac{\beta\lambda}{2\pi\varepsilon_0} \ln \frac{\lambda}{2\pi\varepsilon_0 a E_{min}} . \tag{4.6}$$

The value of β can be related in the following way to the average energy $e\,\Delta V$ required to produce one more electron. The potential difference between $r = a$ and $r = s_{min}$ that corresponds to relation (4.3) is

$$\Phi(a) - \Phi(s_{min}) = \int_a^{s_{min}} E(r)\,dr = \frac{\lambda}{2\pi\varepsilon_0} \ln \frac{s_{min}}{a} = \frac{\lambda}{2\pi\varepsilon_0} \ln \frac{\lambda}{2\pi\varepsilon_0 a E_{min}} . \tag{4.7}$$

It gives rise to a number Z of generations of doubling the electrons in the avalanche:

$$Z = [\Phi(a) - \Phi(s_{min})]/\Delta V , \tag{4.8}$$

$$G = 2^Z , \tag{4.9}$$

$$\ln G = \frac{\ln 2}{\Delta V} \frac{\lambda}{2\pi\varepsilon_0} \ln \frac{\lambda}{2\pi\varepsilon_0 a E_{min}} . \tag{4.10}$$

We recognize that in this model the constant β of (4.6) has the meaning of the inverse of the average potential required to produce one electron in the avalanche multiplied by $\ln 2$.

We write E_{min}, which is proportional to the gas density ρ, in the form

$$E_{min}(\rho) = E_{min}(\rho_0) \frac{\rho}{\rho_0} , \tag{4.11}$$

where ρ_0 is the normal gas density. We finally obtain

$$\ln G = \frac{\ln 2}{\Delta V} \frac{\lambda}{2\pi\varepsilon_0} \ln \frac{\lambda}{2\pi\varepsilon_0 a E_{min}(\rho_0)(\rho/\rho_0)} . \tag{4.12}$$

Relation (4.12) was compared with measurements using proportional counter tubes, and was shown to be better than the older models of Rose–Korff and Curran–Craggs; see [HEN 72, WOL 74, KIS 60] and the literature quoted therein. In a proportional counter tube with inner radius b and wire radius a, the charge density λ is related to the voltage V by

$$\frac{\lambda}{2\pi\varepsilon_0} = \frac{V}{\ln(b/a)} . \tag{4.13}$$

Therefore, (4.12) can also be expressed in the following form (Diethorn's formula):

$$\ln G = \frac{\ln 2}{\ln(b/a)} \frac{V}{\Delta V} \ln \frac{V}{\ln(b/a) a E_{min}(\rho_0)(\rho/\rho_0)} . \tag{4.14}$$

Experimentally, we vary ρ/ρ_0, a and $V/\ln(b/a)$, and measure G. A plot of

$$\frac{\ln G \ln(b/a)}{V} \quad \text{versus} \quad \ln \frac{V}{\ln(b/a) a (\rho/\rho_0)}$$

Table 4.2. Measured Diethorn parameters for various gases

Gas mixture	$E_{min}(\rho_0)$ (kV/cm)	ΔV (V)	Refs.
Ar(90%) + CH$_4$(10%)	48 ± 3	23.6 ± 5.4	HEN 72
Ar(95%) + CH$_4$(5%)	45 ± 4	21.8 ± 4.4	HEN 72
CH$_4$	69 ± 5	36.5 ± 5.0	HEN 72
C$_3$H$_8$	100 ± 4	29.5 ± 2.0	HEN 72
He(96%) + (CH$_3$)$_2$CHCH$_3$(4%)	148 ± 2	27.6 ± 3.0	HEN 72
Ar(75%) + Xe(15%) + CO$_2$(10%)	51 ± 4	20.2 ± 0.3	HEN 72
Ar(69.4%) + Xe(19.9%) + CO$_2$(10.7%)	54.5 ± 4.0	20.3 ± 2.5	HEN 72
Ar(64.6%) + Xe(24.7%) + CO$_2$(10.7%)	60 ± 5	18.3 ± 5.0	HEN 72
Xe(90%) + CH$_4$(10%)	36.2	33.9	WOL 74
Xe(95%) + CO$_2$(5%)	36.6	31.4	WOL 74
CH$_4$(99.8%) + Ar(0.2%)	171	38.3	KIS 60
Ar(92.1%) + CH$_4$(7.9%)	77.5	30.2	DIE 56
CH$_4$(76.5%) + Ar(23.5%)	196	36.2	DIE 56
CH$_4$(90.3%) + Ar(9.7%)	21.8	28.3	DIE 56
Ar(90%) + CH$_3$CH$_2$OH(10%)	62	27.0	DIE 56
CH$_4$	144	40.3	DIE 56

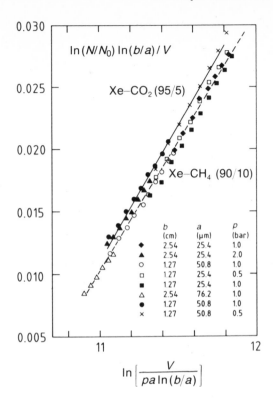

Fig. 4.9. Diethorn plot of amplification measurements using two Xe gas mixtures [HEN 72]; $p = \rho/\rho_0$

must be linear and yields the two constants $E_{min}(\rho_0)$ and ΔV. Figure 4.9 shows such a plot made by Hendricks [HEN 72] for two xenon gas mixtures, and Table 4.2 contains a collection of Diethorn parameters measured in this way, typically for amplification factors between 10^2 and 10^5. They are only in moderate agreement with each other.

We recognize the value of formulae (4.12) and (4.14), discussed above, more from a practical than from a fundamental point of view. Obviously, for a given gas mixture, the Townsend coefficient $\alpha(E)$ must first be known accurately as a function of the electric field. Then the multiplication factor can be calculated using (4.5). See also Charles [CHA 72] and Shalev and Hopstone [SHA 78], who compared and criticized various formulae for the gas gain.

4.4.2 Dependence of the Gain on the Gas Density

The variation of the gain with the gas density is of particular interest since very often the chambers are operated at atmospheric pressure and the gas density changes proportionally to it. From (4.10) and (4.11), a small relative change $d\rho/\rho$

of the density can be seen to result in a change of amplification of

$$\frac{dG}{G} = -\frac{\lambda \ln 2}{\Delta V \, 2\pi\varepsilon_0} \frac{d\rho}{\rho} .$$ (4.15)

In practical cases the factor that multiplies $d\rho/\rho$ ranges between 5 and 8: variation of the gas pressure causes global variations of the gain that are typically 5 to 8 times larger. Since the gas pressure can be easily monitored these variations can be corrected for.

4.4.3 Measurement of the Gain Variation with Sense-Wire Voltage and Gas Pressure

In this section we give some measurements of the variation of the gain with the sense wire voltage and with the atmospheric pressure. These measurements have been obtained using prototypes of the ALEPH TPC that had a cell geometry as described in Sect. 3.2.

Figure 4.10 shows a plot of the gain as a function of the voltage (cf. (4.14)) using the Diethorn parameters for an Argon (90%)–CH_4(10%) mixture (cf. Table 4.2) and assuming atmospheric pressure. The two side-lines show the large uncertainty in the prediction of the gain owing to the large error (20%) in ΔV. The superimposed experimental points have been measured with a chamber under the same operating conditions. The agreement is very good.

Figure 4.11 shows the variation of the atmospheric pressure p over a period of 2 days (a) and the corresponding average pulse height measured by the chamber when detecting the electrons produced in the gas by the absorption of X-rays from a ^{55}Fe source (b). One notices how the measured pulse height follows the variations of the pressure (note the zero-suppressed scales of the two

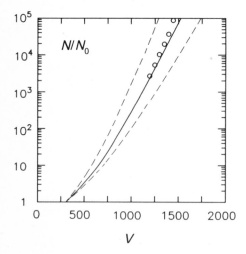

Fig. 4.10. Gain as a function of the sense-wire voltage in a chamber with a cell geometry as described in Sect. 3.2. *Dots* are measurements done with a model of the ALEPH TPC; the *full line* is computed with (4.14) and Table 4.2; *broken lines* show the error margins introduced by the error in Table 4.2

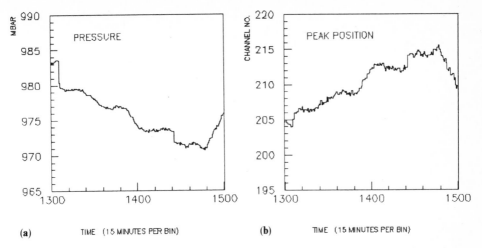

Fig. 4.11. (a) Atmospheric pressure p as a function of the time in a time interval of two days. (b) Average pulse-height measured by the chamber in the same period. Measurements done with the ALEPH TPC

plots). The chamber was operated in an Argon (90%)–CH$_4$(10%) mixture. The sense-wire voltage was 1.4 kV and the field wires were grounded. Using Table 3.2 we can compute the linear charge density λ of the sense wires and from the measured value of ΔV (cf. Table 4.2) using (4.15) we predict that

$$\frac{dG}{G} = -(6.7 \pm 1.5)\frac{dp}{p}.$$

Inspecting Fig. 4.11a we read an overall change of pressure of 0.8% and predict a change of gain of $-(5.5 \pm 1.2)\%$, in agreement with the variation (-4%) shown in Fig. 4.11b.

4.5   Local Variations of the Gain

Equation (4.12) shows that the gain depends on the local charge density of the wire λ. A relative change $d\lambda/\lambda$, as may result from geometrical imperfections of a wire chamber, from fluctuations of the supply voltage, or from space charge near the wire, will change the amplification by the factor (cf. (4.12))

$$\frac{dG}{G} = \left(\ln G + \frac{\lambda \ln 2}{\Delta V 2\pi\varepsilon_0}\right)\frac{d\lambda}{\lambda}. \tag{4.16}$$

In practical cases the two factors in the parentheses of the previous equation are of the same order of magnitude and their sum lies in the range 10–20: the local relative variations of the charge density cause local relative variations of the gain that are typically 10–20 times larger.

In Sect. 3.2 we saw how the linear charge density λ depends on the potentials and on the geometry of the wire chamber, assuming that there is an ideal geometry. Real chambers deviate from the ideal geometry: the wires are not infinite and we can expect changes in the charge density in the region close to the wire supports. Moreover, the mechanical components of the chamber have small construction imperfections, which can induce local changes in the charge density along the wire. The mathematical solution of the electrostatic problem is not simple because as soon as we introduce variations from the ideal geometry we lose the symmetry of the boundary conditions.

In drift chambers that are operated in high-density fluxes of particles another problem arises: the local charge density on the wire seen by ionization electrons when they reach the wire can be modified by the presence of the ions produced in a previous avalanche and still drifting inside the drift cell.

4.5.1 Variation of the Gain Near the Edge of the Chamber

In wire chambers the gain usually drops to zero near the support frame. This behaviour depends on the details of the geometry and extends over a region of the wire comparable to the distance between the sense-wire grid and the cathode plane. Figure 4.12 from [BRA 85] shows the relative variation of the gain near the edge of a wire chamber.

The dependence of the wire gain on the distance from the chamber edge can be modified by the addition of field correction strips [BRA 85, AME 86c], which can be trimmed to an appropriate voltage to modify the electric field in the region of the frame and to obtain a more uniform charge density (see Fig. 4.12).

4.5.2 Local Variation of the Gain Owing to Mechanical Imperfections

Here we want to evaluate the extent to which small deviations from the ideal geometry can affect the charge density on the sense wires, and we refer again to the example discussed in Sect. 3.2. The problem has been extensively studied by Erskine [ERS 72], and we refer to his paper for a general discussion.

The order of magnitude of the charge variation induced by some imperfections of the chamber can be obtained using the general formulae of Sect. 3.2: this is possible for imperfections that can be approximated to a large extent as a global variation of the geometry of the chamber. For example, the effect of a bump in the cathode plane can be evaluated from the global effect of a reduction of the distance between the sense-field grid and the cathode plane.

In the following we refer to the notation of Sect. 3.2 and to the geometry sketched in Fig. 31. Equation (3.8) gives the potentials of the electrodes as a function of their charges and of the matrix of the potential coefficients A:

$$V = A\sigma , \tag{4.17}$$

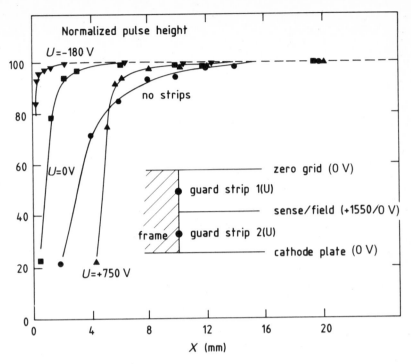

Fig. 4.12. Measured pulse heights of wire signals created by a laser beam near the wire support frame as a function of the distance between frame and laser beam. Guard strips can extend the sensitive length of the wire. The distance between neighbouring grids was 4 mm [BRA 85]

where $\sigma_i = \lambda_i/s_i$ are the linear charge densities on the wires of each grid divided by the pitch of the grid. A change in the geometry of the wire chamber induces a change in the matrix. Differentiating (4.17) with the fixed values of the potentials V, we obtain

$$A\,d\sigma = -\,dA\sigma\ ,$$

where $d\sigma$ is the vector of the charge-density variations and dA is the matrix of the variations in the potential coefficients induced by a particular change in the geometry of the chamber. Using the previous equation we obtain

$$d\sigma = -\,A^{-1}\,dA\sigma\ . \tag{4.18}$$

As a first example we calculate the charge-density variation induced by a displacement of the cathode plane. The matrix dA can be calculated by differentiating (3.10) with respect to z, noticing that the effect of a displacement dz of the cathode plane changes z_1 and z_2 by the same amount (cf. Fig. 3.1):

$$dA = -\frac{dz}{\varepsilon_0}\begin{pmatrix} 1 & 1 & 1 \\ 1 & 1 & 1 \\ 1 & 1 & 1 \end{pmatrix}. \tag{4.19}$$

From (4.18) we obtain

$$d\sigma = \frac{dz}{\varepsilon_0} A^{-1} \begin{pmatrix} \sigma_s + \sigma_F + \sigma_z \\ \sigma_s + \sigma_F + \sigma_z \\ \sigma_s + \sigma_F + \sigma_z \end{pmatrix}. \tag{4.20}$$

In the approximations of Sect. 3.2 the quantity $\sigma_s + \sigma_F + \sigma_z$ is the charge density on the cathode plane. The charge-density variation $d\sigma$ induced by a displacement of the cathode plane vanishes if there are no charges on the plane. In the particular geometry of the example considered in Sect. 3.2, we obtain:

$$d\sigma_s = (70\,\text{m})^{-1} dz(\sigma_s + \sigma_F + \sigma_z).$$

As another example we now compute the charge variation induced by a change of the sense-wire diameter. Differentiating (3.10) with respect to the sense-wire radius r_s, we obtain

$$dA = -\frac{s_1}{2\pi\varepsilon_0} \frac{dr_s}{r_s} \begin{pmatrix} 1 & 0 & 0 \\ 0 & 0 & 0 \\ 0 & 0 & 0 \end{pmatrix}, \tag{4.21}$$

and

$$d\sigma = \frac{s_1}{2\pi\varepsilon_0} \frac{dr_s}{r_s} A^{-1} \begin{pmatrix} \sigma_s \\ 0 \\ 0 \end{pmatrix}. \tag{4.22}$$

For the particular example considered in Sect. 3.2 we obtain

$$\frac{d\sigma_s}{\sigma_s} = 0.16 \frac{dr_s}{r_s}.$$

The effect of the displacement of a single wire is more difficult. In this case we cannot use (3.10) because it was calculated for a set of symmetric grids, and this symmetry is lost when a single wire is displaced. We notice that in first order the displacement of a wire does not affect the charge density of the wire itself (again because of the symmetry), but only that of the other wires close to it. The displacement of a field wire changes the charge density of the two closest sense wires proportionally to its own charge density. If there is no charge density on the field wires, their displacement does not affect the gain of the chamber.

Table 4.3 shows the gain variations induced by mechanical imperfections in the geometry of the example discussed here. We have assumed that the setting of the voltages produces a gain of 10^4, and using (4.16) we compute the relation between the relative variation of the gain and the local relative charge-density variation:

$$\frac{dG}{G} = (15.9 \pm 1.5) \frac{d\sigma_s}{\sigma_s}.$$

Table 4.3. Gain variation dG/G induced by mechanical imperfections in the geometry discussed in Sect. 3.2. A gain 10^4 is assumed and all displacements (Δ) are in mm. σ_F/σ_s is the ratio between the field wires and sense wires charge densities

Imperfection	dG/G
Bump on the pad plane	0.69 $\Delta z(1 + \sigma_F/\sigma_s)$
Displacement of a field wire	1.1 $\Delta x \, \sigma_F/\sigma_s$
Displacement of a sense wire	0.2 Δx
Variation of the sense wire diameter	2.5 $\Delta r/r$

The same gain can be obtained for different settings of the sense-wire voltage (V_s) and of the field wire voltage (V_F) that leave unchanged the charge density on the sense wire σ_s. Using Table 3.2 we compute the condition of constant gain:

$$0.25 \, V_s - 0.11 \, V_F = \text{constant} .$$

The different settings correspond to different values of the ratio σ_F/σ_s. Using again Table 3.2 we deduce that if $V_F/V_s = 0$, then $\sigma_F/\sigma_s = -0.44$, while $\sigma_F/\sigma_s = 0$ when $V_F/V_s = 0.34$. As shown in Table 4.3 the ratio σ_F/σ_s influences the local variations of the gain that are due to mechanical imperfections.

4.5.3   Space Charge Near the Wire

When the avalanche has come to an end, all electrons are collected on the anode and the space near the wire is filled with the remaining positive ions. We may assume that in the regime of proportional wire amplification, the amount of charge in the avalanche is small compared to the charge on the wire (integrated along the wire over the lateral extent of the avalanche). The positive ions are already moving away from the wire at relatively low speed. With a mobility of a few centimetres per second per volt per centimetre, a typical speed near the thin wire ($E \sim 2 \times 10^5$ V/cm) would be several 10^5 cm/s, and a hundred times less at hundred times the radius. The travelling time T between the wire and some point at a radius R is expressed by the mobility μ and the field $E = \lambda/(2\pi\varepsilon_0 r)$ as a function of the radius R as

$$T = \int_a^R \frac{dr}{\mu E(r)} = \frac{R^2 - a^2}{2a\mu E(a)} , \tag{4.23}$$

where the ions are assumed to have started from the wire radius at $t = 0$. In Fig. 4.13 we give an example of the distance travelled by CH_4 ions as a function of the time elapsed since the moment of the avalanche.

The space charge obviously takes hundreds of microseconds to settle on a cathode several millimeters away. During this time, what exactly is the effect on the amplification of any subsequent multiplication process, at the same spot on the wire? This question is relevant to drift chambers that measure events with

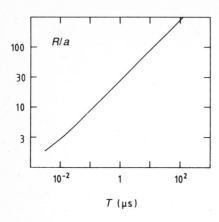

Fig. 4.13. Position of positive ion as it moves away from the wire after the instant of the avalanche: example of a CH_4^+ ion in Ar at a wire radius $a = 10\,\mu m$ and a field of $200\,kV/cm$ on the surface

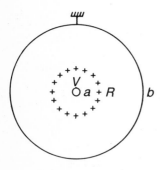

Fig. 4.14. Geometry of a cylindrical cloud of positive ions moving away from the wire of the proportional tube

a high track density. Since the lateral extent L of the avalanche is not smaller than the width of the initial group of electrons that occasion the avalanche, it will be of the order of a millimeter at least, owing to the diffusion of the incoming electrons. If the diffusion is not reduced by a magnetic field (cf. Chap. 2), L may be considerably larger.

In order to estimate the reduction of the amplification factor as a function of the time elapsed since the last avalanche, we approximate the space charge as a thin sheet of charge around the wire, infinitely long with linear charge density λ^*. Let it move according to (4.23).

In a proportional tube with radii a, b (see Fig. 4.14) and with the wire at potential V, the field inside the space-charge cylinder is

$$E^{(i)}(r) = \frac{\lambda'}{2\pi\varepsilon_0 r} \quad (r < R),\tag{4.24}$$

where λ' is the perturbed charge density of the wire. Outside the space-charge cylinder the field is

$$E^{(o)}(r) = \frac{\lambda' + \lambda^*}{2\pi\varepsilon_0 r} \quad (r > R).\tag{4.25}$$

The potential difference V determines the value of λ':

$$V = \frac{\lambda}{2\pi\varepsilon_0} \int_a^b \frac{dr}{r} = \frac{\lambda'}{2\pi\varepsilon_0} \int_a^R \frac{dr}{r} + \frac{\lambda' + \lambda^*}{2\pi\varepsilon_0} \int_R^b \frac{dr}{r} , \qquad (4.26)$$

$$\lambda' = \lambda - \lambda^* \frac{\ln b/R}{\ln b/a} , \qquad (4.27)$$

where λ is the charge density in the unperturbed situation ($\lambda^* = 0$).

Referring to (4.16) and (4.24), we recognize the shielding effect of the ground potentials.

The charge density that determines the multiplication is reduced by

$$\frac{d\lambda}{\lambda} = \eta(T) \frac{\lambda^*}{\lambda} , \qquad (4.28)$$

$\eta(T)$ being the shielding factor that changes with the travelling time,

$$\eta(T) = \frac{\ln b/R(T)}{\ln b/a} . \qquad (4.29)$$

Starting with $\eta = 1$ when the ions are on the wire, η decreases to zero when they have reached the ground electrode. The approximation of an infinitely long charge cylinder breaks down when R becomes of the order of the avalanche width. We note that the cloud of ions is almost all the time outside the multiplication region. The value of λ^* is given by the charge in the first avalanche, divided by its width L:

$$\lambda^* = Ne/L .$$

Assuming a gain of 10^4 and $L = 1$ mm, 100 ionization electrons produce $\lambda^* \approx 1.6 \times 10^{-10}$ C/m, to be compared with a typical value of $\lambda \approx 1.2 \times 10^{-8}$ C/m. Using (4.28) and (4.16) we obtain

$$\frac{dG}{G} \approx 15 \frac{d\lambda}{\lambda} \approx 0.2 \, \eta(T) .$$

4.6 Statistical Fluctuation of the Gain

The gain of an avalanche is equal to the number of ion pairs inside it divided by the number of electrons that started the avalanche. In the regime of proportionality, the average number of ions produced in the avalanche is proportional to the number of initial electrons. We are interested in the fluctuations in the number of ions that are caused by the random nature of the multiplication process.

We may assume that each initiating electron develops its own little avalanche, independent of the presence of the others near by. If we want to know the probability distribution of the number N of the ions in the total avalanche we simply have to sum over the probability distributions $P(n)$ of the numbers n of ions in the individual little avalanches. The first step is therefore to find an expression $P(n)$. This is done in Sect. 4.6.1.

If the number k of initiating electrons is large, the central-limit theorem of statistics applies. It makes a statement about the distribution function $F(N)$ of the sum N:

$$N = n_1 + n_2 + n_3 + \ldots + n_k \,, \tag{4.30}$$

where each of the independent variables n_i has the distribution function $P(n)$. If the mean of $P(n)$ is called $\bar{n}$, and the variance σ^2, then the central-limit theorem states that in the limit of $k \to \infty$, $F(N)$ is a Gaussian,

$$F(N) = \frac{1}{S\sqrt{2\pi}} \exp[(N - \bar{N})^2/2S^2] \,, \tag{4.31}$$

with

$$\bar{N} = k\bar{n} \quad \text{and} \quad S^2 = k\sigma^2 \,.$$

This means that the exact shape of $P(n)$ is not needed for the distribution of the avalanches started by a large number of electrons. On the other hand, $P(n)$ is particularly interesting for drift chambers where individual ionization electrons have to be detected, as in ring-imaging Cherenkov (RICH) counters.

Statistical fluctuations of the gain may be created by chemical reactions associated with the formation of avalanches. Such deterioration of performance occurs either through deposits on old wires but also through the presence of small contaminations of the gas. A discussion of these effects is given in Sect. 11.6; we refer especially to Fig. 11.7. – Here we assume that the proportional wire is in unperturbed working conditions.

4.6.1   Distributions of Avalanches Started by Single Electrons (Theory)

A convenient way of catching the fluctuation phenomenon derives from the concept of the stochastic process: as the avalanche develops, the variation of the number n of ions is followed as a function of a parameter t which can be thought of as the time but which is not identical with the physical time as measured by a clock. Rather, its increase describes the progress of the process in the order given by the causal sequence of the events. Therefore, any monotonic function of time will serve the purpose. For an introduction to stochastic processes, the reader is referred to a suitable textbook on statistics, for example [FIS 58].

The Yule–Furry Process. We begin by describing the stochastic process due to Yule and Furry, who invented it for the phenomena of biological population

growth and of electromagnetic showers. It was applied to proportional avalanches by Snyder and by Frisch (see [BYR 62] for references).

Let $P(n, t)$ be the probability that at some 'time' t there are n ions (or
electrons) in the avalanche. At $t = 0$ we have

$$P(1, 0) = 1 \tag{4.32}$$

because the avalanche was started by a single electron.

The increase is characterized by a constant parameter λ, and we suppose that
the probability for the birth of one electron in any interval Δt is proportional to
n and equal to

$$n\lambda \, \Delta t \, . \tag{4.33}$$

New electrons are created singly, i.e the probability for the birth of two
(three) electrons is quadratically (cubically) small for small Δt. We now express
the probability of having n electrons at the time $t + \Delta t$ by the probability at the
time t. Either nothing happened or one electron was created:

$$P(n, t + \Delta t) = P(n, t)(1 - n\lambda\Delta t) + P(n - 1)(n - 1)\lambda\Delta t \, . \tag{4.34}$$

$P(n, t + \Delta t)$ is the sum of the probabilities that at t there were already n electrons and no birth took place during Δt, plus the probability that at t there were
$n - 1$ electrons and one birth took place during Δt. The statistical independence
of these increments was used here. In the limit of vanishing Δt we have a system
of differential equations in the variable t:

$$\frac{dP(n, t)}{dt} = P(n - 1, t)(n - 1)\lambda - P(n, t)n\lambda \quad (n = 2, 3, \dots) \, . \tag{4.35}$$

For the beginning of the process, this has to be supplemented by

$$\frac{dP(1, t)}{dt} = -\lambda P(1, t) \quad (n = 1) \, , \tag{4.36}$$

which, together with (4.32), implies

$$P(1, t) = e^{-\lambda t} \, .$$

The system (4.35) integrates stepwise to yield the solutions

$$P(n, t) = e^{-\lambda t}(1 - e^{-\lambda t})^{n - 1} \, . \tag{4.37}$$

This represents the probability distribution at a given t, the mean $\bar{n}$ and the
variance σ^2 being

$$\bar{n} = \sum_{n=1}^{\infty} nP(n, t) = e^{\lambda t} \, , \tag{4.38}$$

$$\sigma^2 = \sum_{n=1}^{\infty} n^2 P(n, t) - \bar{n}^2 = e^{\lambda t}(e^{\lambda t} - 1) \, . \tag{4.39}$$

Here we have used the expression for the sum of the geometrical series. Rewriting (4.37) using (4.38), the distribution takes the form

$$P(n) = \frac{1}{\bar{n}}\left(1 - \frac{1}{\bar{n}}\right)^{n-1} , \qquad (4.40)$$

from which the 'time' variable t has formally vanished. In the limit of $\bar{n} \to \infty$, which is appropriate for avalanches,

$$P(n) = \frac{1}{\bar{n}} e^{-n/\bar{n}} , \qquad (4.41)$$

$$\sigma^2 = \bar{n}^2 . \qquad (4.42)$$

Equations (4.41) and (4.42) tell us that the Yule–Furry process has an exponential signal distribution, small signals being the most probable; the r.m.s. width is equal to the mean.

Let us remark in passing that a Poisson distribution is obtained if the increment in probability (4.33) is not proportional to n but is a constant. Then one deals with a 'Poisson process'; see an appropriate textbook on statistics.

Ever since the first careful measurements of signal distribution on proportional wires were done, it has been recognized that the simple expression (4.41) had to be modified in order to have more flexibility in reproducing the observed shapes of the signal distributions. One such generalization is due to Byrne [BYR 62].

The Byrne Process. In addition to the constant λ, Byrne introduces a second one, b, and writes the probability for the birth of an electron as

$$n\lambda\theta(r)\left(b + \frac{1-b}{n}\right)\Delta r \qquad (4.43)$$

in lieu of (4.33). Here the process parameter is the radius r, or the space distance between the wire and the point of development of the avalanche. The probability for ionization per unit path length depends on r through the unspecified function $\theta(r)$, but it decreases as the avalanche develops, through the dependence of (4.43) on n. The ansatz (4.43) represents the idea that a fluctuation to larger n in the first part of the avalanche reduces the rate of development in the second part. There is no better justification of this ansatz than the fact that it leads to a class of distributions that include the monotonic Furry distribution ($b = 1$) and the strongly peaked Poisson distribution ($b = 0$) as limiting cases. See the remarks following (4.42).

The difference equations corresponding to (4.34) now take the form

$$P(1, r - \Delta r) = (1 - \lambda\theta(r)\Delta r)P(1, r) ,$$

$$P(n, r - \Delta r) = [1 - n\lambda\theta(r)(b + (1 - b)/n)\Delta r]P(n, r) \qquad (4.44)$$

$$+ (n - 1)\lambda\theta(r)(b + (1 - b)/(n - 1))\Delta r\, P(n - 1, r)$$

$$(n = 2, 3, \ldots) .$$

At this point Byrne writes these equations in term of the transformed variable x and the number $n' = n - 1$ of secondary electrons:

$$t = \int_r^{r_0} \theta(r)\,dr\,,$$

$$1 + \lambda bx = e^{\lambda bt}\,, \quad r \le r_0; \quad x = 0, \quad r > r_0\,. \tag{4.45}$$

Taking the limit, the differential equations for $P(n', x)$ are obtained:

$$\frac{dP(0, x)}{dx} = -\frac{\lambda}{1 + \lambda bx}\,P(0, x)\,,$$

$$\frac{dP(n', x)}{dx} = \frac{\lambda[1 + b(n' - 1)]}{1 + \lambda bx}\,P(n' - 1, x) - \frac{\lambda[1 + bn']}{1 + \lambda bx}\,P(n', x) \tag{4.46}$$

$$(n' = 1, 2, \ldots)\,.$$

These are the differential equations satisfied by the 'Polya' or 'negative binomial' distribution functions. Their solutions under the boundary conditions for a single initial electron (leaving out the prime of the n),

$$P(n, 0) = \delta_{n0}\,,$$

are given by the functions

$$P(n, x) = \left(\frac{\lambda x}{1 + \lambda bx}\right)^n \frac{1[1 + b][1 + 2b]\ldots[1 + (n - 1)b]}{n!}(1 + \lambda bx)^{-1/b}$$

$$(n = 1, 2, \ldots)\,. \tag{4.47}$$

They have mean $\bar{n}$ and variance σ^2 equal to

$$\bar{n} = \lambda x\,, \tag{4.48}$$

$$\sigma^2 = \bar{n}(1 + b\bar{n})\,. \tag{4.49}$$

Whereas the variance of the Furry distribution (4.41) was given by the square of the mean, the variance of the Polya distribution is governed by its mean and the additional parameter b. The parameter x formally disappears when we replace $\bar{n}$ for λx in (4.47). For $b \to 0$, (4.47) approaches the Poisson distribution.

Another way of expressing the Polya distribution is by making use of (4.48) and the abbreviation $k = 1/b - 1$:

$$P(n) = \frac{n!}{k!(n - k)!}\left(\frac{b\bar{n}}{1 + b\bar{n}}\right)^n \left(\frac{1}{1 + b\bar{n}}\right)^{k+1}\,. \tag{4.50}$$

In order to reach the limit of large $\bar{n}$ that is appropriate for proportional wire avalanches, we express the factorials in (4.50) using Stirling's formula. The asymptotic form of (4.47) or (4.50) is

$$P(n) = \frac{1}{b\bar{n}}\frac{1}{k!}\left(\frac{n}{b\bar{n}}\right)^k e^{-n/b\bar{n}}\,. \tag{4.51}$$

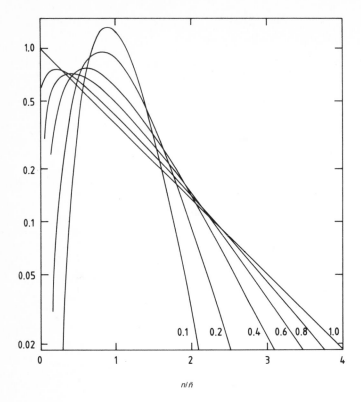

$n/\bar{n}$

Fig. 4.15. Polya distributions according to (4.51) for various values of the parameter b

We see that for $b = 1$ (which implies that $k = 0$), (4.41) is recovered. Figure 4.15 contains graphs of (4.51) for various values of b. Using (4.49), the variance of (4.51) is

$$\sigma^2 = b\bar{n}^2 . \tag{4.52}$$

4.6.2 Distributions of Avalanches Started by Single Electrons (Measurements)

It was shown by Raether and his school [RAE 64] that it is the ratio

$$\chi = \frac{E}{\alpha(E)\,U_{ion}}$$

that decides the nature of the distribution function. Here, E is the electric field, α the first Townsend coefficient and U_{ion} the ionization potential of the gas. Due to the dependence of α on E (Fig. 4.8) χ usually decreases with increasing E.

We go back to the Yule–Furry process and (4.33), where the probability for the creation of a new electron was proportional to the number already present and did not depend on their previous history; it was independent of how much energy the electrons had picked up since their own creation. This amount of energy is eE/α, because the average distance the electrons travel between two successive ionizing collisions is, by definition, equal to $1/\alpha$, and it has to be large compared with the ionization energy eU_{ion}, if the probability for the creation of a new electron is independent of it.

According to Raether and Schlumbohm, deviations from the simple exponential distribution (4.41) are observed unless $\chi \gg 1$. Figure 4.16 shows measurements taken in parallel-plate geometry in methylal vapour for various values of χ [RAE 64], [SCH 58]. For large χ ($\chi = 26$) there is a purely exponential distribution; for $\chi = 10.5$, a depletion at small numbers of ions is visible and causes a shallow maximum at $\approx 20\%$ of the average; for $\chi = 4.1$ the depletion is very deep and the peak is near 85% of the average. All these distributions have approximately exponential behaviour at the upper end.

These curves show that the Yule–Furry process is quite adequate for describing avalanches at low field strengths, but that it is too simple for the high field strength characterized by low values of χ. The extent to which the Polya distributions describe avalanches correctly has, to our knowledge, not been studied in a very systematic way. As an indication, we have given in Fig. 4.16e the Polya distribution of (4.51) for $b = 0.4$, which has approximately the same variance as the measured curve. From this one example and from more material that can be found in [GEN 73], it appears that the Polya distribution can describe very well the avalanches created by a single electron. Let us note in passing that the Polya distributions have recently been used very successfully to fit multiplicity distributions of high-energy multiple-particle production [ALN 85].

The cylindrical geometry of the wires involves a full range of field strength for every avalanche. It can be expected that the shape of the distributions is determined mainly by the weaker fields at the beginning of the avalanche, because the shape is influenced more by the statistics of the small numbers than by the intensity variations in the fully developed avalanche. Various distributions for wire counters are treated by Alkhazov [ALK 70].

Single-electron spectra and avalanche fluctuations in proportional counters have also been measured by Hurst et al., using laser resonant ionization spectroscopy (RIS) [HUR 78]. A review of single-electron spectra is given by Genz [GEN 73].

4.6.3 The Effect of Avalanche Fluctuations on the Wire Pulse Heights

Let each electron create an avalanche of average size h (a large number), distributed according to a density function $e_h(k)$ which has some variance σ_e^2. As an example, we may imagine e_h to be the exponential distribution $e_h(k) = (1/h)\exp(-k/h)$ that has $\sigma_e = h$, or one of the Polya distributions discussed in Sect. 4.6.1. In any case, σ_e will vary together with h, as larger

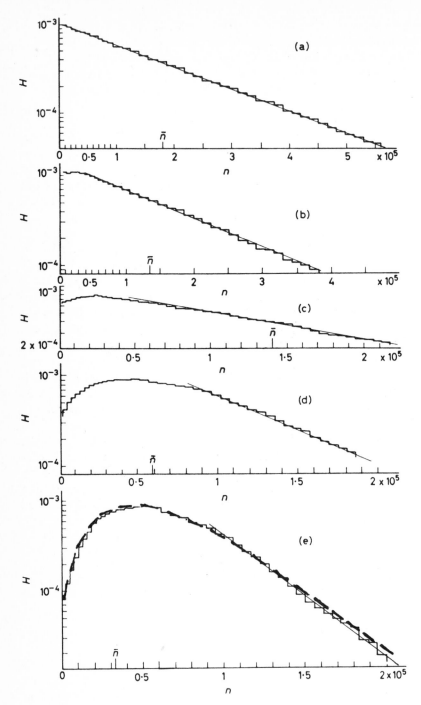

Fig. 4.16a–e. Distributions of numbers of ions in avalanches started by single electrons as measured by Schlumbohm [SCH 59] in methylal, in parallel-plate geometry, for various values of χ: $\chi = 26$ (**a**), 22.6 (**b**), 10.5 (**c**), 5.3 (**d**) and 4.1 (**e**). Superimposed in (**e**) is the Polya distribution with $b = 0.4$

avalanches can be expected to have larger fluctuations. If there is a fixed number n of electrons to create an avalanche, the resulting average avalanche size on a proportional wire is nh, the variance being $n\sigma_e^2$.

If the number n is not fixed but itself subject to fluctuations around a mean m with variance σ_m^2, then the mean and the variance of the resulting number of electrons at the wire are

$$N_e = mh \,, \tag{4.53}$$

$$\sigma^2 = m\sigma_e^2 + h^2\sigma_m^2 \,. \tag{4.54}$$

4.6.4   A Measurement of Avalanche Fluctuations Using Laser Tracks

A simple method that is close to the practice of recording particle tracks is to compare the pulse heights of many wires that record a laser track. The ionization of the gas by a beam of pulsed laser light is described in more detail in Chap. 1. The electrons along the beam are created independently; hence the numbers that arrive on each wire are randomly distributed according to Poisson's law, with a mean common to all the wires (provided the light beam is sufficiently uniform). The basic idea is to observe how much the pulse heights fluctuate more owing to the presence of the term containing σ_e^2 in (4.54).

For the case at hand, in which the number of ionization electrons that start the avalanche is Poisson-distributed, we have $\sigma_m^2 = m$. The measured pulse height P is proportional to the avalanche size; let the constant of proportionality be called a – it could have the dimension volts per electron. Then the mean and the variance of the pulses created by these electrons are

$$\langle P \rangle = aN_e = amh \,, \tag{4.55}$$

$$\sigma_P^2 = a^2\sigma^2 = a^2 m\sigma_e^2 + a^2 mh^2 \,. \tag{4.56}$$

In order to determine σ_e^2 experimentally, we measure the ratio $R = \sigma_P^2/\langle P\rangle^2$ for various m:

$$R = \frac{1 + \sigma_e^2/h^2}{m} \,. \tag{4.57}$$

This can be achieved by varying the laser intensity using grey filters. The track from one or several shots of the same intensity that lead to the same m are measured on a number K of wires (measured pulse heights P_i ($i = 1, 2, \ldots, K$)). The average pulse height is given by

$$\langle P \rangle = \sum_i \frac{P_i}{K} \tag{4.58}$$

and the above ratio R by

$$R = \sum_i \frac{(P_i - \langle P\rangle)^2}{K\langle P\rangle} \,. \tag{4.59}$$

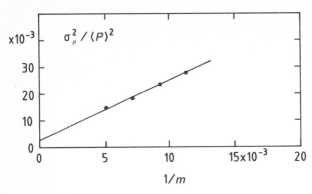

Fig. 4.17. Square of the measured relative pulse-height variation as a function of the inverse number of primary electrons

The value of m can be obtained from the pulse height by calibration with a ^{55}Fe radioactive source. Since the peak of the spectrum is at $m = 227$ in argon, a pulse height P corresponds to

$$m = 227 \frac{P}{P_{\text{peak}}} \ .$$

In an experiment using the TPC 90 model of the ALEPH TPC with approximately 100 wires, the ratio R was measured as a function of $1/m$ (Fig. 4.17), and a straight-line fit gave

$$R = 2.2 \times \frac{1}{m} + 0.003 \ .$$

The constant term represents an intrinsic pulse-height variation from wire to wire due to the apparatus, whereas the measured slope implies a value of

$$\sigma_e^2 = 1.2 \, h^2 \quad \text{or} \quad \sigma_e = 1.1 \, h \ ,$$

using (4.57). This determination of σ_e is roughly 10% accurate.

Within the limits of this simple experiment, we may conclude that the variance of the single-electron avalanche was equal to that of the Yule–Furry process.

References

[ALE 80]   G.D. Alekseev, N.A. Kalinina, V.V. Karpukhin, D.M. Khazins and V.V. Kruglov, Investigation of self-quenching streamer discharge in a wire chamber, *Nucl. Instrum. Methods* **177**, 385 (1980)

[ALE 82]   G.D. Alekseev, V.V. Kruglov, D.M. Khazins, Self-quenching streamer discharge in a wire chamber, *Sov. J. Part. Nucl.* **13**, 293 (1982)

[ALK 70]   G.D. Alkhazov, Statistics of electron avalanches and ultimate resolution of proportional counters, *Nucl. Instrum. Methods* **89**, 155 (1970)

[ALN 85]   G.J. Alner et al. (UA5 Collab.), A new empirical regularity for multiplicity distributions in place of KNO scaling, *Phys. Lett.* **160B**, 193 and 199 (1985)

[AME 86c]   S.R. Amendolia et al., Studies of the wire gain and track distortions near the sector edges of the ALEPH TPC, *Nucl. Instrum. Methods A* **252**, 399 (1986)

[ATA 82]   M. Atac, A.V. Tollestrup and D. Potter, Self-quenching streamers, *Nucl. Instrum. Methods* **200**, 345 (1982)

[BAT 85]   G. Battistoni et al., Sensitivity of streamer mode to single ionization electrons. *Nucl. Instrum. Methods A* **235**, 91 (1985)

[BRA 85]   C. Brand et al., Improvement of the wire gain near the frame of a time projection chamber. *Nucl. Instrum. Methods A* **237**, 501 (1985)

[BYR 62]   J. Byrne, Statistics of the electron multiplication process in proportional counters, *Proc. R. Soc. Edinburgh*, **XVI A 33** (1962)

[CHA 72]   M.W. Charles, Gas gain measurements in proportional counters, *J. Phys. E* **5**, 95 (1972)

[CUR 58]   S.C. Curran, The proportional counter as detector and spectrometer, in *Handbuch der Physik*, ed. by S. Flügge, Vol. 45 (Springer, Berlin Heidelberg 1958)

[DIE 56]   W. Diethorn, A methane proportional counter system for natural radiocarbon measurements, USAEC Report NY06628 (1956); also doctoral dissertation, Carnegie Inst. of Technology, 1956

[DWU 83]   A. Dwurazny, K. Jelen, E. Rulikowska-Zabrebska, inorganic gas mixtures for proportional counters, *Nucl. Instr. Methods* **217**, 301 (1983)

[ENG 56]   A. von Engel, Ionization in gases by electrons in electric fields, in *Handbuch der Physik*, ed. by S. Flügge, Vol. 21 (Springer, Berlin Heidelberg, 1956)

[ERS 72]   G.A. Erskine, Electrostatic problems in multiwire proportional chambers, *Nucl. Instrum. Methods* **105**, 565 (1972)

[FIS 58]   M. Fisz, Wahrscheinlichkeitsrechung und mathematische Statistik (VEB Verlag Deutscher Wissenschaften, Berlin, 1954) (translated from Polish)

[GEN 73]   H. Genz, Single electron detection in proportional gas counters, *Nucl. Instrum. Methods* **112**, 83 (1973)

[GRO 89]   J. Groh et al., Computer simulation of the electron avalanche in cylindrically symmetric electric fields, *Nucl. Instrum. Methods* A**283**, 730 (1989)

[HAN 81]   *Handbook of Chemistry and Physics*, ed. by R.C. Weast (C.R.C. Press, Bica Raton, Fl. 1981/82), p. E205

[HEN 72]   R.W. Hendricks, The gas amplification factor in xenon-filled proportional counters, *Nucl. Instrum. Methods* **102**, 309 (1972)

[HUR 78]   G.S. Hurst et al., Resonant ionization studies of the fluctuation of proportional counters, *Nucl. Instrum. Methods* **155**, 203 (1978)

[IAR 83]   E. Iarocci, Plastic streamer tubes and their applications to high-energy physics, *Nucl. Instrum. Methods* **217**, 30 (1983)

[KIS 60]   R.W. Kiser, Characteristic parameters of gas-tube proportional counters. *Appl. Sci. Res. B* **8**, 183 (1960)

[KNO 79]   G.F. Knoll, Radiation detection and measurements (Wiley, New York 1979)

[KOR 55]   S.A. Korff, Electron and nuclear counters, 2nd ed. (Van Nostrand, Princeton, NJ 1955)

[MAT 85]   M. Matoba, T. Hirose, T. Sakae, H. Kametani, H. Ijiri and T. Shintake, Three dimensional Monte Carlo simulation of the electron avalanche around an anode wire of a proportional counter, *IEEE Trans. Nucl. Sci.* NS-**32**, 541 (1985)

[OKU 79]   H. Okuno, J. Fisher, V. Radeka and A.H. Walenta, Azimuthal spread of the avalanche in proportional chambers, *IEEE Trans. Nucl. Sci.* NS-**26**, 160 (1979)

[RAE 64]   H. Raether, Electron avalanches and breakdown in gases (Butterworth, London 1964)

[SCH 58]   H. Schlumbohm, Zur Statistik der Electronenlawinen in ebenen Feld, *Z. Phys.* **151**, 563 (1958)

[SCH 62]   R.I. Schoen, Absorption, ionization and ion-fragmentation cross-section of hydro-carbon vapors under vacuum-ultraviolet radiation, *J. Chem. Phys.* **17**, 2032 (1962)

[SHA 78]   S. Shalev and P. Hopstone, Empirical expression for gas multiplication in ^{3}He proportional counters, *Nucl. Instrum. Methods* **155**, 237 (1978)

[TAY 90]   F.E. Taylor, A model of the limited streamer process, *Nucl. Instrum. Methods A* **289**, 283 (1990)

[WOL 74]   R.S. Wolff, Measurements of the gas constants for various proportional counter gas mixtures, *Nucl. Instrum. Methods* **115**, 461 (1974)

5. Creation of the Signal

The moving charges between the electrodes of a chamber give rise to electrical signals that can be picked up by amplifiers connected to the electrodes. Every avalanche will create a signal in the wire, and we treat this case first, emphasizing the development with time of the signal (Sect. 5.1). If, for the purpose of coordinate measurement, the cathode is subdivided into several pieces, the signal created in each part will have the same time dependence in a first approximation, but it will depend on the distance from the avalanche. This dependence is derived for the important class of plane cathode strips in Sect. 5.2. We also derive Ramo's theorem for the most general case in Sect. 5.3.

5.1 Signal Generation on the Wire

In order to understand how the electric signal develops with time, we imagine a proportionl tube with the wire set at positive high voltage via a resistor R_1. Let the radius of the wire be a and the inner radius of the grounded tube be b. Once the ion pairs of the avalanche separate, the electrons travel the short remaining distance to the wire surface, and the positive ions travel in the opposite direction towards the wall of the tube.

A small charge q that travel between two points 1 and 2 under the influence of the field E reduces the electric energy ε of the condenser by the amount

$$\Delta\varepsilon = \int_1^2 q\boldsymbol{E}\cdot\mathrm{d}\boldsymbol{r} = q(\varPhi_1 - \varPhi_2)\,, \qquad (5.1)$$

proportional to the potential difference between the two points. This change in energy is the source of the signal.

In comparison with the potential difference ΔV needed to double the number of ion pairs in the avalanche (Sect. 4.4.1), the electrons do not have much potential difference to travel before reaching the wire. One half of them have ΔV, one quarter have $2\Delta V$, and so forth, roughly $2\Delta V$ on the average. In contrast the positive ions have almost the full potential difference between the electrodes, about two orders of magnitude more under typical conditions (see Table 4.2). Therefore, almost all the energy and all the signal in a proportional counter are due to the motion of the positive ions.

The energy difference, as a function of time, may be calculated from the motion of the positive ions starting at radius a. Following Wilkinson [WIL 50], we write

$$\Delta\varepsilon = \int_a^{R(t)} qE \cdot dr = \int_a^{R(t)} \frac{q\lambda \, dr}{2\pi\varepsilon_0 r} = \frac{q\lambda}{2\pi\varepsilon_0} \ln \frac{R(t)}{a} \,. \tag{5.2}$$

The radius R reached at a time t can be calculated using the expression for the drift velocity of the ions $u = \mu E$. Although the mobility μ is not really constant over the whole range of values of the electric field traversed by the ions, we may see in Fig. 2.4 that some effective constant value of μ will be a fairly good approximation. Compare also (4.23). When we insert it we find that

$$\frac{R(t)}{a} = \left(1 + \frac{2\mu E(a)}{a} t\right)^{1/2} \tag{5.3}$$

and obtain

$$\Delta\varepsilon(t) = q\frac{\lambda}{4\pi\varepsilon_0} \ln\left(1 + \frac{t}{t_0}\right). \tag{5.4}$$

Here, the characteristic time

$$t_0 = \frac{a}{2\mu E(a)} \tag{5.5}$$

is expressed in terms of the field on the wire surface, the wire radius, and the ion mobility. For a typical counter, t_0 is of the order of one or a few nanoseconds. The energy change continues until the ions have reached the wall at radius b, the total energy change being

$$(\Delta\varepsilon)_{\text{tot}} = \frac{q\lambda}{2\pi\varepsilon_0} \ln \frac{b}{a} = qV \,. \tag{5.6}$$

Therefore, we may write the change in energy as function of the time as

$$\Delta\varepsilon = qVF(t) \,, \tag{5.7}$$

with

$$F(t) = \frac{\ln(1 + t/t_0)}{2\ln(b/a)} \,, \tag{5.8}$$

where $F(t)$ is a function that rises from 0 to 1. It depends on the counter dimensions and on the ion drift velocity at the wire. An example for a typical counter is plotted in Fig. 5.1 on a logarithmic time scale: $F(t)$ is seen to reach the value 1 after 320 µs, and the value of 0.5 after 1 µs.

For the generation of the electric signal, we distinguish two limiting cases:

(i) The potential of the wire is re-established during the development of the pulse. This requires charges to flow quickly enough into the counter, which then acts as a "current source".

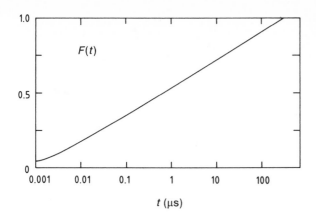

Fig. 5.1. Plot of $F(t)$ from (5.8) on a logarithmic time scale. The example was calculated for $t_0 = 1.25$ ns and $b/a = 500$

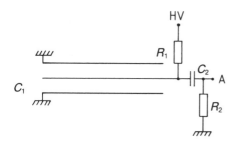

Fig. 5.2. Wiring diagram of a typical connection of a proportional counter tube to the amplifier at A

(ii) Charges are prevented from flowing into the counter during the development of the pulse. This results in a drop in potential across the counter, which then acts as a "voltage source". The wire potential is eventually reestablished on a longer time scale.

Figure 5.2 shows the wiring diagram of a typical connection to the amplifier input at A: R_1 connects the wire to the positive high voltage and is assumed to be very large; C_1 describes the capacitance between the wire and the tube, as well as any other parallel capacitance present in a technical realization of the counter. The resistor R_2 summarizes the internal (perhaps active) and external impedances of the amplifier, and C_2 decouples the amplifier from the high voltage.

In the limit where the time constants R_2C_2 and R_2C_1 are small compared with the pulse rise-time (or, more precisely, with the inverse frequencies of the pulse Fourier spectrum), we have case (i), and the signal is the current $I(t)$ that flows through R_2. If the field energy in C_1 is QV, its change is

$$\Delta \varepsilon = \Delta Q(t) V \quad \text{(case (i))}$$

and, using (5.7),

$$I(t) = \frac{\mathrm{d}}{\mathrm{d}t} \Delta Q(t) = q \frac{\mathrm{d}}{\mathrm{d}t} F(t) = \frac{q}{\ln(b/a)} \frac{1}{t + t_0} . \tag{5.9}$$

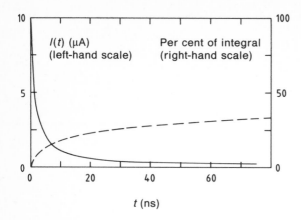

Fig. 5.3. Current signal according to (5.9) (*full line*, left-hand scale) for $t_0 = 1.25$ ns, $b/a = 500$, and $q = 106$ elementay charges. Time integral of this pulse as a percentage of the total (*broken line*, right-hand scale)

The current signal involves the derivative of $F(t)$. An example, calculated as for Fig. 5.1, is plotted in Fig. 5.3. We observe that the peak of the signal is extremely narrow (width t_0), which allows precise time information to be derived from it, but the total signal is b^2/a^2 times longer than the width.

In the other limit, where the time constants $R_2 C_2$ and $R_2 C_1$ are large compared with the pulse rise-time, we have case (ii), and there the signal is the voltage $\Delta V(t)$ that is apparent at A. The change in energy is then

$$\Delta \varepsilon = Q \, \Delta V(t) \quad (\text{case (ii)}) \, ,$$

where $Q = C_1 V$ is the charge contained in the counter. Using Eq. (5.7), we have

$$\Delta V(t) = \frac{q}{C_1} F(t) \, . \tag{5.10}$$

The voltage signal is proportional to $F(t)$, depicted in Fig. 5.1, and inversely proportional to the capacity C_1; as before, q is the positive charge in the avalanche.

If the dimensions of C_1, C_2, and R_2 fall between the two limiting cases (i) and (ii), the evolving signal is more complicated.

In this section we have calculated the time development of the electric pulse created by a proportional-wire avalanche. Depending on the geometry of the drift situation, the ionization electrons may arrive at the wire staggered in time. In this case, each electron causes its own avalanche when it arrives, and the resulting pulse is a superposition of many pulses displaced in time.

5.2   Signal Generation on Cathode Strips

Some drift chambers have the cathode subdivided into separate strips with independent signal readouts for the purpose of localizing the avalanche. We must therefore discuss proportional wires that are opposite to several electrodes, and we must understand how electric signals are created in the electrodes.

These signals can be quite different, depending on whether the moving charge q travels to the electrode in question and settles down there, or whether it settles down on a neighbouring electrode. In the first case the time integral over the current signal equals q, in the other case it vanishes. However, a normal electronic amplifier produces its signal in a short time – usually much less than a microsecond – by using only the early part of the counter signal and of the ion motion when the ions are still in the vicinity of the wire. In this case the signals on neighbouring electrodes have the same time dependence to good approximation.

5.2.1   Pad Response Function

We treat the case of a plane geometry where the proportional wires (perhaps interspaced with field wires) form a plane between two grounded cathode planes, one of which has a strip separated from ground by the small resistor R_3 (Fig. 5.4).

In order to derive the current that flows through R_3 when there are avalanches on many wires, we concentrate on a series of avalanches that come from a straight particle track. At first we neglect the width of the avalanches. It is then appropriate to think of the positive charges created by such a track as a line of charge with density λ which is located in the middle of a parallel-plate condenser. The influence of the wires is neglected. The task is to calculate the

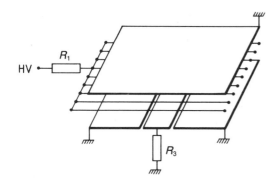

Fig. 5.4. Proportional wire plane between two grounded parallel cathode planes, one of which has an isolated strip connected to ground through a small resistor R_3

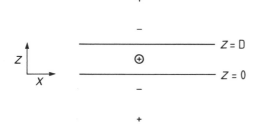

Fig. 5.5. Charge images created by a linear charge in the centre of the parallel-plate condenser

surface charge density $\sigma(x)$ induced in the cathode plate. This is achieved using the *method of images*.

The images of the positive line of charge are alternatively negative and positive, and are situated at $z_k = \pm (2k + 1)D/2, (k = 1, 2 \ldots)$ and $z_0 = -D/2$, half the plate distance (Fig. 5.5). Each pair of lines of charge at $\pm z_k$ produces an electric field which, at $z = 0$, is normal to the surface and equal to

$$E_k^{(n)}(x) = -\frac{\lambda}{\pi\varepsilon_0}\frac{z_k}{x^2 + z_k^2} .$$ (5.11)

The surface charge density equivalent to this field is given by Gauss's law:

$$\sigma = \varepsilon_0 E^{(n)} .$$

The total charge density is obtained by summation over all the pairs:

$$\sigma(x) = -\frac{\lambda}{\pi}\sum_{k=0}^{\infty}(-1)^k\frac{(2k + 1)D/2}{x^2 + (2k + 1)^2 D^2/4}$$

$$= \frac{-\lambda}{2D}\cdot\frac{1}{\cosh(\pi x/D)} .$$ (5.12)

This function is plotted in Fig. 5.6. Its integral is half the charge on the wire, with the opposite polarity:

$$\int_{-\infty}^{+\infty}\sigma(x)dx = -\lambda/2 .$$ (5.13)

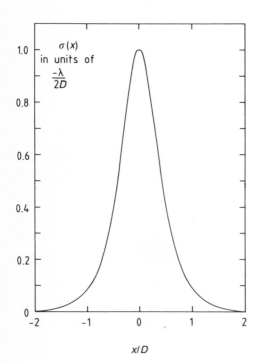

Fig. 5.6. Charge density on one plate of the parallel-plate condenser induced by a line charge λ in the centre, according to (5.12)

The 'pad response function' is obtained by integrating $\sigma(x)$ over the area of the strip. The important case is the one in which the ionization track is parallel to the strip. (The case when the track is at an angle is mentioned below, after (5.17).) The pad response function takes the form

$$P_0(x) = \int_{x-W/2}^{x+W/2} \sigma(x')\,dx', \tag{5.14}$$

where W is the strip width. It represents the amount of induced charge that has to flow into the strip, and x now denotes the distance between the track and the centre of the strip; x is the coordinate to be determined for a track by making a comparison of the pulses on neighbouring strips.

The pad response function of (5.14) depends on the width of the strip. For the construction of drift chambers with strip readout, it is important to choose the strip width W in relation to the cathode distance D, so that the signal ratio on neighbouring strips does not make excessive demands on the dynamic range of the electronics. In Fig. 5.7 these different pad response functions are shown for three different values of W/D. For example, a track over the centre of one strip will make a signal 6.1 times smaller on the neighbouring strips if $W/D = 1$, and 2.1 times smaller if $W/D = 0.5$.

It has been found experimentally [FAN 79] that $P_0(x)$ can be approximated by a Gaussian curve to within a few per cent of its maximum value:

$$P_0 \approx e^{-x^2/2s_0^2}. \tag{5.15}$$

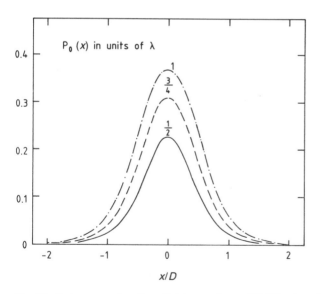

Fig. 5.7. Pad response functions $P_0(x)$ according to (5.14) for three different values of strip width W. The area and the width become smaller as W decreases

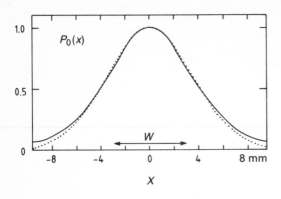

Fig. 5.8. Pad response function $P_0(x)$ according to (5.14) calculated for $W = 6$ mm and $D = 8$ mm, and compared with a Gaussian (*broken line*). The avalanche is in the centre of the strip

A typical case, using $W = 6$ mm, $D = 8$ mm, is compared with a Gaussian fit in the interval $-W < x < W$ in Fig. 5.8, giving $s_0 = 3.5$ mm. The Gaussian approximation (5.15) falls more quickly for large x than does (5.14), but for most practical cases (5.15) is quite sufficient.

For the determination of coordinates from pulse-height ratios, the original pad response function $P_0(x)$ has to be modified because of the spread of the avalanche along the coordinate. The spread is caused mainly by diffusion of the primary-ionization electrons on their drift path to the counting wire. This diffusion is also a limiting factor of the accuracy of measurement and is discussed in more detail in Chap. 6. Here we note that the displacement of an electron along x is also distributed like a Gaussian and has a characteristic width s_{diff} that depends on the drift conditions, the drift length z, and the magnetic field B. Using (5.15), the response function becomes

$$P(x) \approx e^{-x^2/2s^2} , \tag{5.16}$$

$$s^2 = s^2(z, B) = s_0^2 + s_{diff}^2(z, B) . \tag{5.17}$$

The pad response function in the case where the track is at an angle α with the strip may be treated by replacing the integral (5.14) by the appropriate two-dimensional integral over the strip area, which involves again the strip width W, but also the wire pitch l, which defines the length of the track sampled by one wire. Alternatively, one may use the approximate relation (5.16) and add an effective width to (5.17), which for small α is changed into

$$s^2 = s_0^2 + s_{diff}^2(z, B) + \alpha^2 l^2/12 . \tag{5.18}$$

When in some chambers the coordinates are measured using cathode strips, the determination of the pad response function and its dependence on the track and gas parameters is the important first step towards precision measurements. The systematic errors that can be associated with this method were discussed by Piuz et al. [PIU 82].

5.2.2 Principle of the Measurement of the Coordinate along the Wire Direction Using Cathode Strips

The measurement of the coordinate of the avalanche along the wire direction is made by interpolating the pulse height recorded on adjacent cathode strips with the help of (5.16). The strip width W and the cathode distance D have to be chosen such that the typical pulse height induced on 2 or 3 adjacent strips falls in the dynamic range of the readout electronics. When the coordinate of the avalanche is exactly in between two strips, it induces an equal signal on them. If the avalanche is produced in correspondence with the centre of one strip, it induces a large signal on it and two smaller signals on the two adjacent strips. Typically one records two signals in the first case and three in the second.

Let us call the coordinates of the centres of three adjacent strips $x_i - W$, x_i and $x_i + W$ and the pulse heights recorded by them p_{i-1}, p_i and p_{i+1}. These pulse heights are correlated to the x coordinate of the avalanche by (5.16):

$$p_{i-1} = A e^{-(x-(x_i-W))^2/2s^2} ,$$

$$p_i = A e^{-(x-x_i)^2/2s^2} , \qquad (5.19)$$

$$p_{i+1} = A e^{-(x-(x_i+W))^2/2s^2} ,$$

where A is proportional to the total charge of the avalanche.

In this set of equations we have, in principle, three unknowns A, x and s. Solving for s we obtain

$$s^2 = - W^2 \ln \frac{p_{i-1} p_{i+1}}{p_i^2} . \qquad (5.20)$$

Using this equation one can measure the pad response function directly from the recorded pulse heights. This formula was used for the measurements shown in Fig. 2.22 where the pad response function was measured as a function of z and B to derive the dependence of the diffusion of the drift length in presence of magnetic field (cf. (5.17)).

In principle, (5.19) can be also used to derive a measurement of x *independent* of the a priori knowledge of the pad response function s. In practical cases this is not convenient since one obtains a better resolution if one uses a known parameterization of s^2. In this case x is given by the equation

$$x = \frac{1}{w_1 + w_2} \left[w_1 \left(x_i - \frac{W}{2} + \frac{s^2}{W} \ln \frac{p_i}{p_{i-1}} \right) + w_2 \left(x_i + \frac{W}{2} + \frac{s^2}{W} \ln \frac{p_{i+1}}{p_i} \right) \right] .$$

$$(5.21)$$

The coordinate of the avalanche x is given by the weighted average of two measurements, one done with strips $i - 1$ and i and the other with strips i and $i + 1$: the weights are w_1 and w_2. Since the measurement error is roughly inversely proportional to the recorded pulses on the side strips (the ones that

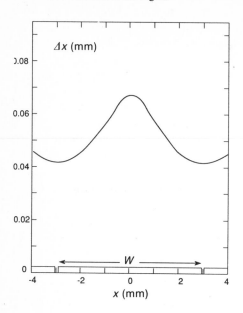

Fig. 5.9. The error in the coordinate measured interpolating the pulse heights of adjacent strips (5.21) is shown as a function of the position of the avalanche. $\Delta p/A = 1\%$, $W = 6$ mm and $s = 3.5$ mm are assumed

have smaller pulse heights) one can use as weights p_{i-1}^2 and p_{i+1}^2 instead of w_1 and w_2. If the signal is recorded only on two strips then the weights are 1 and 0.

Figure 5.9 shows the error induced on the measurement of the avalanche coordinate by an error on the measured pulse heights $\Delta p/A = 1\%$. A strip width W of 6 mm and a pad response function of 3.5 mm are assumed and the coordinate is calculated using (5.21) with weights p_{i-1}^2 and p_{i+1}^2. The error is smaller when the avalanche is between two strips ($x = -3$ mm and $x = 3$ mm) and is larger when it is at the centre of the strip ($x = 0$).

5.3   Ramo's Theorem

The most general way to calculate signal currents induced in electrodes by wire avalanches is provided by Ramo's theorem. It describes the situation where a charge q is moving in the space between several electrodes, causing charges to flow into and out of the electrodes, each of which is connected at some potential, to an infinite reservoir of charge. In practice, this could be a sufficiently large condenser. Figure 5.10 symbolizes the situation. The box representing the 'earth' could be infinitely far away and does not have to be closed – this would not alter the results.

We would like to compute the current I_k that flows between the electrode k and its reservoir as a function of the movement of the charge q in the inter-electrode space. This corresponds to case (i) in Sect. 5.1. The drift of ions in gases is slow enough for a quasi-electrostatic calculation. We follow Ramo

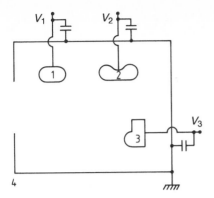

[RAM 39] almost entirely, although he treated grounded electrodes. The solution $\Phi(x)$ of the electrostatic problem, with the charge q at some point x_1, is the sum of solutions Φ_k ($k = 0, 1, 2, \ldots$) to the following problems:

Φ_i ($i = 1, 2, \ldots$): all electrodes grounded, but the ith electrode at V_i, q removed.
Φ_0: all electrodes grounded, q at x_1;

In the space between electrodes and charges q, Laplace's equation holds for all Φ_k and Φ:

$$\nabla^2 \Phi_k(x) = 0, \quad \nabla^2 \Phi = 0 . \tag{5.22}$$

Now Green's theorem says that for every i,

$$\int_V (\Phi \nabla^2 \Phi_i - \Phi_i \nabla^2 \Phi) \, dV = \int_S \left(\Phi \frac{\partial \Phi_i}{\partial n} - \Phi_i \frac{\partial \Phi}{\partial n} \right) dS . \tag{5.23}$$

The surface integral is taken over all the electrodes plus the surface of a small sphere around the point x_1. The volume integral is taken over the space delimited by the surfaces; the integral vanishes because of (5.22). The partial derivative of Φ_i with respect to the normal n to the surface is the electric field there, and, according to Gauss's theorem, every surface integral equals the charge inside, divided by ε_0:

$$\int_{\text{elect } j} \frac{\partial \Phi}{\partial n} \, dS = \frac{Q_j}{\varepsilon_0} , \tag{5.24}$$

$$\int_{\text{elect } j} \frac{\partial \Phi_i}{\partial n} \, dS = \frac{Q_j^{(i)}}{\varepsilon_0} , \tag{5.25}$$

$$\int_{\text{sphere } x_1} \frac{\partial \Phi}{\partial n} \, dS = \frac{q}{\varepsilon_0} , \tag{5.26}$$

$$\int_{\text{sphere } x_1} \frac{\partial \Phi_i}{\partial n} \, dS = 0 , \tag{5.27}$$

where $Q_j^{(i)}$ is the charge induced on electrode j by the potential of electrode .

i alone, and Q_i is the charge whose time rate of change we want to know. We can now perform the surface integrals, and (5.23) becomes

$$0 = \Phi_i(x_1)q - V_iQ_i + \sum_j V_jQ_j^{(i)} \ . \tag{5.28}$$

The time derivative of the first term comes from the movement of the charge q in the inter-electrode space, and is equal to the scalar product of the velocity v of the charge with the electric field created by solution i at the place where the charge is

$$\frac{d}{dt}\Phi_i(x_1)q = q\frac{dx_1}{dt}\cdot \nabla\Phi_i(x_1) = -qv\cdot E_i(x_1) \ . \tag{5.29}$$

The time derivative of the second term is the current in question, multiplied by the potential of the ith electrode:

$$\frac{d}{dt}V_iQ_i = V_iI_i \ . \tag{5.30}$$

The third term does not vary with time because it does not depend on the moving charge. So from (5.28–5.30) we finally obtain

$$I_i = -q\frac{v\cdot E_i(x_1)}{V_i} \ . \tag{5.31}$$

This holds for arbitrary V, and therefore we have Ramo's theorem: the current I that flows into one particular electrode i under the influence of a charge q moving at x_1 with velocity v can be calculated using (5.31) from the field $E_i(x_1)$ created by raising this electrode to potential V_i and grounding all others, in the absence of the charge.

A simple application of Ramo's theorem to the sense-wire grid of a drift chamber explains the well-known fact that the signal induced in the wire that carried the avalanche has a sign opposite to that of the signal induced in its neighbouring electrodes.

References

[FAN 79]   D.L. Fancher and A.C. Schaffer, Experimental study of the signals from a segmented cathode drift chamber, *IEEE Trans. Nucl. Sci.* **NS-26**, 150 (1979)

[PIU 82]   F. Piuz, R. Rosen, J. Timmersmans, Evaluation of systematic errors in the avalanche localization along the wire with cathode strips read-out MWPC, *Nucl. Instrum. Methods* **196**, 451 (1982)

[RAM 39]   S. Ramo, Currents induced in electron motion, *Proc. IRE* **27**, 584 (1939)

[WIL 50]   D.H. Wilkinson, Ionization Chambers and Counters, (Cambride University Press, Cambridge 1950)

6. Coordinate Measurement and Fundamental Limits of Accuracy

6.1 Methods of Coordinate Measurement

After the passage of a charged particle through the sensitive gas volume, the electrons produced in the ionization process along a trajectory segment drift toward the sense wire, where they are collected and amplified in avalanches. There are essentially four different methods to determine particle coordinates in drift chambers:

1. *Measurement of the drift time.* Using the known drift velocity of the ionization electrons along their drift trajectory, this determines the distance along the drift trajectory between wire and track.
2. *Measurement of the pulse-height ratios on pick-up electrodes* (strips, "pads", wires) near the sense-wire avalanche. With electrode response known this determines the coordinate of the avalanche between the pick-up electrodes. If the drift trajectory is known, the corresponding coordinate is determined. In this way the coordinate along the sense wire is measurable.
3. *Measurement of the pulse-height ratios at the two ends of a sense wire* (charge division). The position of the avalanche along the wire is determined using the known damping of the pulse as it propagates along the wire.
4. *Measurement of the difference of arrival times at the two ends of a sense wire* (time difference). This has been used for the same purpose, applying the known propagation time of the signal on the wire.

Of these, methods 3 and 4 measure the coordinate along the wire direction and method 1 measures the coordinate in the drift direction. The coordinate direction of method 2, finally, is given by the arrangement of the pick-up electrodes. A combination of method 1 with one of the others allows a three-dimensional position measurement of the track segment to be made, provided the drift trajectories are known.

Methods 1 and 2 are the most powerful in terms of the achievable accuracy because it turns out that the electronic measurement of the time and of the pulse-height ratio on neighbouring electrodes can often be made better than the fundamental limits imposed by the properties of the ionization that arrives on the wire.

Comparing methods 2 and 3 we notice that in both cases the coordinate is obtained from a known function of a pulse-height ratio times a length: the distance D between the pick-up electrodes in method 2 and the wire length L in the other case. It is clear that the response (i.e. the change in pulse-height ratio per unit length of avalanche displacement) with method 2 is better than the response with method 3 by the order of magnitude of the ratio L/D. The designer of the chamber has to pay for this advantage with a correspondingly larger number of electronic channels.

Comparing methods 1 and 4, both based on a measurement of time, method 4 is less favourable because the response (i.e. the change in measured time per unit length of coordinate displacement) is inversely proportional to the velocity involved, this being the drift velocity u in the gas for method 1 and the signal propagation velocity v (a good fraction of the speed of the light) along the sense wire for method 4. Therefore the response with method 1 is better than the response with method 4 by the order of magnitude of the ratio v/u, which has typical values of 10^3 to 10^4.

It is not our goal to review the various technical aspects of these methods, such as pulse propagation, amplifier noise or discriminator behaviour. Our interest is focussed on the fundamental processes that create limitations which cannot be overcome by technical improvements. In methods 1 and 2 we have techniques to measure the coordinate such that the error resulting from imperfections of the electronics can be made much smaller than that deriving from the fundamental processes.

If all the technical problems have been solved, which means that the wire positions are under control, the drift-velocity field is known well enough (implying the absence of unknown field distortions of the electric and magnetic fields) and if the electronics is capable of measuring the drift time or the pulse-height ratio of the induced signals with sufficient accuracy, then we are ultimately faced with the statistical fluctuations of the finite number of electrons involved in the measuring process. These fluctuations are strongly influenced by the clustering of the ionization process itself. They are the fundamental limitation of any coordinate measurement in drift chambers and become effective because of the presence of mechanisms which spread the ionization cloud as it reaches the wire. In this chapter we will study the most important mechanisms of spread: the diffusion in the drift toward the sense wire, the drift path, and time variations due to the inhomogeneous electric field near the wire and those due to an angle the track might have with the wire. In the presence of a magnetic field, the cylindrical electric field near the wire can combine with the magnetic field and create a spread of ionization as if the track had an angle with the wire.

The problem of accuracy can be described in the following terms. The ionization electrons from the track element in question arrive at the wires; they are spread both in arrival time and along the wires. What are the best estimates for the two coordinates and what is their statistical variation?

In order to understand this better, we will first analyse the spreading on a single wire and how it depends on the track angles, the wire geometry, the magnetic and electric fields and the diffusion. Then we have to discuss how much we gain from the fact that there is not one electron but many electrons arriving. Here the nature of the ionization, which is clustered, plays a role.

Anticipating the result of this chapter, we summarize the benefit of having many electrons as follows. When diffusion is the main source of spreading, the increase in accuracy is a function of the total number N of electrons, in the sense that the r.m.s. measurement error decreases as $1/\sqrt{N}$. When any of the other mechanisms is the main source of spreading, the accuracy is a function only of N_{eff}, a number which is usually much smaller than N and depends on the length of the track segment as well as on diffusion.

6.2 Basic Formulae for a Single Wire

For the coordinate measurements along the wire direction and along the drift direction there are some common aspects that we want to underline. In both cases the accuracy is limited by a track angle which spreads the ionization at the wire and by diffusion which amplifies this spread.

The most general orientation a track might have with the wire plane is given by two angles θ and α. For the measurement along the wire, only the angle θ between the track projected onto the wire plane and the wire is relevant, and for the measurement along the drift direction it is the angle α between the track projected onto the plane orthogonal to the wire and the wire plane.

We consider the simplified scheme of a drift chamber given in Figs. 6.1 and 6.2. The sense wire is parallel to the x axis and contains the point of origin $y = z = 0$. We show two regions: the "drift region", where the ionization electrons drift and diffuse, and a "wire region" where they are collected on the wire. The drift-velocity vector is assumed constant in the drift region and the diffusion in the wire region is neglected.

In Fig. 6.1 we have drawn the case relevant for the coordinate along the wire, where the track may be assumed to be parallel to the x–y plane. In Fig. 6.2 we

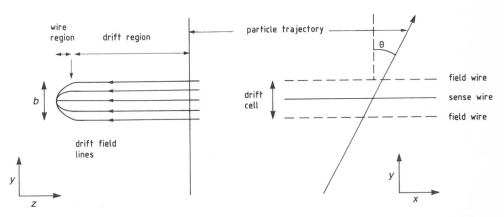

Fig. 6.1. Scheme of a drift cell. For a study of the ionization spread on the wire, we consider a track parallel to the x–y plane

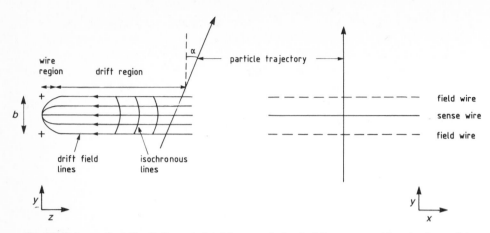

Fig. 6.2. Scheme of a drift cell. For a study of the spread of arrival times we consider a track parallel to the z–y plane

have drawn a different track, parallel to the x–z plane, relevant for the coordinate along the drift direction. In the following we discuss those aspects of the problem that are common to the determination both coordinates.

The ionization clusters are randomly distributed along the particle trajectory with uniform distribution. An electron is collected by the wire if it arrives at $-b/2 < y < b/2$ at the entrance of the wire region after its drift and its random diffusion. Because of diffusion, electrons produced at $y < -b/2$ or at $y > b/2$ may also be collected by the wire and contribute to the measurement.

Since we neglect the diffusion in the wire region, the arrival position and time of one electron on the wire are completely determined by its arrival coordinates and time at the entrance to the wire region. We discuss first the distribution of these variables in the case of a single electron, and later we treat the statistical problems connected with the ensemble of collected electrons.

6.2.1 Frequency Distribution of the Coordinates of a Single Electron at the Entrance to the Wire Region

This distribution is determined by the trajectory of the particle and by the random diffusion of the drifting electron in the drift region.

We begin with the particular case shown in Fig. 6.1: a charged particle parallel to the x–y plane produces ionization clusters at coordinates $x = x_0 + s \sin \theta$, $y = s \cos \theta$ and $z = z_0$, where s is the coordinate along the track and θ is the angle between the perpendicular to the wire and the projection of the trajectory on the wire plane.

The arrival position of an electron produced at coordinate s along the track is distributed according to a Gaussian diffusion around the production coordinate, while s is uniformly distributed along the trajectory. Introducing the symbol G,

$$G(x|\langle x \rangle, \sigma) = \frac{1}{\sqrt{2\pi}\sigma} \exp\left(-\frac{(x - \langle x \rangle)^2}{2\sigma^2}\right),$$

to denote the generic Gaussian function of average $\langle x \rangle$ and r.m.s. σ we can write down the frequency distribution of the arrival positions and time of the electron at the entrance of the wire region ($z = z_w$) as

$$F(x, y, t)\mathrm{d}x\mathrm{d}y\mathrm{d}t = \frac{\mathrm{d}x\,\mathrm{d}y\,\mathrm{d}t}{R} \int_{-R/2}^{+R/2} \mathrm{d}s\, G(x|x_0 + s \sin\theta, \sigma_x)$$

$$\times G(y|s \cos\theta, \sigma_y)\, G(t|(z_0 - z_w)/u, \sigma_t)\,. \tag{6.1}$$

Here σ_x, σ_y, $\sigma_t = \sigma_z/u$ denote the diffusion r.m.s. in the three directions (in the most general case we should introduce the diffusion tensor) and u is the drift velocity assumed along z. R is the length of a track arbitrarily chosen so large that $R \cos\theta \gg b$ and $R \cos\theta \gg \sigma_i$ ($i = x, y, z$). In the following we will always take the limit $R \to \infty$.

This frequency distribution has been written assuming that just one ionization electron is produced along the track segment R. This electron is collected by the wire and contributes to the coordinate measurement only if its arrival position at the entrance to the wire region satisfies the condition $-b/2 < y < b/2$. The probability that this electron is collected by the wire is

$$\int_{-\infty}^{+\infty} \mathrm{d}x \int_{-\infty}^{+\infty} \mathrm{d}t \int_{-b/2}^{+b/2} \mathrm{d}y\, F(x, y, t) = \frac{b}{R \cos\theta} \tag{6.2}$$

and is equal to the ratio between the length of the portion of the track between the planes $y = -b/2$ and $y = b/2$ and R.

If we are interested in the measurement of the coordinate along the wire direction (x) alone, the readout electronics will integrate electrons arriving at different times: the probability distribution of the arrival positions at the entrance of the wire region for the electrons that are collected by the wire is obtained integrating the distribution $F(x, y, t)$ given by (6.1) in the variable t and normalizing it using (6.2):

$$f_x(x, y)\mathrm{d}x\,\mathrm{d}y = \mathrm{d}x\mathrm{d}y\frac{\cos\theta}{b} \int_{-R/2}^{+R/2} \mathrm{d}s\, G(x|x_0 + s \sin\theta, \sigma_x)\, G(y|s \cos\theta, \sigma_y)\,.$$

$$\tag{6.3}$$

Assuming that the x coordinate on the wire is the same as at the entrance to the wire region we can compute the average and the variance of the arrival position at the wire as

$$\langle x \rangle = \int_{-b/2}^{b/2} \mathrm{d}y \int_{-\infty}^{+\infty} x f_x(x, y)\mathrm{d}x = x_0\,, \tag{6.4}$$

$$\langle x^2 \rangle - \langle x \rangle^2 = \sigma_x^2 + \sigma_y^2 \tan^2\theta + \frac{b^2}{12} \tan^2\theta\,. \tag{6.5}$$

The variance of x is the sum of three terms: the first two depend on diffusion, the third is the projection of the track segment on the wire; this latter we call the *angular wire effect*. The factor $1/12$ that divides the square of the width b of the drift cell comes from the variance of the uniform distribution of the electrons along the track.

6.2.2   Frequency Distribution of the Arrival Time of a Single Electron at the Entrance to the Wire Region

We now turn to the particular case shown in Fig. 6.2: a charged particle parallel to the y–z plane produces ionization clusters at coordinates $x = x_0$, $y = s \cos \alpha$ and $z = z_0 + s \sin \alpha$, where s is the coordinate along the track and α is the angle between the perpendicular to the drift-velocity direction (z) and the projection of the trajectory on the wire z–y plane.

The frequency distribution of the arrival positions and time of the electron at the entrance of the wire region ($z = z_w$) is

$$F(x, y, t)\mathrm{d}x\,\mathrm{d}y\,\mathrm{d}t = \frac{\mathrm{d}x\,\mathrm{d}y\,\mathrm{d}t}{R} \int\limits_{-R/2}^{+R/2} \mathrm{d}s\, G(x\,|\,x_0, \sigma_x)$$

$$\times\, G(y\,|\,s \cos \alpha, \sigma_y)\, G(t\,|\,(z_0 + s \sin \alpha - z_w)/u, \sigma_t)\,. \qquad (6.6)$$

If we are interested in the measurement of the coordinate along the drift direction z alone, the readout electronics will not distinguish among electrons arriving at different wire positions: the probability distribution of the arrival time at the entrance to the wire region for the electrons that are collected by the wire is obtained by integrating the distribution given by (6.6) in the variable x and normalizing it using (6.2):

$$f_z(t, y)\mathrm{d}t\,\mathrm{d}y = \mathrm{d}t\,\mathrm{d}y\, \frac{\cos \alpha}{b} \int\limits_{-R/2}^{+R/2} \mathrm{d}s\, G(t\,|\,(z_0 + s \sin \alpha - z_w)/u, \sigma_t)$$

$$\times\, G(y\,|\,s \cos \alpha, \sigma_y)\,. \qquad (6.7)$$

Assuming for the moment that the arrival time on the wire is equal to the time at the entrance to the wire region, plus the constant shift z_w/u, we can compute the average and the variance of the arrival time at the wire;

$$\langle t \rangle = \int\limits_{-b/2}^{b/2} \mathrm{d}y \int\limits_{-\infty}^{+\infty} t f_z(t, y)\mathrm{d}t = z_0/u\,, \qquad (6.8)$$

$$\langle t^2 \rangle - \langle t \rangle^2 = \sigma_t^2 + \frac{\sigma_y^2}{u^2} \tan^2 \alpha + \frac{b^2}{12u^2} \tan^2 \alpha\,. \qquad (6.9)$$

Comparing (6.3) and (6.7) we observe that one can obtain one distribution from the other with the substitutions

$$f_x \leftrightarrow f_z\,,$$

$$x \leftrightarrow z = tu \, ,$$

$$x_0 \leftrightarrow z_0 - z_w \, ,$$ \hfill (6.10)

$$\sigma_x \leftrightarrow \sigma_z = \sigma_t u \, ,$$

$$\theta \leftrightarrow \alpha \, .$$

In the following, we discuss how the cluster fluctuations influence the resolution in the coordinate measurement along the wire direction (x). This will be done using the probability distribution (6.3) and assuming that the x coordinate at the wire is the same as at the entrance of the wire region. Analogously, the same conclusions can be drawn with regard to the coordinate measurement in the drift direction, using the formal substitutions (6.10). Drift-path variations are introduced in Sect. 6.4.

6.2.3 Influence of the Cluster Fluctuations on the Resolution – the Effective Number of Electrons

Up to this point we have calculated the average and the variance of the position x of a single electron (see (6.4) and (6.5)). For the measurement of the coordinate we average over all electrons that are collected by the wire in a given time interval, defined by the readout electronics, and the variance of the coordinate measurement will be reduced. As it turns out, this reduction is not given by a single factor. This is because of the ionization clustering.

We measure the x coordinate averaging the arrival positions of N electrons:

$$X_{AV} = \frac{x_1 + x_2 + \ldots + x_N}{N} \, . \hfill (6.11)$$

The variance of X_{AV} depends on the characteristics of the ionization process because the electrons are produced in clusters and therefore the x are, in general, not independent.

Assume that the N electrons are grouped in M sets of n electrons, each set containing the electrons that were produced in the same cluster and after drift and diffusion are collected by the wire:

$$\sum_{j=1}^{M} n_j = N \, . \hfill (6.12)$$

In general not all the electrons produced in the same cluster are collected by the wire because of the random diffusion along the drift path.

The variance of X_{AV} is by definition

$$\sigma_{X_{AV}}^2 = \langle X_{AV}^2 \rangle - \langle X_{AV} \rangle^2 \, , \hfill (6.13)$$

where the average $\langle \; \rangle$ is taken over the probability distribution f_x:

$$\langle X_{AV} \rangle = \frac{\sum \langle x_i \rangle}{N} = \int_{-b/2}^{b/2} dy \int_{-\infty}^{+\infty} x f_x(x, y) dx = x_0 \hfill (6.14)$$

Table 6.1. Classification of the pairs of all electrons from a track that were collected on one wire

Cluster		1				2		3 ....M
Electron		1 2 3 4				5 6		7 ....N
1	1	× · · ·				· ·		· ·
	2	· × · ·				· ·		· ·
	3	· · × ·				· ·		· ·
	4	· · · ×				· ·		· ·
2	5	· · · ·				× ·		· ·
	6	· · · ·				· ×		· ·
3	7	· · · ·				· ·		× ·
⋮	⋮	· · · ·				· ·		· ×
M	N							

and

$$\langle X^2_{\text{AV}} \rangle = \frac{1}{N^2} \sum_{ij} \langle x_i x_j \rangle . \tag{6.15}$$

The sum in the last equation contains three different kinds of terms; refer to Table 6.1 for a graphical representation. When $i = j$ then $\langle x_i x_j \rangle = \langle x^2 \rangle$. There are N terms of this kind; they are marked by crosses in the table. When $i \neq j$ and the electrons do not belong to the same cluster, then $\langle x_i x_j \rangle = \langle x \rangle^2$, because two different ionization processes are not correlated. There are $N^2 - \sum n_j^2$ terms of this kind, they are represented by the dots in the rectangular boxes of the table. When $i \neq j$ and the electrons do belong to the same cluster (the dots in the square boxes), then x_i and x_j are correlated. We define the average $\langle x_i x_j \rangle = \langle xx \rangle$. There are $\sum n_j^2 - N$ terms of this kind.

Using (6.13–15) one finds that the variance is equal to

$$\sigma^2_{X_{\text{AV}}} = \frac{1}{N} (\langle x^2 \rangle - \langle xx \rangle) + \frac{\sum n_j^2}{N^2} (\langle xx \rangle - \langle x \rangle^2) . \tag{6.16}$$

In the absence of diffusion, $\langle xx \rangle$ is equal to $\langle x^2 \rangle$ and the first term of (6.16) is zero, so we obtain the result derived in Sect. 1.2, where the quantity $\sum n_j^2/N^2$ was defined as $1/N_{\text{eff}}$. If all the clusters consisted of one electron, the term $\sum n_j^2/N^2$ there became equal to $1/N$ and we obtained the standard formula of the variance of the mean.

We may therefore distinguish the following limiting cases of (6.16):

● Limit of no diffusion, using (6.5):

$$\sigma^2_{X_{\text{AV}}} \to \frac{1}{N_{\text{eff}}} (\langle x^2 \rangle - \langle x \rangle^2) = \frac{1}{N_{\text{eff}}} \frac{b^2}{12} \tan^2 \theta . \tag{6.17}$$

● Limit of no clustering, using (6.5) and $\sigma_x = \sigma_y = \sigma$:

$$\sigma^2_{X_{AV}} \to \frac{1}{N}(\langle x^2 \rangle - \langle x \rangle^2) = \frac{1}{N}\left(\frac{\sigma^2}{\cos^2\theta} + \frac{b^2}{12}\tan^2\theta\right). \qquad (6.18)$$

● Limit $\theta = 0$, in the presence of clustering:

$$\sigma^2_{X_{AV}} \to \frac{\sigma}{N}. \qquad (6.19)$$

In (6.17–19), N_{eff} is the effective number and N is the total number of electrons on the track segment delimited by the cell width; their values are θ-dependent through the length of the track segment.

An analytical expression of the covariance $\langle xx \rangle$ in (6.16) is derived under certain assumptions in the appendix to this chapter, resulting in formula (6.72) for $\sigma^2_{X_{AV}}$. This can be rewritten in the form

$$\sigma^2_{X_{AV}} \to \frac{1}{N'}\frac{\sigma^2}{\cos^2\theta} + \frac{1}{N'_{eff}}\frac{b^2}{12}\tan^2\theta, \qquad (6.20)$$

where the quantities N' and N'_{eff} depend on θ as well as on σ and b.

Let us first deal with the term proportional to $(b^2/12)\tan^2\theta$. We may regard N'_{eff} as the effective number of electrons that would cause the same fluctuation as the combined action of ionization plus diffusion, thus extending the original meaning of N_{eff} defined in Sect. 1.2 and used in (6.17).

The quantity N'_{eff}, for which an analytic expression is given in the appendix to this chapter, can also be calculated with a numerical simulation of the ionization and diffusion process. In order to compute it as a function of the diffusion parameter σ we evaluate the variance V_{AV} of the average of the coordinates along the track direction of those electrons that, after diffusion σ along the track direction, are contained in a track segment of length l. We write

$$\frac{1}{N'_{eff}} = \frac{1}{N_{eff}(\sigma)} = \frac{l^2}{12}\frac{1}{V_{AV}(\sigma)}. \qquad (6.21)$$

Figure 6.3 shows how $N_{eff}(\sigma)$ varies with diffusion. We have plotted on the vertical axis the quantity

$$N_{eff}(\sigma)\frac{\lambda}{l},$$

where the ratio λ/l is equal to the average number of clusters on the track segment l (λ was defined in Sect. 1.1.1 as the mean distance between two clusters). Therefore $N_{eff}(\sigma)\,\lambda/l$ is the ratio of the effective number of electrons over the number of clusters in the track length l.

On the horizontal scale we have the diffusion expressed in units of the length of the track segment σ/l. This is suggested by (6.72) and is a very natural choice because the change of $N_{eff}(\sigma)$ is caused by electrons that have diffused so much

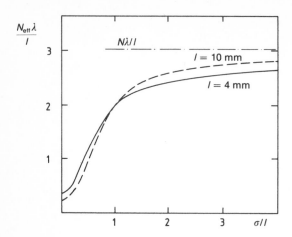

Fig. 6.3. Declustering through diffusion: effective number of electrons divided by the total number of clusters, as a function of σ/l, for two values of l. In the Monte Carlo calculation the cluster-size distribution of Fig. 1.7 and a value of $1/\lambda = 2.7$ clusters/mm were used

that they cross over from one track segment to the neighbouring one, thus breaking the cluster correlation. It can therefore be expected that $N_{\text{eff}}(\sigma)$ goes up when σ exceeds some distance comparable to the length of the track segment, i.e. the distance between the wires. And this is in fact what we see.

The curves in Fig. 6.3 where computed using the argon cluster-size distribution of Fig. 1.7. At zero diffusion $N_{\text{eff}}(0)$ is only 0.3 ($l = 0.4$ cm) or 0.2 ($l = 1$ cm) of the number of clusters. We know from Fig. 1.19 that this number must go down with increasing l. Now we switch on the diffusion, and before σ has reached half the length l of the segment, $N_{\text{eff}}(\sigma)$ is as large as the number of clusters. For very large σ/l all the clustering is destroyed by the diffusion, and $N_{\text{eff}}(\sigma)$ approaches the total number N of electrons. The increase of $N_{\text{eff}}(\sigma)$ with σ/l was first observed in a TPC and termed *declustering through diffusion* [BLU 86].

We notice that the wire angular term in (6.20) can be written in the following way:

$$\frac{1}{N_{\text{eff}}(\sigma)} \frac{(b \tan \theta)^2}{12} = \frac{1}{N_{\text{eff}}(\sigma)} \frac{1}{12} \left(\frac{b}{\cos \theta}\right)^2 \sin^2 \theta . \tag{6.22}$$

It represents the variance of the average of the position along the track direction of the electrons sampled by the wire, projected onto the wire direction. The first term is the variance of a flat distribution of width $b/\cos \theta$, which is the length of the track segment sampled by the wire. The factor $\sin^2 \theta$ projects this variance onto the wire direction.

After this discussion of N'_{eff} as contained in the second term of (6.20), we deal now with the first term. The simple form into which we have cast the terms directly proportional to σ^2 (cf. (6.72)) hides the complications in the symbol N'. Although it is true that in the limiting cases of (6.18) and (6.19) N' is exactly equal to N, the total number of electrons on the track segment delimited by the cell width, N' changes as a function of θ, σ and b so as to make the variance

$\sigma^2_{X_{AV}}$ larger. The physical cause of this is the rare larger clusters outside the cell that send some electrons via diffusion to the edge of the cell, and more so when θ is large. We have two reasons not to go into these details here. When we omit the primes in (6.20) the omissions amount to less than 25% in the quantity $\sigma_{X_{AV}}$ (except in extreme cases). The omitted parts, which pull measurements to one side in our cell, have a tendency to pull the corresponding measurement in the neighbouring cell to the opposite side. Therefore it is better to leave them out when combining several cells. The contributions of several wires are discussed in Sects. 6.3 and 6.4.

In conclusion we simplify (6.20) to read

$$\sigma^2_{X_{AV}} = \frac{1}{N} \frac{\sigma^2}{\cos^2\theta} + \frac{1}{N_{\text{eff}}} \frac{b^2}{12} \tan^2\theta . \tag{6.23}$$

It represents the square of the accuracy with which a track coordinate X_{AV} (6.11) can be measured along a single wire in the absence of a magnetic field.

6.3 Accuracy in the Measurement of the Coordinate in or near the Wire Direction

6.3.1 Inclusion of a Magnetic Field Perpendicular to the Wire Direction: the Wire $E \times B$ Effect

Drift chambers with precise measurements of the track coordinates along the wire direction are often operated with a magnetic field perpendicular to the direction of the wires. With this configuration one can obtain a precise determination of the curvature induced on the particle trajectory by the magnetic field, and the momentum of the particle can be determined (see Chap. 7 for details).

The presence of a magnetic field perpendicular to the wire direction modifies the angular wire term of the variance of the arrival position of the electrons (6.5). It becomes asymmetric and, on average, larger because the track segment is projected onto the wire in a more complicated way. The electrons that after their drift are collected in the cylindrical field of the wire have to move transverse to the magnetic field (Fig. 6.4). This produces an $E \times B$ force according to (2.6) and causes the electrons to drift under an effective angle ψ toward the wire. The angle ψ is such that $\tan\psi = \omega\tau$, where ω is the cyclotron frequency and τ is the time between two electron collisions suitably averaged. Details have been treated in Sect. 2.1.

The arrival position x_w of an electron entering the region close to the wire with coordinates x and y is $x_w = x - y\tan\psi$; the variance of x_w on the frequency distribution (6.3) is

$$\langle x_w^2 \rangle - \langle x_w \rangle^2 = \frac{\sigma^2}{\cos^2\theta} + \frac{b^2}{12}(\tan\theta - \tan\psi)^2 . \tag{6.24}$$

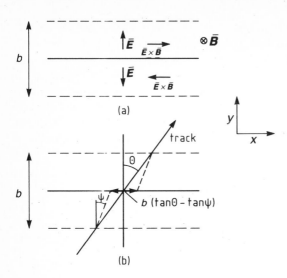

Fig. 6.4. (a) Directions of the electric and magnetic fields in the region close to the sense wire. The magnetic field points into the page. **(b)** Direction of the drift velocity of the electrons in this region projected onto the x–y plane

The electrons produced by a track at $\theta = 0$ are spread over a region $b \tan \psi$, and those of a track at an arbitrary angle are smeared over a distance $b(\tan \theta - \tan \psi)$. The width of the charge distribution on the wire has a minimum at $\theta = \psi$ and not at $\theta = 0$.

This effect was discovered with the TRIUMF-TPC [HAR 84] and is typically the most important limitation of the measuring accuracy in all TPC-like detectors using a gas with high values of $\omega\tau$. In the TRIUMF-TPC the angle ψ was 29°, and 32° in the ALEPH TPC in standard running conditions. The broadening of the avalanche width has been systematically studied in [BLU 86].

Using (6.24) we can rewrite (6.23) to include this effect:

$$\sigma^2_{X_{AV}} = \frac{1}{N} \frac{\sigma^2}{\cos^2\theta} + \frac{1}{N_{eff}} \frac{b^2}{12} (\tan \theta - \tan \psi)^2 . \qquad (6.25)$$

Equation (6.25) is the general expression for the accuracy $\sigma_{X_{AV}}$ with which a track segment can be measured on one wire along the wire direction. Summarizing our findings up to here, we see that this accuracy depends on the projected angle θ the track has with respect to the normal to the wire direction as well as on the angle ψ of the wire $E \times B$ effect. N_{eff} is roughly 6 for 1 cm of argon NTP and is more accurately obtained in Figs. 6.3 and 1.19. The diffusion term is characterized by the width σ of the single-electron diffusion transverse to the drift and the total number N of electrons, typically 100/cm in argon NTP.

The size and relative importance of the two terms in (6.25) depend on the electron drift length L, because σ^2 is proportional to L and N_{eff} depends on L through the declustering effect (Fig. 6.3). Also the angle θ has some influence on N and N_{eff} as the length of the track segment varies with θ. If one wants to be more specific, one has to take into account the details of a particular chamber.

6.3.2   Case Study of the Explicit Dependence of the Resolution on L and θ

Writing (6.25) in the form

$$\sigma_{\bar{X}_{AV}}^2 = \frac{1}{n_0}\left(\frac{C^2 L}{b \cos\theta} + \frac{b}{12}\frac{(\tan\theta\tan\psi)^2\cos\theta}{g(L)}\right) \qquad (6.26)$$

we have introduced the L dependence of the diffusion and of N_{eff}:

$$\sigma^2 = C^2 L, \qquad (6.27)$$

$$N_{eff} = N g(L), \qquad (6.28)$$

as well as the θ dependence of N:

$$N = n_0 b/\cos\theta. \qquad (6.29)$$

Here n_0 is the number of collected electrons per unit track length. (We recall that $g(L)$ is not a universal function but depends to some extent on N, b and θ.)

Figure 6.5 shows $\sigma_{\bar{X}_{AV}}^2$ calculated as a function of L at $\theta = 0$ for $C^2 = 3.4 \times 10^{-3}$ mm, $\psi = 32°$ (corresponding to a gas mixture of 80% argon and 20% methane in a field of 0.85 T), $b = 4$ mm, $n_0 = 8.1$ electrons/mm. We notice that the resolution goes through a shallow minimum at small L and is dominated by diffusion at large L.

Figure 6.6 shows $\sigma_{\bar{X}_{AV}}^2$ as a function of the angle θ for two different values of L and for the same choice of the other parameters.

6.3.3   The General Situation – Contributions of Several Wires, and the Angular Pad Effect

The most general situation is the one where the coordinate direction along the pad row is inclined with respect to the wire direction, and several wires are

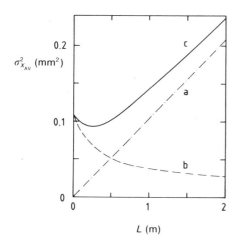

$\sigma_{\bar{X}_{AV}}^2$ (mm²)

L (m)

Fig. 6.5a–c. Variance of the average arrival position as a function of the drift length at $\theta = 0$ – the special case described in the text. (a) Contribution of the diffusion; (b) contribution of the angular wire effect; (c) sum of the two

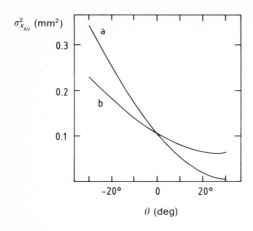

Fig. 6.6. Variance of the average arrival position of the electrons as a function of the angle θ – the special case described in the text. (a) $L = 0$, (b) $L = 50$ cm

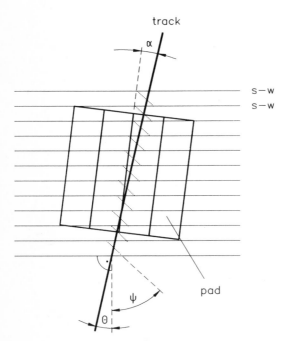

Fig. 6.7. Scheme of a chamber with cathode-strip readout. The *bold lines* indicate the arrival positions of the electrons on the wire; the *dotted lines* indicate the direction of the drift velocity in the region close to the sense wires

located opposite the same cathode strips, as seen in Fig. 6.7. Apart from θ, the angle between the track and the wire normal, and ψ, the effective angle of approach caused by the wire $\boldsymbol{E} \times \boldsymbol{B}$ effect, we also have α, the angle between the normal to the direction of the pad row and the track; all angles are measured in the wire plane. The angles θ and α are the same when the pad row follows the wire direction.

The signals induced in a cathode strip are from k wires, where

$k = $ pad length h/wire pitch b .

These signals are added in the measuring process. We want to compute the achievable accuracy. As the situation is quite complicated we wish to start with a simple case. The basic statistical relation will be derived for the special geometry given by $\alpha = 0$, $\theta = 0$; also we will assume $\sigma = 0$. Later on we will generalize our findings step by step.

The simplified geometry is sketched in Fig. 6.8. The ionization clusters along the track are redistributed onto the sense wires, where they occupy sections of length $b \tan \psi$. The situation is quite similar to the one in Sect. 1.2.6, where we treated the problem of charge localization along a track. The coordinate to be determined by the pads in one measurement is

$$X_{\text{AVP}} = \frac{\sum\limits_{i=1}^{m_1} x_i n_i + \sum\limits_{i=m_1+1}^{m_2} x_i n_i + \ldots + \sum\limits_{i=m_{k-1}+1}^{m_k} x_i n_i}{\left(\sum\limits_{i=1}^{m_k} n_i\right)^2}, \qquad (6.30)$$

where each x_i is the position of a cluster with n_i electrons; there are m_1 clusters on the first, and $m_j - m_{j-1}$ on the jth sense wire ($j = 1, \ldots, k$). Averaging over the positions x_i – which are distributed between $-(b/2) \tan \psi$ and $+(b/2) \tan \psi$ – we find for the average and the variance

$$\langle X_{\text{AVP}} \rangle = 0$$

and

$$\sigma^2_{\text{AVP}} = \langle X^2_{\text{AVP}} \rangle - \langle X_{\text{AVP}} \rangle^2$$

$$= \frac{b^2 \tan^2 \psi}{12} \frac{\sum\limits_{i=1}^{m_k} n_i^2}{\left(\sum\limits_{i=1}^{m_k} n_i\right)^2} \qquad (6.31)$$

$$\sigma^2_{\text{AVP}} = \frac{b^2 \tan^2 \psi}{12} \frac{1}{N_{\text{eff}}(h)} \quad (\alpha = \theta = \sigma = 0).$$

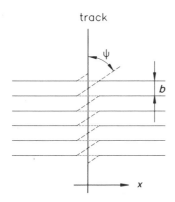

Fig. 6.8. Track and sense wires for (6.31)

Here we have denoted by $1/N_{\text{eff}}(h)$ the statistical factor which is equal to the sum of the squared cluster size divided by the square of the summed cluster sizes, taken over all the clusters on the piece of track that is delimited by the pad length h. Let us repeat (cf. Sect. 1.2.9) that the effective number N_{eff} is a measure of fluctuation which is smaller than the number of clusters because of the fluctuations of the cluster size. It is not proportional to h; for argon it scales according to Fig. 1.19. Comparing $N_{\text{eff}}(b)$, the corresponding number that belongs to a wire cell, we have for that case

$$N_{\text{eff}}(h) \sim N_{\text{eff}}(b)(h/b)^{0.54} \ .$$

For this reason the achievable accuracy of pads does not improve with the square root but (for argon) more with the fourth root of the pad length.

In the next step we let θ and σ be different from zero, but keep $\sigma \ll b$ and $\alpha = 0$. The geometry is sketched in Fig. 6.9. The angle θ the track makes with the wire normal has two consequences: it changes the length on the wire occupied by charges, from $b \tan \psi$ to $b(\tan \psi - \tan \theta)$, and it introduces the projection factor $\cos \theta$ between this length and the coordinate direction. The diffusion, finally, contributes to the variance the same term as in (6.23) but with two changes. Firstly, the projection factor onto the coordinate direction is 1 in the present case. Secondly, N now represents the total number of electrons from the track segment delimited by h; we write $N(h)$, which is of course proportional to h. The variance of X_{AVP} therefore assumes the form

$$\sigma^2_{X_{\text{AVP}}} = \frac{\sigma^2}{N(h)} + \frac{1}{N_{\text{eff}}(h)} \frac{b^2}{12} (\tan \psi - \tan \theta)^2 \cos^2\theta \quad (\alpha = 0, \ \sigma \ll b) \ . \quad (6.32)$$

Next we let α assume a non-zero value. This has the consequence that the two projection factors must be referred to the new coordinate direction, that is the diffusion term receives in its denominator the factor $\cos^2\alpha$, and the projection factor of the angular term changes into $\cos^2(\theta - \alpha)$. The variance (6.32) of X_{AVP} receives another contribution, which describes the fluctuation of the

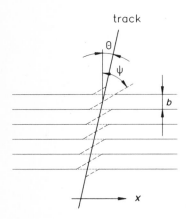

track

Fig. 6.9. Track and sense wires for (6.32)

position of the centre of charge along the track segment, because a non-zero angle α projects this fluctuation onto the coordinate direction. The size of this term is proportional to the squared length of the pad,

$$\frac{1}{N_{\text{eff}}(h)} \frac{(h \tan \alpha)^2}{12}. \tag{6.33}$$

This is the *angular pad effect*. We recognize the similarity with the corresponding angular wire term in (6.23); in (6.33) the fluctuation is controlled by the effective number of electrons on the track segment defined by the pad.

This contribution to the variance can be suppressed if the pulse height on the relevant wires is recorded, because one way of looking at this fluctuation is that it is caused by the differences in the charge deposited on the wires (see Fig. 6.10). It has been demonstrated in practice [AME 83, BAR 82], that the measured track position can be corrected using the wire pulse heights, leaving only a small residual error.

Finally we have to lift the restriction that σ^2 should be small compared to b^2. In order for declustering to occur, the diffusion must reach the value of a length parameter which reorganizes the charges in such a way that the cluster correlation is broken, thus making N_{eff} larger. For the angular terms in (6.23) and (6.25) it was b that set the scale whereas in the angular pad term (6.33) it is h that sets the scale. (It is irrelevant on which wire the charge is collected – only new charges from outside the pad region improve the cluster statistics.)

Although the declustering effect often does not play a decisive role in practical drift chambers – because the diffusion is not large enough – we want to be specific in distinguishing the respective scales of the declustering.

For this purpose we denote by

$$N_{\text{eff}}\left(h, \frac{\sigma}{b}\right)$$

the effective number of electrons on the track segment delimited by h and declustered by diffusion on the scale b. In this sense the variance of X_{AVP} takes

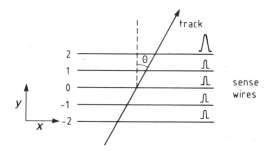

Fig. 6.10. Displacement of the measured coordinate owing to charge fluctuations between the sense wires that contribute to a pad signal. The large pulse height collected by wire 2 moves the centroid towards positive x

the form

$$\sigma_{\text{AVP}}^2 = \frac{1}{N(h)}\frac{\sigma^2}{\cos^2\alpha} + \frac{b^2(\tan\theta - \tan\psi)^2\cos^2(\theta-\alpha)}{12N_{\text{eff}}\left(h,\dfrac{\sigma}{b}\right)} + \frac{(h^2-b^2)\tan^2\alpha}{12N_{\text{eff}}\left(h,\dfrac{\sigma}{h}\right)}.$$

(6.34)

This is the principal limit of the coordinate measuring accuracy using pads; θ, ψ, and α were defined in Fig. 6.7. In the usual case of no declustering, the values of N_{eff} for argon can be taken from Fig. 1.19.

6.3.4 Consequences of (6.34) for the Construction of TPCs

The relative size of the three terms in (6.34) depends, firstly, on the diffusion of the electrons during their drift. In the presence of a typical longitudinal magnetic field, the magnetic anisotropy of diffusion will make the first term the smallest one on average, for any practical choice of b and h.

In the usual case, a pad extends over several wires, and h is much larger than b; therefore the third term will dominate even for a small α. The most critical coordinate measurements are those for a determination of large track momenta on tracks with a small curvature. The pad rows should therefore be oriented at right angles to these critical tracks, so that, for them, $\alpha \approx 0$. For this reason, the ALEPH and the DELPHI TPCs have circular pad rows.

With the choice of h the designer of the TPC wants to take advantage of the fact that $N(h)$ increases in proportion to h and $N_{\text{eff}}(h)$ approximately in proportion to $\sqrt{h}$. Also, the larger the collected amount of charge on a pad the less critical is the noise of the electronic amplifier. On the other hand the third term in (6.34) can be kept small only within a range of α which becomes smaller and smaller as h is increased. Even the modest magnetic bending of a high-momentum track may exceed this range if h is made too large. It can be shown that the consequence is a *constant* limiting relative-momentum-measuring accuracy $\delta p/p$ rather than one which decreases as p is reduced.

Problems of double-track resolution are discussed in Sect. 10.7.1.

6.3.5 A Measurement of the Angular Variation of the Accuracy

The second term in (6.34) can be studied in detail by measuring σ_{AVP}^2 as a function of θ, keeping α fixed at 0. This has been done by Blum et al. [BLU 86] using a small TPC with rotatable wires in a beam of 50 GeV muons. Each track had its transverse position measured at four points using the pulse heights on the pads (see Fig. 6.11). The pad length h was 3 cm and the wire pitch b was 4 mm. The variance was determined for every track as one half the sum of the squared residuals of the straight-line fit. The result is σ_{m}^2, the measured variance

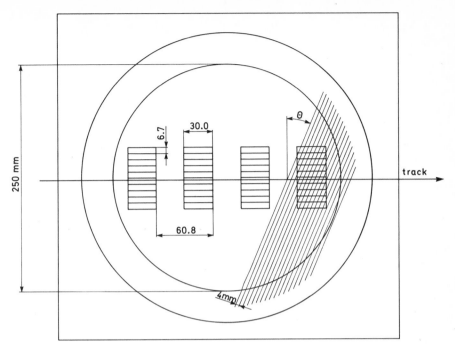

Fig. 6.11. Layout of TPC with rotatable wires [BLU 86]

averaged over many tracks, as a function of the angle θ between the tracks and the normal to the wires. According to (6.34) σ_m^2 should vary as

$$\sigma_m^2 = \sigma_0^2 + \frac{b^2}{12 N_{\text{eff}}\left(h, \dfrac{\sigma}{b}\right)} (\tan\theta - \tan\psi)^2 \cos^2\theta$$

$$= \sigma_0^2 + \frac{b^2}{12 N_{\text{eff}}\left(h, \dfrac{\sigma}{b}\right)} \frac{\sin^2(\theta - \psi)}{\cos^2\psi}, \tag{6.35}$$

where σ_0^2 contains the effects of diffusion and, perhaps, of the electronics and any remaining contribution of the angular pad effect.

The results are depicted in Fig. 6.12. The curves represent fits to (6.35) on adjusting the three free parameters ψ, σ_0 and $N_{\text{eff}}(h, \sigma/b)$. A reversal of the magnetic field changes the sign of ψ so that two curves are seen in Fig. 6.12a. The values of the fitted parameters are shown in Table 6.2.

Keeping in mind that a magnetic field of 1.5 T reduces the transverse diffusion coefficient by a factor 50 (cf. Sect. 2.4.5) we expect for the mean square width of the diffusion cloud after 4 cm of drift a value of 1.6 mm² ($B = 0$) or 3.2×10^{-2} mm² ($B = 1.5$ T). Since on a pad length of 3 cm the number of

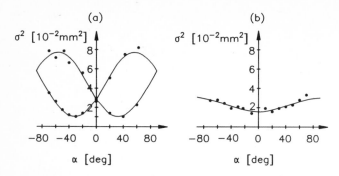

Fig. 6.12a, b. The square of the measuring accuracy σ determined by [BLU 86] as a function of the track angle θ, (a) in a magnetic field of ± 1.5 T, (b) without magnetic field

Table 6.2. Parameters fitted to the measured accuracy σ_m^2 according to (6.35) for two values of the magnetic field (gas: Ar (90%) + CH$_4$ (10%))

B (T)	σ_0^2 (mm^2)	$N_{\text{eff}}\left(h, \dfrac{\sigma}{b}\right)$	ψ (deg)
1.5	$(1.08 \pm 0.03)10^{-2}$	27.8 ± 0.6	32.3 ± 1.5
0	$(1.58 \pm 0.06)10^{-2}$	83 ± 8	-1 ± 2

ionization electrons is approximately 300, the expected change in the uncertainty of the centre is 0.5×10^{-2} mm^2, equal to the measured difference in σ_0^2 appearing in Table 6.2.

The increase by a factor of three in the parameter N_{eff}, connected with the removal of the B field, is attributed to the declustering effect: as the width of the diffusion cloud increases it becomes comparable to b, more electrons from the same cluster arrive on different wires, and N_{eff} goes up. The increase does not occur when the tracks are produced by a laser ray rather than a particle.

The angle ψ of electron drift near the wire was 32° at 1.5 T in gas composed of argon (90%) and methane (10%). This value depends strongly on the gas and on the electric field near the sense wires. The wire and field configuration employed here is characteristic of TPCs and is described in the case studies in Sect. 3.1.2. The tangent of ψ is found to be roughly proportional to B.

6.4 Accuracy in the Measurement of the Coordinate in the Drift Direction

In Sect. 6.2 we pointed out that there is a symmetry between the two situations symbolized in Figs. 6.1 and 6.2, the coordinate measurement along the wire and along the drift direction. The variances (6.5) and (6.9) of the single-electron coordinate are basically the same if one employs the substitution rule (6.10); this holds under the assumption that the time of arrival of the drifting electron at the entrance of the wire region is the same, up to a constant, as that recorded by the electronics.

We must now complement this argument by the observation that for most drift chambers this last approximation is too crude. Figure 6.2 shows that, owing to the shape of the velocity field lines in the wire region, even electrons produced by a track at $\alpha = 0$ have different paths before reaching the wire: electrons collected at the edge of the cell have to drift more than those collected in the middle.

The isochronous lines in the wire region have almost a parabolic shape and can be approximated by

$$z = z_0 - a\frac{y^2}{b},$$

where a is a dimensionless constant of the order of unity, which depends on the shape of the field lines in this region. In the example of Fig. 6.2, $a = 0.8$.

The arrival time t on the wire of a single electron entering the wire region at y at the time t is then

$$t = t' + \frac{1}{u}\left(z_w + a\frac{y^2}{b}\right). \tag{6.36}$$

The average and variance of this last expression on the distribution (6.7) are

$$\langle t \rangle = \frac{1}{u}\left(z_0 + a\frac{b}{12}\right), \tag{3.37}$$

$$\langle t^2 \rangle - \langle t \rangle^2 = \frac{1}{u^2}\left(\sigma_z^2 + \sigma_y^2\tan^2\alpha + \frac{b^2}{12}\tan^2\alpha + \frac{a^2 b^2}{180}\right), \tag{6.38}$$

where the last and new term is the contribution of the drift-path variations in the wire region.

6.4.1 Inclusion of a Magnetic Field Parallel to the Wire Direction: the Drift $E \times B$ Effect

Drift chambers with precise measurements of the track coordinates along the drift direction are often operated with a magnetic field perpendicular to the drift direction and parallel to the wires. With this configuration one can obtain a precise determination of the curvature induced on the particle trajectory by the magnetic field, and the momentum of the particle can be determined (see Chap. 7 for details).

If a magnetic field perpendicular to the drift electric field is present, the drifting electrons have to move transverse to it. This produces an $E \times B$ force according to (2.6) and causes the electrons to drift under an effective angle

ψ with respect to the direction of the electric field. The angle ψ is such that $\tan \psi = \omega\tau$, where ω is the cyclotron frequency and τ is the time between two collision suitably averaged.

Electrons produced by a track at $\alpha = 0$ do not arrive simultaneously at the entrance to the wire region, but their arrival positions depend linearly on y with a slope $\tan \psi$.

The variance of the arrival time of a single electron depends now on the angle ψ:

$$\langle t^2 \rangle - \langle t \rangle^2 = \frac{1}{u^2}\left(\sigma_z^2 + \sigma_y^2 \tan^2\alpha + \frac{b^2}{12}(\tan \alpha - \tan \psi)^2 + \frac{a^2 b^2}{180} \right), \qquad (6.39)$$

where the diffusion parameter σ_z refers to the new drift direction, and σ_y to the direction perpendicular to it and to the magnetic field. The two parameters are not necessary equal, owing to the electric anisotropy of diffusion (cf. Sect. 2.2.5), but for simplicity we will replace them below by a common value.

Up to now we have dealt with the drift-time variations of a single electron. When a measurement is performed, the whole swarm arrives. There are various ways to extract the time information. In terms of electronics, one may either employ discriminators of various types in order to trigger on the first electrons, or on the centroid of the pulse or on some other value based on a threshold-crossing time. Or one may apply a very fast sampling technique in order to obtain a numerical representation of the pulse shape, which may be used to derive a suitable estimator. We discuss here two estimators of the arrival time of the swarm: the average and the Mth electron, assuming for simplicity $\psi = 0$.

6.4.2 Average Arrival Time of Many Electrons

When we average over the arrival times of all the electrons, the statistical reduction of the different terms of (6.39) is not the same because of the correlation between electrons produced in the same cluster. Following the lines of Sect. 6.2 we can write the variance of the average arrival time as

$$\sigma_{\text{TAV}}^2 = \frac{1}{u^2}\left[\frac{1}{N}\frac{\sigma^2}{\cos^2\alpha} + \frac{1}{N_{\text{eff}}}\left(\frac{b^2}{12}\tan^2\alpha + \frac{a^2 b^2}{180} \right) \right], \qquad (6.40)$$

where N represents the total ionization collected by the wire, and N_{eff} is the effective number of independent electrons.

The contribution of the drift-path variation of the variance of the average is always present, even for tracks in good geometry ($\alpha = 0$) and close to the wire ($\sigma \approx 0$). It is quite large and increases with b more than linearly, since N_{eff} increases with the track length more slowly than in proportion.

Obviously the average over all arrival times fluctuates so badly because of the late-arriving electrons, those that have to take the longest path when they approach the wire. If one derives the time signal from the few electrons that arrive first, the latecomers do not contribute. In fact, a discriminator with a low threshold will do this automatically.

6.4.3   Arrival Time of the Mth Electron

Let the time signal be defined by the moment that the Mth electron arrives. Here M can either be a fixed number (this is the case when a fixed-threshold discriminator is used) or a given fraction of the total number of electrons contributing (this is the case of a constant-fraction discriminator).

The probability $p(t)$ of one electron arriving before the time t is calculable using (6.7) and (6.36):

$$p(t) = \frac{\int\limits_{-\infty}^{t} f(\tau)\,d\tau}{\int\limits_{-\infty}^{+\infty} f(\tau)\,d\tau} \,, \tag{6.41}$$

where

$$f(\tau)\,d\tau = \int\limits_{-b/2}^{+b/2} dy\, f_z(\tau - (z_w + ay^2/b)/u, y)\,d\tau \,. \tag{6.42}$$

Next we have to write out the probability $R(t)$ that the Mth electron out of a group of N arrives in the interval dt. It is equal to the probability that any one of the N electrons arrives precisely in dt (equal to $N(dp/dt)$ or $N\,dp$) multiplied by the probability that any of $(M-1)$ out of the $(N-1)$ remaining ones arrives before t and the rest after $t + dt$:

$$R(t)\,dt = N\binom{N-1}{M-1} p^{M-1}(1-p)^{N-M} \frac{dp}{dt}\,dt$$

$$= M\binom{N}{M} p^{M-1}(1-p)^{N-M}\,dp = P(p)\,dp \,. \tag{6.43}$$

In the last expression we have defined a probability density $P(p)$ for the variable p. It is easy to show that

$$\int\limits_{-\infty}^{+\infty} R(t)\,dt = \int\limits_{0}^{1} P(p)\,dp = 1 \,.$$

Here and in the following steps we make use of the well-known expression for the definite integral

$$\int\limits_{0}^{1} x^k(1-x)^j\,dx = \frac{j!\,k!}{(j+k+1)!} \,.$$

The variance of the arrival time of the Mth electron is given by

$$\sigma_{TM}^2 = \langle t^2 \rangle - \langle t \rangle^2 \,,$$

with

$$\langle t^2 \rangle = \int_{-\infty}^{+\infty} t^2 R(t) dt = \int_0^1 t^2(p) P(p) dp$$

and

$$\langle t \rangle = \int_{-\infty}^{+\infty} t R(t) dt = \int_0^1 t(p) P(p) dp \ . \tag{6.44}$$

The function $t(p)$ is obtained by inverting (6.41).

We consider now separately the two contributions of the difference of the drift paths and of the diffusion and study the particular case of a track perpendicular to the drift velocity direction ($\alpha = 0$).

6.4.4   Variance of the Arrival Time of the Mth Electron: Contribution of the Drift-Path Variations

In order to study the effect of the drift-path variation alone we go to the limit $\sigma_y \to 0$, $\sigma_t \to 0$ of (6.42) at $\alpha = 0$, and obtain

$$f(t) dt = \frac{u}{R} \frac{1}{\sqrt{\dfrac{a}{b}(tu - z_0)}} dt, \quad 0 \le (tu - z_0) \le \frac{ab}{4} \tag{6.45}$$

and $f(t) = 0$ outside of this time interval. Using this expression we can evaluate $p(t)$ from (6.41) and invert it:

$$t(p) = \frac{1}{u}\left(\frac{ab}{4} p^2 + z_0 \right) . \tag{6.46}$$

Now we compute the integrals (6.44) and use this expression for $t(p)$:

$$\sigma_{\text{TPM}}^2 = \left(\frac{ab}{4u} \right)^2 \frac{M(M+1)}{(N+1)(N+2)} \left(\frac{(M+2)(M+3)}{(N+3)(N+4)} - \frac{M(M+1)}{(N+1)(N+2)} \right) . \tag{6.47}$$

This is the variance in the arrival time of the Mth electron among N independent ones when only the drift-path variation is taken into account.

If we want to apply this formula to a practical case we have to use for N the number of clusters, and we realize that because of the cluster-size fluctuations it describes correctly only the case $M = 1$, since the threshold can only be set at a certain number of electrons and not of clusters.

In order to take into account the cluster-size fluctuations we have to use a Monte Carlo program. Figure 6.13 shows the result of a simulation of the distribution of the arrival times, assuming the conditions valid for 1 cm of argon gas at NTP (and 27 clusters/cm). In both parts of the figure the r.m.s. of the arrival time is normalized to the width of the collection time interval:

$$\Delta T = ab/4u \ .$$

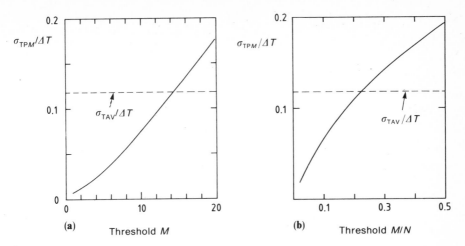

Fig. 6.13a, b. R.m.s. variation of the arrival time of the Mth electron in 1 cm of argon, caused by differences in drift-path length, (**a**) as a function of M, (**b**) when M is a fixed fraction of the total number of electrons N. The r.m.s. of the average time is also shown. (See text for explanations.)

Also visible is the r.m.s. of the average (6.40), evaluated for $\alpha = 0$, $\sigma = 0$, $N_{eff} = 6$ and normalized to the same quantity. We observe that the limit of the accuracy due to the drift-path variation can be made very small by lowering the threshold. For example, σ_{TM} comes down to 30% of σ_{TAV} when triggering on a value of M equal to 5% of N. But by gaining on the drift-path variations, one loses on the diffusion term.

6.4.5   Variance of the Arrival Time of the Mth Electron: Contribution of the Diffusion

Next we study the contribution of diffusion. The probability density $f(t)$ of (6.42) with $a = 0$ and $\alpha = 0$ takes the form

$$f(t)dt = \frac{b}{R} \frac{\exp\left(-\dfrac{\left(t - \dfrac{z_0}{u}\right)^2}{2\sigma_t^2}\right)}{\sqrt{2\pi}\sigma_t} \, dt \ , \tag{6.48}$$

and the function $p(t)$ of (6.41) is the function E:

$$p(t) = E\left(\frac{t - z_0/u}{\sigma_t}\right) = \int_{-\infty}^{\frac{t - z_0/u}{\sigma_t}} dq \, \frac{\exp(-q^2/2)}{\sqrt{2\pi}} \ . \tag{6.49}$$

To compute the variance of the arrival time of the Mth electron we should solve the integrals (6.44) using the function $t(p)$ defined by the inverse of (6.41).

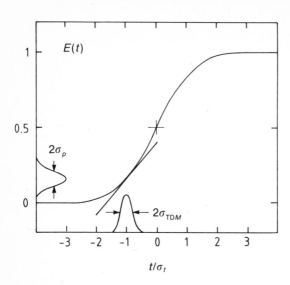

Fig. 6.14. Illustration of (6.51): the variance of p is projected onto the time axis using the tangent to the error function erf(t) in order to obtain σ_{TDM}

Since we cannot invert the expression (6.49) analytically, we linearize it in the vicinity of the average value of the function $P(p)$ of (6.43). This procedure is justified by the fact that the variance of the function $P(p)$ is small when M is small and N large. It is illustrated in Fig. 6.14. The variance of the arrival time is the variance of p on the probability distribution $P(p)$ projected onto the t axis using the derivative

$$\frac{dt}{dp} = \sigma_t \frac{dq}{dE(q)} = \sigma_t \sqrt{2\pi} \exp\left(+\frac{\langle q \rangle^2}{2} \right),$$

evaluated at $\langle q \rangle$ defined by $\langle p \rangle = E(\langle q \rangle)$.

The variance of p is given by

$$\langle p \rangle = \int_0^1 pP(p)\,dp = \frac{M}{N+1}$$

and

$$\langle p^2 \rangle = \int_0^1 p^2 P(p)\,dp = \frac{M(M+1)}{(N+1)(N+2)},$$

from which we obtain

$$\langle p^2 \rangle - \langle p \rangle^2 = \frac{M}{(N+1)(N+2)}\left(1 - \frac{M}{N+1} \right) \approx \frac{M}{N^2}\left(1 - \frac{M}{N} \right). \tag{6.50}$$

Projecting now this variance onto the time axis (see Fig. 6.14) we obtain

$$\sigma^2_{TDM} \approx \frac{\sigma_t^2}{N} 2\pi \exp(\langle q \rangle^2) \frac{M}{N}\left(1 - \frac{M}{N} \right), \tag{6.51}$$

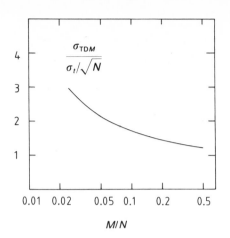

Fig. 6.15. Errors of time measurement caused by diffusion: r.m.s. variation σ_{TDM} of the arrival time of the Mth electron, in units of the r.m.s. variation of the average arrival time of N electrons, $\sigma_t/\sqrt{N}$, plotted as a function of M/N

where $\langle q \rangle$ is the normalized average arrival time, given by the condition that the corresponding integral probability p takes on the value $M/(N + 1)$, or in good approximation, M/N. In other words, $\langle q \rangle$ is given by the inverse of the function E at its argument M/N:

$$\langle q \rangle = E^{-1}(M/N) . \tag{6.52}$$

Expression (6.51) with (6.52) represents the answer to the question how much the arrival time of the Mth electron fluctuates when each single electron has a Gaussian distribution with a width σ_t; σ_{TDM} goes down with the total number N of electrons and is equal to $\sigma_t/\sqrt{N}$ times a factor which depends only on M/N. The factor describes the loss of accuracy when the arrival-time measurement is based on the Mth electron rather than on the average of all N. (The variance σ_{TAV}^2 of the average arrival time caused by diffusion alone was σ_t^2/N as implied by (6.40).) The loss factor is plotted in Fig. 6.15 and is seen to be always larger than 1. For example the loss is a factor 2 if one triggers on the first 5% of the swarm.

An often-quoted formula according to which

$$\sigma_{\text{TDM}} = \frac{\sigma_t}{\sqrt{2 \log N}} \sum_{r=M}^{\infty} \frac{1}{r^2} \tag{6.53}$$

instead of (6.51) has been discussed by Cramer [CRA 51]. The right-hand side of (6.53) is the first term of a power series in $1/\log N$. It turns out that for $M = 1$ (6.53) is more accurate than (6.51), but it cannot reproduce the variation of σ_{TDM} with M.

We conclude this chapter with a comment on the various ways to estimate the time of arrival of the electron swarm. It is clear from the previous pages that this is a question of the relative importance of the two main contributions to the measurement accuracy. Where the drift-path variations are relatively large, one gains a lot by triggering on a low threshold. Where the diffusion is relatively large, one gains, but not so much, by using the average. The contribution from the angular effect and from the effect of the drift-path variations quickly increase as the angle α increases. Since the diffusion increases with the drift length, the relative importance of the two contributions changes through their dependence on σ_t and N_{eff}, and the best estimator is a function of the drift length. Depending on how far one wants to go in the optimization, the fast sampling technique offers the greatest freedom.

Appendix. Influence of the Cluster Fluctuations on the Measurement Accuracy of a Single Wire

An analytical form of (6.16) can be derived under certain assumptions as follows. The symbols used here were defined in Sect. 6.2.

We consider the N_c clusters produced along the track segment R. N_c^i are the clusters of cluster size i, where $N_c^i = N_c P(i)$ and $P(i)$ is the cluster-size distribution defined in Sect. 1.2.2 (1.35).

The N_c^i clusters produce in total iN_c^i electrons and on average m_i of them,

$$m_i = iN_c P(i) \frac{b}{R \cos \theta} = iN_c^i \frac{b}{R \cos \theta} , \tag{6.54}$$

are collected by the wire. The total number N of electrons collected by the wire is

$$N = \sum_i m_i ;$$

here the sum on i is performed on the cluster size i.

As we have shown in Sect. 6.2.3, N does not depend on diffusion, and we assume in the following that also each m_i does not depend on it.

Defining $\bar{x}_i$ to be the average position of the m_i electrons collected by the wire and produced in clusters of original cluster size i, the quantity X_{AV} defined by (6.11) can be rewritten as

$$X_{AV} = \frac{\sum m_i \bar{x}_i}{\sum m_i} = \frac{\sum m_i \bar{x}_i}{N} . \tag{6.55}$$

Its variance is simply given by

$$\sigma_{X_{AV}}^2 = \text{Var}(X_{AV}) = \frac{\sum m_i^2 \, \text{Var}(\bar{x}_i)}{N^2} , \tag{6.56}$$

since the various $\bar{x}_i$ are not correlated because they are by definition averages of arrival positions of electrons produced in different clusters.

The variance of $\bar{x}_i$, $\mathrm{Var}(\bar{x}_i)$, depends on the distribution of the number k ($k = 0, \ldots, i$) of collected electrons when i are produced in the same cluster. Referring to (6.1) and (6.3) and assuming for simplicity $x_0 = 0$, we find that this probability that k electrons among i are collected is given by

$$\frac{1}{R} \int\limits_{-R/2}^{+R/2} ds \left[\int\limits_{-\infty}^{+\infty} dx\, G(x \mid s \sin \theta, \sigma_x) \right]^i \left[\int\limits_{-b/2}^{+b/2} dy\, G(y \mid s \cos \theta, \sigma_y) \right]^k$$

$$\times \left[1 - \int\limits_{-b/2}^{+b/2} dy\, G(y \mid s \cos \theta, \sigma_y) \right]^{i-k} \binom{i}{k}, \qquad (6.57)$$

where the term $\binom{i}{k}$ takes into account the different combinations of k electrons among i.

Since we have assumed that $x_0 = 0$, then we have also that

$$\langle x \rangle = \langle \bar{x}_i \rangle = \langle X_{\mathrm{AV}} \rangle = 0 . \qquad (6.58)$$

The integrals in dx of (6.57) are equal to 1. Defining

$$\Delta(s \cos \theta) = \int\limits_{-b/2}^{+b/2} dy\, G(y \mid s \cos \theta, \sigma_y) \qquad (6.59)$$

as the probability that one electron is collected by the wire ($0 < \Delta < 1$), we rewrite the probability that k among i electrons produced in the same cluster are collected by the wire as

$$\Pi_k^i = \frac{1}{R} \int\limits_{-R/2}^{+R/2} \Delta^k (1 - \Delta)^{i-k} \binom{i}{k} ds . \qquad (6.60)$$

We note that

$$\sum_{k=0}^{} i\, \Pi_k^i = 1 .$$

If along the track segment R we produce N_c^i clusters of i electrons, on the wire we collect on average

- $N_c^i \, \Pi_1^i$ single electrons
- $N_c^i \, \Pi_2^i$ pairs of electrons
- $N_c^i \, \Pi_3^i$ triplets of electrons
- . . .
- $N_c^i \, \Pi_i^i$ ith of electrons

and in total we have

$$N_c^i \sum_{k=0}^{i} k \Pi_K^i = m_i$$

electrons. This last equality can be proved using (6.60) and (6.59) and the properties of the binomial distribution.

The numbers N_{pairs}^i of different *pairs* of electrons, produced in the same cluster of cluster size i and collected by the wire, are given by

$$N_{\text{pairs}}^i = N_c^i \sum_{k=2}^i \binom{k}{2} \Pi_k^i = \frac{N_c^i}{2} \left(\sum_{k=0}^i k^2 \, \Pi_k^i - \sum_{k=0}^i k \Pi_k^i \right)$$

$$= \frac{N_c^i}{2} \sum_{k=0}^i k^2 \, \Pi_k^i - \frac{m_i}{2} . \tag{6.61}$$

We are now in a position to compute the variance $\text{Var}(\bar{x}_i)$ of the average position $\bar{x}_i$ of the m_i electrons produced by clusters of cluster size i and collected by the wire: we have m_i variance terms $\langle x^2 \rangle$ and $2 \times N_{\text{pairs}}^i$ covariance terms $\langle xx \rangle$. The average $\langle \bar{x}_i \rangle$ is zero because we have assumed that $x_0 = 0$ in (6.57) (see also (6.58)). We obtain

$$\text{Var}(\bar{x}_i) = \frac{m_i \langle x^2 \rangle + N_{\text{pairs}}^i \langle xx \rangle}{m_i^2}$$

$$= \frac{m_i \langle x^2 \rangle + \left(N_c^i \sum_{k=0}^i k^2 \, \Pi_k^i - m_i \right) \langle xx \rangle}{m_i^2}$$

$$= \frac{1}{m_i} \left[(\langle x^2 \rangle - \langle xx \rangle) + \frac{\sum_{k=0}^i k^2 \, \Pi_k^i}{\sum_{k=0}^i k \Pi_k^i} \langle xx \rangle \right]. \tag{6.62}$$

Using (6.60) and the properties of the binomial distribution one can show that

$$\sum_{k=0}^i k^2 \, \Pi_k^i = \frac{i^2 - i}{R} \int_{-R/2}^{+R/2} \Delta^2 \, ds + \frac{i}{R} \int_{-R/2}^{+R/2} \Delta \, ds ,$$

$$\sum_{k=0}^i k \Pi_k^i = \frac{i}{R} \int_{-R/2}^{+R/2} \Delta^2 \, ds .$$

From these equations we obtain

$$\frac{\sum_{k=0}^i k^2 \, \Pi_k^i}{\sum_{k=0}^i k \Pi_k^i} = 1 + (i - 1) I_2 ,$$

where

$$I_2 = \frac{\int_{-R/2}^{+R/2} \Delta^2 \, ds}{\int_{-R/2}^{+R/2} \Delta \, ds} , \tag{6.63}$$

and we can rewrite (6.62) as

$$\text{Var}(\bar{x}_i) = \frac{1}{m_i}(\langle x^2 \rangle - I_2 \langle xx \rangle + iI_2 \langle xx \rangle) . \tag{6.64}$$

Finally we substitute the variance $\text{Var}(\bar{x}_i)$ from this last equation into (6.56) and obtain

$$\text{Var}(X_{\text{AV}}) = \frac{\sum m_i(\langle x^2 \rangle - I_2 \langle xx \rangle + iI_2 \langle xx \rangle)}{N^2}$$

$$= \frac{\langle x^2 \rangle - I_2 \langle xx \rangle}{N} + \frac{\sum i\, m_i}{N^2} I_2 \langle xx \rangle . \tag{6.65}$$

Using (6.54) we can show that

$$\frac{\sum\limits_i im_i}{N^2} = \frac{\sum\limits_i i^2 N_c P(i)[b/(R\cos\theta)]}{N^2} = \frac{\sum\limits_i i^2 N_c P(i)[b/(R\cos\theta)]}{\left(\sum\limits_i i N_c P(i)[b/(R\cos\theta)]\right)^2}$$

$$= \frac{\sum\limits_j n_j^2}{\left(\sum\limits_j n_j\right)^2} \frac{1}{b/(R\cos\theta)}$$

$$= \frac{1}{N_{\text{eff}}}, \tag{6.66}$$

where the sum over i is performed on the cluster size i, while the sum over j is performed on the N_c clusters produced by the particle in the track segment R and n_j is the number of electrons in the cluster j.

N_{eff} was defined in Sect. 1.2.5 as the effective number of electrons computed along the track segment R; here it is scaled by the normalization factor $b/R\cos\theta$ derived in (6.2). In Sect. 1.2.5 we showed that N_{eff} does not scale linearly with the length of track sampled, while (6.66) implies a linear dependence of N_{eff} on R. This is an implicit consequence of the assumption that the m_i are constant, since doing so we do not take into account the effect of the rare and large clusters present in the tails of the cluster-size distribution, and this effect is a limitation of our derivation. We will comment on this point at the end of this appendix.

The quantity I_2 defined in (6.63) can be computed using (6.59):

$$I_2 = \text{Erf}\left(\frac{b}{2\sigma_y}\right) - \frac{1}{\sqrt{\pi}} \frac{2\sigma_y}{b}\left[1 - \exp\left(-\frac{b^2}{4\sigma_y^2}\right)\right], \tag{6.67}$$

where Erf is the error function [ABR 64],

$$\text{Erf}(z) = \frac{2}{\sqrt{\pi}} \int_0^z \exp(-t^2)\mathrm{d}t .$$

In the limit $\sigma_y \ll b$, $I_2 \to 1$, since $\text{erf}(\infty) = 1$. When $\sigma_y \gg b$, then $I_2 \to b/(2\sqrt{\pi}\sigma_y) \to 0$.

In order to compute explicitly the variance $\text{Var}(X_{AV})$ of (6.65) we have still to evaluate the correlation term $\langle xx \rangle$.

The covariance of the arrival position of two electrons produced in the same cluster is evaluated from the distribution function of the arrival position of two electrons. Starting from (6.1) and (6.3) we write it as

$$F_{xx}(x_1, y_1, x_2, y_2) = \frac{1}{R} \int_{-R/2}^{+R/2} ds \, G(x_1 | s \sin\theta, \sigma_x) G(y_1 | s \cos\theta, \sigma_y)$$

$$\times G(x_2 | s \sin\theta, \sigma_x) G(y_2 | s \cos\theta, \sigma_y) \qquad (6.68)$$

where the two indices 1 and 2 refer to the two electrons.

This distribution is not normalized. Its normalization is the probability that two electrons produced in the same cluster are both collected by the wire. It can be shown that

$$\int_{-\infty}^{+\infty} dx_1 \int_{-b/2}^{b/2} dy_1 \int_{-\infty}^{+\infty} dx_2 \int_{-b/2}^{b/2} dy_2 \, F_{xx}(x_1, y_1, x_2, y_2) = I_2 \frac{b}{\cos\theta}, \qquad (6.69)$$

where I_2 is defined by (6.63). Since in (6.65) we have the product $I_2 \langle xx \rangle$, we can compute directly

$$I_2 \langle xx \rangle = \frac{\cos\theta}{b} \int_{-\infty}^{+\infty} dx_1 \int_{-b/2}^{b/2} dy_1$$

$$\times \int_{-\infty}^{+\infty} dx_2 \int_{-b/2}^{b/2} dy_2 \, x_1 x_2 \, F_{xx}(x_1, y_1, x_2, y_2), \qquad (6.70)$$

and after some tedious integration we obtain

$$I_2 \langle xx \rangle = \tan^2\theta \left[\left(\sigma_y^2 + \frac{b^2}{12} \right) \left(\text{Erf}\left(\frac{b}{2\sigma_y} \right) - \frac{2}{\sqrt{\pi}} \frac{\sigma_y}{b} \left(1 - \exp\left[-\frac{b^2}{4\sigma_y^2} \right] \right) \right) \right.$$

$$+ b^2 \frac{1}{3\sqrt{\pi}} \left(4 \frac{\sigma_y^3}{b^3} \left(1 - \exp\left[-\frac{b^2}{4\sigma_y^2} \right] \right) - \frac{\sigma_y}{b} \right)$$

$$\left. - \sigma_y^2 \frac{1}{\sqrt{\pi}} \left(\frac{\sigma_y}{b} \left(1 - \exp\left[-\frac{b^2}{4\sigma_y^2} \right] \right) \right) \right]. \qquad (6.71)$$

Now we have all the information we require and we can write down the final formula for the variance $\text{Var}(X_{AV})$. Using (6.65), (6.66), (6.5) and (6.71) we obtain

$$\sigma_{X_{AV}}^2 = \text{Var}(X_{AV}) = \sigma_x^2 \frac{1}{N} + \tan^2\theta \, \sigma_y^2 \left[\frac{1}{N} + \left(\frac{1}{N_{eff}} - \frac{1}{N} \right) F_\sigma\left(\frac{\sigma_y}{b} \right) \right]$$

$$+ \tan^2\theta \frac{b^2}{12} \left[\frac{1}{N} + \left(\frac{1}{N_{eff}} - \frac{1}{N} \right) F_b\left(\frac{\sigma_y}{b} \right) \right], \qquad (6.72)$$

where the two functions F_σ and F_b are given by

$$F_\sigma = \mathrm{Erf}\left(\frac{b}{2\sigma_y}\right) - \frac{3}{\sqrt{\pi}}\frac{\sigma_y}{b}\left(1 - \exp\left[-\frac{b^2}{4\sigma_y^2}\right]\right) \tag{6.73}$$

$$F_b = \mathrm{Erf}\left(\frac{b}{2\sigma_y}\right) - \frac{2}{\sqrt{\pi}}\frac{\sigma_y}{b}\left(1 - \exp\left[-\frac{b^2}{4\sigma_y^2}\right]\right)\left(1 - 8\frac{\sigma_y^2}{b^2}\right) - \frac{4}{\sqrt{\pi}}\frac{\sigma_y}{b}. \tag{6.74}$$

These are equal to 1 when σ_y/b is zero and go to zero when $\sigma_y/b \to \infty$. Inspecting (6.72) we notice that this behaviour corresponds to a smooth transition between the coefficients $1/N_{\mathrm{eff}}$ and $1/N$ in the angular terms. In the absence of diffusion there is a complete correlation and the reduction factor of the angular term in the variance of the average (6.11) is only N_{eff}; when the diffusion is large compared to the width b of the drift cell, there is no correlation and the reduction factor of the angular term is N, the total number of collected electrons. The functions F_σ and F_b are shown in Fig. 6.16.

If we had calculated the plot of Fig. 6.3 with (6.72) (instead of using the numerical simulation, see Sect. 6.2.3) we would have found a similar behaviour but a somewhat steeper increase of $N_{\mathrm{eff}}(\sigma)$ with σ/l. This is a consequence of the assumption introduced: we have assumed that the average number m_i of electrons originated by clusters of size i and collected by the wire is constant. If we do not make this assumption, then (6.56) contains an additional term,

$$\mathrm{Var}(X_{\mathrm{AV}}) = \frac{\sum m_i^2\,\mathrm{Var}(\bar{x}_i)}{N^2} + \frac{\sum(\bar{x}_i - X_{\mathrm{AV}})^2\,\mathrm{Var}(m_i)}{N^2}. \tag{6.75}$$

The second term plays a role when the rare large clusters are present, since they dominate the average. The numerical simulation introduced in Sect. 6.2.3 and used to produce Fig. 6.3 takes into account all the various effects due to the

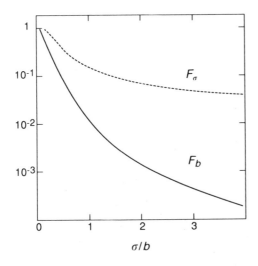

Fig. 6.16. Declustering as a function of σ/b. Plot of the functions defined in (6.73) and (6.74)

peculiar shape of the cluster-size distribution and then correctly produces a variance larger than that evaluated in (6.72) and consequently a smaller $N_{\mathrm{eff}}(\sigma)$.

References

[ABR 64]   M. Abramovitz and I. Stegun (eds.), *Handbook of Mathematical Functions* (Dover, New York 1964) p. 297.

[AME 83]   S.R. Amendolia et al. $E \times B$ and angular effects in the avalanche localization along the wire with cathode pad readout, *Nucl. Instrum. Methods* **217**, 317 (1983)

[BAR 82]   A. Barbaro Galtieri, Tracking with the PEP-4 TPC, TPC-LBL 82-24, Berkeley preprint 1982 (unpublished)

[BLU 86]   W. Blum, U. Stiegler, P. Gondolo and L. Rolandi, Measurement of the avalanche broadening caused by the wire $E \times B$ effect, *Nucl. Instrum. Methods A* **252**, 407 (1986)

[CRA 51]   H. Cramer, *Mathematical Methods of Statistics* (Princeton Univ. Press 1951), Sec. 28.6

[HAR 84]   C.K. Hargrove et al., The spatial resolution of the time projection chamber at TRIUMF, *Nucl. Instrum. Methods* **219**, 461 (1984)

7. Geometrical Track Parameters and Their Errors

Now we come to the point to which all the efforts of drift-chamber design lead up – the determination of the track parameters. The point of origin of a track, its angles of orientation, and its curvature in a magnetic field are the main geometric properties that one aims to measure. The proof of the pudding is in the eating, and the proof of the drift chamber is in the track parameters.

In a particle experiment a track is often measured by several detectors, and the accuracy of its parameters depends on the integrity of the whole ensemble, and in particular on the knowledge of the relative detector positions. Likewise, the track parameters determined with a drift chamber depend on its overall geometry, the electrode positions, fields and electron drift paths. If one wants to calculate the achievable accuracy of a drift chamber in all parts of its volume, one needs to have quantitative knowledge of these factors in addition to the point-measuring accuracy. Such knowledge is often difficult to get, and then one may want to work the other way around: starting from a measurement of the achieved accuracy of track parameters, one can compare it to that expected from the point-measuring accuracy alone. If they agree, then the other factors make only a small contribution, if they do not, the contribution of the other factors is larger, or perhaps dominant.

The achieved accuracy can be measured in a number of ways. The common methods include the following: vertex localization by comparing tracks from the same vertex, momentum resolution by measuring tracks with known momentum, a combination of momentum and angular precision with a measurement of the invariant mass of a decaying particle.

It is obviously our first task to ascertain the accuracy that can be achieved in a given geometry when considering the resolution of each measuring point alone, ignoring all the other factors by assuming they are negligible or have been corrected for. We will do this in two sections, one for the situation without magnetic field, using straight-line fits, the other for the situation inside a magnetic field, where a quadratic fit is appropriate. One section is devoted to accuracy limitations due to multiple Coulomb scattering in parts of the apparatus. Finally, the results on spectrometer resolution are summarized.

A general treatment of the methods of track fitting using different statistical approaches is presented in the book of Böck, Grote, Notz and Regler [BOC 90]. Here our aim is more modest and more specific: on the basis of the least-squares method we would like to understand how the point-measuring errors of a drift chamber propagate into the track parameters.

7.1   Linear Fit

Referring to Fig. 7.1, there are $N + 1$ points at positions x_i $(i = 0, 1, \ldots, N)$ where the coordinates y_i of a given track were measured. We ask for the accuracy with which the coordinate and the direction of the track are determined at $x = 0$. If there is a magnetic field, the track is curved. As a first step we imagine the curvature to be well measured outside the vertex detector so that we are allowed to consider a straight-line extrapolation; the influence of an error of curvature will be estimated in Sect. 7.2.

A least-squares fit to the straight line

$$y = a + bx \tag{7.1}$$

is obtained by minimizing

$$\chi^2 = \sum (y_i - a - bx_i)^2/\sigma_i^2 \ , \tag{7.2}$$

where the σ_i are the uncorrelated point-measuring errors whose reciprocal squares are used as weights. The two conditions $\partial \chi^2/\partial a = 0$ and $\partial \chi^2/\partial b = 0$ lead to the following linear equations:

$$aS_1 + bS_x = \sum y_i/\sigma_i^2 \ ,$$

$$bS_x + bS_{xx} = \sum x_i y_i/\sigma_i^2 \ ,$$

where $S_1 = \sum 1/\sigma_i^2$, $S_x = \sum x_i/\sigma_i^2$, $S_y = \sum y_i/\sigma_i^2$, $S_{xy} = \sum x_i y_i/\sigma_i^2$, $S_{xx} = \sum x_i x_i/\sigma_i^2$ (all sums from 0 to N); also $D = S_1 S_{xx} - S_x^2$. The two linear equations are solved for the best estimates of the coefficients a and b by

$$
\begin{aligned}
a &= (S_y S_{xx} - S_x S_{xy})/D \ , \\
b &= (S_1 S_{xy} - S_x S_y)/D \ .
\end{aligned}
\tag{7.3}
$$

Let us now consider the fluctuations of a and b. They are caused by the fluctuations of the y_i, which can be characterized by the $(N + 1)^2$ covariances $[y_i y_k]$ defined as the average over the product minus the product of the averages:

$$[y_i y_k] = \langle y_i y_k \rangle - \langle y_i \rangle \langle y_k \rangle \ ; \tag{7.4}$$

they vanish when $i \neq k$ because the y_i fluctuate independently of one another, and $[y_i^2]$ is what was called σ_i^2 before.

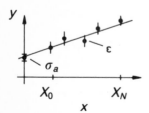

Fig. 7.1. Straight-line fit to the origin

The fluctuations of a and b are described by their covariance matrix,

$$\begin{pmatrix} [a^2] & [ab] \\ [ab] & [b^2] \end{pmatrix},$$

and we have to determine how it depends on the quantities $[y_i^2]$.

To compute $[a^2] = \langle a^2 \rangle - \langle a \rangle^2$ we use (7.3):

$$[a^2] = (S_{xx}^2 \sum (\langle y_i^2 \rangle - \langle y_i \rangle^2)/\sigma_i^4 + S_x^2 (\sum x_i^2 \langle y_i^2 \rangle - x_i^2 \langle y_i \rangle^2)/\sigma_i^4$$
$$- 2 S_x S_{xx} \sum x_i (\langle y_i^2 \rangle - \langle y_i \rangle^2)/\sigma_i^4)/D^2$$
$$= (S_{xx} \sum [y_i^2]/\sigma_i^4 + S_x^2 \sum x_i^2 [y_i^2]/\sigma_i^4 - 2 S_x S_{xx} \sum x_i [y_i^2]/\sigma_i^4)/D^2$$
$$= S_{xx}/D . \tag{7.5}$$

The other two elements are calculated in a similar way, and the covariance matrix of the fitted parameters is

$$\begin{pmatrix} [a^2] & [ab] \\ [ab] & [b^2] \end{pmatrix} = \begin{pmatrix} S_{xx} & -S_x \\ -S_x & S_1 \end{pmatrix} \frac{1}{D} . \tag{7.6}$$

Since the fit is linear, the result (7.5) and (7.6) depends only on the covariances of the y_i and not on their averages.

7.1.1 Case of Equal Spacing Between x_0 and x_N

We evaluate (7.4) for $N + 1$ equally spaced points with equal point errors ε^2. In this case $S_1 = (N + 1)/\varepsilon^2$, $S_x = (N + 1)(x_0 + x_N)/(2\varepsilon^2)$ and $S_{xx} = (N + 1)[x_0 x_N + (x_N - x_0)^2(2 + 1/N)/6]/\varepsilon^2$. We see that the result will depend on the number of points and on the ratio of the two extreme distances x_N and x_0. Let us introduce the ratio r, which expresses how far – in units of the chamber length – the centre of the chamber is away from the origin:

$$r = \frac{x_N + x_0}{2(x_N - x_0)} . \tag{7.7}$$

Inserting the sums into (7.5) and taking the square root we obtain the vertex localization accuracy

$$\sigma_a = \frac{\varepsilon}{\sqrt{(N + 1)}} Z(r, N) , \tag{7.8}$$

with

$$Z(r, N) = \left(\frac{12r^2 + 1 + \dfrac{2}{N}}{1 + \dfrac{2}{N}} \right)^{1/2} .$$

Values of Z have been computed for some r and N; results are listed in Table 7.1. We observe that the strongest functional variation is with the ratio r; this is understandable since it has the meaning of a lever–arm ratio. For any given r,

Table 7.1. Values of the factor $Z(r, N)$ in (7.6) for various distances r of the vertex from the centre of the chamber, in units of the chamber length, and for various numbers of $N + 1$ equally spaced sense wires

N	r:	5.0	3.0	2.0	1.5	1.0	0.75	0.60	0.50
1		10.1	6.08	4.12	3.16	2.24	1.80	1.56	1.41
2		12.3	7.42	5.00	3.81	2.65	2.09	1.78	1.58
5		14.7	8.84	5.94	4.50	3.09	2.41	2.02	1.77
9		15.7	9.45	6.35	4.81	3.29	2.55	2.13	1.86
19		16.5	9.94	6.67	5.04	3.44	2.67	2.22	1.93
infin.		17.3	10.4	7.00	5.29	3.61	2.78	2.31	2.00

the factor Z is more favourable for small N; this means that the dependence on N is weaker than the $1/\sqrt{(N + 1)}$ rule. The error in the direction is given by

$$\sigma_b = \frac{\varepsilon}{x_N - x_0} \frac{1}{\sqrt{N + 1}} \sqrt{\frac{12N}{N + 2}} . \tag{7.9}$$

7.2   Quadratic Fit

In a magnetic spectrometer the particle momenta are determined from a measurement of the track curvature. We recall that in a homogeneous magnetic field B the trajectory is a helix with radius of curvature

$$R = p_T/eB , \tag{7.10}$$

where e is the electric charge of the particle and p_T its momentum transverse to the magnetic field. A remark concerning units: relation (7.10) is in m, if e is in As, B in T (i.e. $V s/m^2$) and p_T in Ns (i.e. $V A s^2/m$). If one wants to express the momentum in units of GeV/c, then one must multiply p_T by the factor f that says how many GeV/c there are to one N s. This factor is

$$f = \frac{1\,GeV/c}{1\,Ns} = \frac{10^9\,e(A\,s)(V)}{(V\,A\,s^2/m)\,2.9979 \times 10^8\,(m/s)} , \tag{7.11}$$

so that (7.10), to better than one tenth of a percent, takes on the simple form

$$R = \frac{10}{3}\left(\frac{T\,m}{GeV/c}\right)\frac{p_T}{B} . \tag{7.12}$$

This radius has to be measured in a plane perpendicular to B. Our interest is in the large momenta because they are the most difficult to measure precisely. Therefore we approximate the helix trajectory, which is actually a circle in a plane perpendicular to its axis, by a parabola in such a plane,

$$y = a + bx + (c/2)x^2 , \tag{7.13}$$

up to terms of the order of x^3/R^2, where $c = 1/R$. The measured y_i at positions x_i are fitted to (7.13) with the aim of determining a, b and c.

7.2.1 Error Calculation

One is interested in the errors of a, b and c. Our procedure is similar to that of Gluckstern [GLU 63], but is more general in that it includes the extrapolation to a vertex point at an arbitrary distance.

The linear equations for a, b and c are derived from the conditions of a minimal χ^2 and read as follows (the weighting factor at each point is denoted by f_n):

$$aF_0 + bF_1 + (c/2)F_2 = \sum f_n y_n \, ,$$
$$aF_1 + bF_2 + (c/2)F_3 = \sum f_n x_n y_n \, , \qquad (7.14)$$
$$aF_2 + bF_3 + (c/2)F_4 = \sum f_n x_n^2 y_n \, ,$$

where

$$F_j = \sum f_n (x_n)^j \, . \qquad (7.15)$$

All sums run from $n = 0$ to $n = N$. The solution of (7.14) for our three quantities is

$$a = \frac{\sum y_n G_n}{\sum G_n} \, ,$$
$$b = \frac{\sum y_n P_n}{\sum x_n P_n} \, , \qquad (7.16)$$
$$\frac{c}{2} = \frac{\sum y_n Q_n}{\sum x_n^2 Q_n} \, ,$$

where the G_n, P_n and Q_n are expressed using determinants of the F_j:

$$G_n = f_n \begin{vmatrix} F_2 & F_3 \\ F_3 & F_4 \end{vmatrix} - f_n x_n \begin{vmatrix} F_1 & F_2 \\ F_3 & F_4 \end{vmatrix} + f_n x_n^2 \begin{vmatrix} F_1 & F_2 \\ F_2 & F_3 \end{vmatrix} \, ,$$

$$P_n = -f_n \begin{vmatrix} F_1 & F_3 \\ F_2 & F_4 \end{vmatrix} + f_n x_n \begin{vmatrix} F_0 & F_2 \\ F_2 & F_4 \end{vmatrix} - f_n x_n^2 \begin{vmatrix} F_0 & F_1 \\ F_2 & F_3 \end{vmatrix} \, , \qquad (7.17)$$

$$Q_n = f_n \begin{vmatrix} F_1 & F_2 \\ F_2 & F_3 \end{vmatrix} - f_n x_n \begin{vmatrix} F_0 & F_1 \\ F_2 & F_3 \end{vmatrix} + f_n x_n^2 \begin{vmatrix} F_0 & F_1 \\ F_1 & F_2 \end{vmatrix} \, .$$

They have the property that

$$\sum Q_n = \sum x_n Q_n = \sum P_n = \sum x_n^2 P_n = \sum x_n G_n = \sum x_n^2 G_n = 0 \qquad (7.18)$$

and

$$\sum x_n^2 Q_n = \sum x_n P_n = \sum G_n \, . \qquad (7.19)$$

Using (7.18), the variance of the first coefficient is given by

$$[a^2] = \sum \sum G_m G_n [y_m y_n] / (\sum G_n)^2 \, , \qquad (7.20)$$

where $[y_m y_n]$ denotes the covariance of the measurements at two points x_m and x_n. The other elements are formed in the same manner.

The point-measurement errors are equal and uncorrelated. We write

$$[y_n \quad y_m] = \varepsilon^2 \delta_{mm} \tag{7.21}$$

and obtain

$$
\begin{aligned}
[a^2] &= \varepsilon^2 \sum G_n^2 / (\sum G_n)^2 , \\
[ab] &= \varepsilon^2 \sum G_n P_n / (\sum G_n)(\sum x_n P_n) , \\
[b^2] &= \varepsilon^2 \sum P_n^2 / (\sum x_n P_n)^2 , \\
[ac] &= 2\varepsilon^2 \sum G_n Q_n / (\sum G_n)(\sum x_n^2 Q_n) , \\
[bc] &= \varepsilon^2 2 \sum Q_n P_n / (\sum x_n^2 Q_n)(\sum x_n P_n) , \\
[c^2] &= \varepsilon^2 4 \sum Q_n^2 / (\sum x_n^2 Q_n)^2 .
\end{aligned}
\tag{7.22}
$$

7.2.2 Origin at the Centre of the Track – Uniform Spacing of Wires

In order to facilitate the evaluation of (7.22) we begin by setting the origin of the x coordinates at the centre of the track. Let

$$q_n = x_n - x_c, \quad \text{with } x_c = \sum x_n/(N + 1) ,$$

and

$$L = x_N - x_0 . \tag{7.23}$$

For uniform spacing and uniform weights $f_n = 1$, the subdeterminants are easily calculated to be

$$
\begin{aligned}
G_n &= F_2 F_4 - q_n^2 F_2^2 , \\
P_n &= q_n(F_0 F_4 - F_2^2) , \\
Q_n &= - F_2^2 + q_n^2 F_0 F_2 ,
\end{aligned}
\tag{7.24}
$$

with $F_j = \sum q^j$. Note that $F_1 = F_3 = 0$ for symmetry. The other F_j are

$$F_0 = N + 1 ,$$

$$F_2 = L^2 \frac{(N + 1)(N + 2)}{12N} ,$$

$$F_4 = L^4 \frac{(N + 1)(N + 2)(3N^2 + 6N - 4)}{240N^3} ,$$

$$S = F_0 F_4 - F_2^2 = L^4 \frac{(N - 1)(N + 1)^2(N + 2)(N + 3)}{180N^3} . \tag{7.25}$$

A combination S, needed later, has been included here. Inserting (7.24) into

(7.22) we find the six elements of the covariance matrix, valid at the origin, that is the centre of the track:

$$[a^2] = \varepsilon^2 F_4/(F_0 F_4 - F_2^2),$$
$$[ab] = 0,$$
$$[b^2] = \varepsilon^2/F_2,$$
$$[ac] = -2\varepsilon^2 F_2/(F_0 F_4 - F_2^2),$$
$$[bc] = 0,$$
$$[c^2] = 4F_0/(F_0 F_4 - F_2^2).$$

(7.26)

Note that there is no correlation between the direction b and the offset a or the curvature c in the middle of the track. The elements of the covariance matrix (7.26) will assume different values outside the centre, and this will be treated in Sect. 7.2.4. The variance of the curvature, however, is the same all along the track, and we can evaluate it by inserting (7.25) into the last of the equations (7.26). The variance of the curvature is equal to

$$[c^2] = \frac{\varepsilon^2}{L^4} \frac{720 N^3}{(N-1)(N+1)(N+2)(N+3)} = \frac{\varepsilon^2}{L^4} A_N,$$

(7.27)

where we have defined a factor A_N, some values of which are enumerated in Table 7.2.

7.2.3 Sagitta

The maximum excursion of a piece of a circle over the corresponding chord is called its sagitta. The sagitta s of a track with length L and radius of curvature $R \gg L$ is proportional to the square of the track length, as can be seen from the

Table 7.2. Values of the factors A_N in (7.27) and $\sqrt{(A_N)}/8$ in (7.29) for various values of N. (The number of measuring points is $N + 1$.)

N	A_N	$\sqrt{(A_N)}/8$
2	96	1.22
3	81	1.12
4	73.1	1.07
5	67.0	1.02
6	61.7	0.982
8	53.2	0.912
10	46.6	0.853
infin.	$\dfrac{720}{N+5}$	$\dfrac{3.35}{\sqrt{N+5}}$

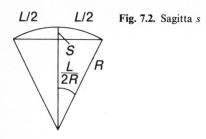

$L/2 \qquad L/2 \qquad$ **Fig. 7.2.** Sagitta s

sketch of Fig. 7.2 when developing the cosine of the small angle $(L/2R)$ into powers of $(L/2R)$:

$$s = L^2/8R , \qquad (7.28)$$

so its r.m.s. error becomes

$$\delta s = \varepsilon \frac{\sqrt{A_N}}{8} . \qquad (7.29)$$

This ratio between the errors of the sagitta and the individual point coordinates is also contained in Table 7.2.

7.2.4 Covariance Matrix at an Arbitrary Point Along the Track

Now we know the full covariance matrix of a, b and c at the centre of the chamber, and we are curious to know the accuracy with which the measured track can be extrapolated to some point x outside the centre. For this we must express the position and the slope of the track as a function of the independent variable which defines the distance from the track centre to the point in question:

$$a(x) = a_0 + b_0 x + (c_0/2)x^2 ,$$
$$b(x) = b_0 + c_0 x , \qquad (7.30)$$
$$c(x) = c_0 .$$

a_0, b_0 and c_0 represent the values of the coefficients at the centre, for which the covariance matrix was given in (7.26). The distance of the point x from the centre x_c of the chamber will be described by the ratio

$$r = \frac{x - x_c}{L} ,$$

which has the same physical meaning as the r of (7.7) used in the linear fit.

For example, to compute the covariance $[a(x) \quad c(x)]$, we form

$$[(a_0 + b_0 x + (c_0/2)x^2) \quad c_0],$$

which is equal to $[a_0 c_0] + x[b_0 c_0] + (x^2/2)[c_0^2]$. In this way the full covariance matrix is constructed, and the result is

$$[a^2(r)] = \varepsilon^2 \left[\frac{F_4}{S} + r^2 \left(\frac{L^2}{F_2} - \frac{2L^2 F_2}{S} \right) + r^4 \frac{L^4 F_0}{S} \right] = \varepsilon^2 \frac{1}{N+1} B_{aa}^2(r, N) ,$$

$$[a(r)b(r)] = \frac{\varepsilon^2}{L} \left[r \left(\frac{L^2}{F_2} - \frac{2L^2 F_2}{S} \right) + 2r^3 \frac{L^4 F_0}{S} \right] = \frac{\varepsilon^2}{L} \frac{1}{N+1} B_{ab}^2(r, N) ,$$

$$[b^2(r)] = \frac{\varepsilon^2}{L^2} \left[\frac{L^2}{F_2} + 4r^2 \frac{L^4 F_0}{S} \right] = \frac{\varepsilon^2}{L^2} \frac{1}{N+1} B_{bb}^2(r, N) ,$$

$$[a(r)c_0] = \frac{\varepsilon^2}{L^2} \left[-\frac{2L^2 F_2}{S} + 2r^2 \frac{L^4 F_0}{S} \right] = \frac{\varepsilon^2}{L^2} \frac{1}{N+1} B_{ac}^2(r, N) ,$$

$$[b(r)c_0] = \frac{\varepsilon^2}{L^3} \left[4r \frac{L^4 F_0}{S} \right] = \frac{\varepsilon^2}{L^3} \frac{1}{N+1} B_{bc}^2(r, N) ,$$

$$[c_0^2] = \frac{\varepsilon^2}{L^4} \left[4 \frac{L^4 F_0}{S} \right] = \frac{\varepsilon^2}{L^4} \frac{1}{N+1} B_{cc}^2(N) .$$

(7.31)

Apart from the dimensional factors in front of the square brackets, these covariances depend only on r and N; they contain a common denominator $N + 1$, the number of measurements. The remaining factors have been called $B(r, N)$; they tend to a constant as $N \to \infty$, and they rapidly increase with r. The factors B_{aa} and B_{bb} belonging to the diagonal elements $[a^2(r)]$ and $[b^2(r)]$ were calculated for some values of N and r using (7.31) and (7.25). They are listed in Tables 7.3 and 7.4. The factor B_{cc} belonging to $[c^2]$ does not vary with r and was already contained in Table 7.2 in the form of $A_N = B_{cc}^2(N)/(N + 1)$.

7.2.5 Comparison Between the Linear and Quadratic Fits in Special Cases

Let us consider the extrapolation of a measured track to a vertex. The accuracy of vertex determination is very different whether it happens inside a magnetic field with unknown track momentum or outside (or momentum known). We

Table 7.3. Values of the factor $B_{aa}(r, N)$ in (7.31) at various points along the track at distances r from the centre of the chamber (in units of its length), for various numbers of $N + 1$ equally spaced sense wires

N r:	5	3	2	1.5	1	0.75	0.60	0.50
2	211	75.3	32.9	18.1	7.63	4.10	2.65	2.05
3	224	80.2	35.2	19.5	8.29	4.48	2.84	2.07
5	250	89.5	39.4	21.9	9.39	5.10	3.20	2.26
10	282	101	44.6	24.8	10.7	5.84	3.65	2.54
19	304	109	48.1	26.8	11.6	6.32	3.95	2.72
infin.	335	120	53.0	29.5	12.8	7.00	4.37	3.00

Table 7.4. Values of the factor $B_{bb}(r, N)$ in (7.31) at various points along the track at distances r from the centre of the chamber (in units of its length), for various numbers of $N + 1$ equally spaced sense wires

N	r:	5	3	2	1.5	1	0.75	0.60	0.50
2		84.9	51.0	34.0	25.6	17.1	13.0	10.5	8.83
3		90.0	54.0	36.1	27.1	18.2	13.8	11.1	9.39
5		100	60.2	40.2	30.2	20.3	15.3	12.4	10.4
10		113	68.0	45.4	34.1	22.9	17.3	14.0	11.8
19		122	73.2	48.8	36.7	24.6	18.6	15.0	12.6
infin.		134	80.6	53.8	40.4	27.1	20.4	16.5	13.9

have a direct comparison in Tables 7.1 and 7.3. It appears that a chamber whose front end is only a quarter of its length away from the vertex ($r = 0.75$) extrapolates half as well when inside a field. This gets quickly worse when the vertex is further away – if the factor of comparison is around 2.3 at $r = 0.75$, it is near 5 at $r = 1.5$. This depends only weakly on the number of wires. It is obviously essential for a vertex chamber that it be supported by a momentum measuring device with precision much better than that of the vertex chamber alone.

The loss of accuracy owing to the presence of a magnetic field is also important for the measurement of direction. We want to compute the precision with which a chamber can measure the direction of a track at the first of $N + 1$ regularly spaced wires. In the quadratic fit we may use the third of equations (7.31) in conjunction with Table 7.4 (last column), or we may evaluate the equation directly for $r = 1/2$, which yields

$$\left[b^2\left(\frac{1}{2}\right)\right] = \frac{\varepsilon^2}{L^2} \frac{12(2N + 1)(8N - 3)N}{(N - 1)(N + 1)(N + 2)(N + 3)} \quad \text{(quadratic fit)} . \tag{7.32}$$

In the linear fit we use (7.9), where L was called $x_N - x_0$:

$$\sigma_b^2 = \frac{\varepsilon^2}{L^2} \frac{12N}{(N + 1)(N + 2)} \quad \text{(linear fit)} . \tag{7.33}$$

The ratio of the errors is

$$\frac{\text{with bending}}{\text{without bending}} = \left(\frac{(2N + 1)(8N - 3)}{(N - 1)(N + 3)}\right)^{1/2} , \tag{7.34}$$

which tends to 4 for infinite N and is 3.60 for $N = 2$. This ratio is independent of the magnetic field and of the particle momentum as long as the curvature can be as large as the uncertainty with which it is measured. If there is a priori knowledge that the sagitta of the track is much smaller than the measurement error, the curvature can be neglected, and (7.33) applies.

The fact that the directional error is dominated by the error in curvature is also visible in their correlation coefficient: with some algebra, we form the ratio

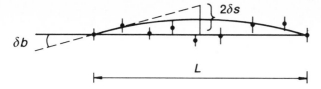

Fig. 7.3. Error δb in direction, created by the sagitta error δs. Case of zero average curvature

$$\frac{[b(\tfrac{1}{2})c]}{\sqrt{[c^2][b^2(\tfrac{1}{2})]}} = \frac{0.968}{\sqrt{\left(1 + \dfrac{1}{2N}\right)\left(1 - \dfrac{3}{8N}\right)}} \tag{7.35}$$

and find it almost equal to 1. This means the error in curvature produces almost the entire error in direction. Figure 7.3 shows a picture of the case $\langle c \rangle = 0$. Since the sagitta is half as long as the corresponding distance between the chord and the end-point tangent, we may estimate the directional error with

$$\delta b = \frac{2\delta s}{L/2}. \tag{7.36}$$

Using (7.29), (7.27) and (7.32) one may verify that (7.36) is correct within 4% for all N.

7.2.6 Optimal Spacing of Wires

With a given number $N + 1$ of wires and a given chamber length L the elements of the covariance matrix can be minimized by varying the wire positions. Summarizing Gluckstern [GLU 63] we mention that the optimal spacing is different for different correlation coefficients, and that for a minimal $[c^2]$ the best wire positions are ideally the following:

$(N + 1)/4$ wires at $x = 0$, $(N + 1)/2$ wires at $x = L/2$,

$(N + 1)/4$ wires at $x = L$.

In practice the clustering of wires will use some of the chamber length. For the ideal case this was set to zero, and he calculates at $r = -1/2$ that

$$\left[b^2\left(\frac{1}{2}\right)\right] = \frac{\varepsilon^2}{L^2}\frac{72}{N + 1},$$

$$\left[b\left(\frac{1}{2}\right)c\right] = -\frac{\varepsilon^2}{L^3}\frac{128}{N + 1}, \tag{7.37}$$

$$[c^2] = \frac{\varepsilon^2}{L^4}\frac{256}{N + 1},$$

and this, at large N, is an important factor of 1.7 better in the momentum resolution than (7.31).

7.3   A Chamber and One Additional Measuring Point Outside

It happens quite often that tracks are measured not only in the main drift chamber but in addition at some extra point which is located some distance away from the the main chamber. Typical examples are a miniature vertex detector near the beam pipe, or a single layer of chambers surrounding the central main chamber. In an event with tracks coming from the same vertex, the vertex constraint serves a similar purpose, especially when the vertex position must be the same for many events. We suppose that all the measurements happen inside a magnetic field.

Since we have developed the covariance matrix of the variables a, b and c at an arbitrary point r along the track, we have the tools to add the constraint of the one additional space-coordinate measurement. The scheme to do this is the following.

Let the full covariance matrix at r (7.31) be abbreviated by

$$C_{ik}(r) = \begin{pmatrix} [a^2(r)] & [a(r)b(r)] & [a(r)c] \\ [a(r)b(r)] & [b^2(r)] & [b(r)c] \\ [a(r)c] & [b(r)c] & [c^2] \end{pmatrix}. \qquad (7.38)$$

This matrix has to be inverted to obtain the χ^2 matrix $C_{ik}^{-1}(r)$. The corresponding matrix D_{ik}^{-1} of the additional measuring point has only the one element that belongs to the space-coordinate non-zero; let us call it $1/\varepsilon_v^2$ (ε_v is the accuracy of the additional measuring point):

$$D_{ik}^{-1} = \begin{pmatrix} \varepsilon_v^{-2} & 0 & 0 \\ 0 & 0 & 0 \\ 0 & 0 & 0 \end{pmatrix}. \qquad (7.39)$$

Now the covariance matrix E_{ik} of the total measurement, $N + 1$ regularly spaced wires plus this one point, a distance rL away from their centre, is given by the inverse of the sum of the inverses of the individual covariances:

$$(E_{ik})^{-1} = (C_{ik}(r))^{-1} + (D_{ik})^{-1} . \qquad (7.40)$$

Now there is a theorem (e.g. [SEL 72]) which states that if the inverse of a first matrix is known, then the inverse of a second matrix, which differs from the first in only one element, can be calculated according to a simple formula. In our case the first matrix is C^{-1}, and its known inverse is C; the second matrix, E^{-1}, differs from the first only in the (1, 1) element which is larger by $1/\varepsilon_v^2$. The inverse of the second matrix is given by

$$E_{ik} = C_{ik} - \frac{C_{1i}C_{1k}}{\varepsilon_v^2 + C_{11}}. \tag{7.41}$$

This means that our covariance matrix (7.31) changes by the addition of one point at r according to the following scheme (we leave out the argument r):

$$[a^2] \to [a'^2] = [a^2] - \frac{[a^2]^2}{\varepsilon_v^2 + [a^2]},$$

$$[ab] \to [a'b'] = [ab] - \frac{[a^2][ab]}{\varepsilon_v^2 + [a^2]},$$

$$[b^2] \to [b'^2] = [b^2] - \frac{[ab^2]^2}{\varepsilon_v^2 + [a^2]},$$

$$[ac] \to [a'c'] = [ac] - \frac{[a^2][ac]}{\varepsilon_v^2 + [a^2]}, \tag{7.42}$$

$$[bc] \to [b'c'] = [bc] - \frac{[ab][ac]}{\varepsilon_v^2 + [a^2]},$$

$$[c^2] \to [c'^2] = [c^2] - \frac{[ac]^2}{\varepsilon_v^2 + [a^2]}.$$

7.3.1 Comparison of the Accuracy in the Curvature Measurement

We are particularly interested in the gain that can be obtained for the curvature measurement. Introducing the ratio $f^2 = \varepsilon_v^2/[a^2(r)]$ (how much better the extra measurement is compared to the extrapolated point measuring accuracy of the chamber), we write the gain factor as

$$\kappa^2(f, r) = \frac{[c^2] - \dfrac{[a(r)c]^2}{\varepsilon_v^2 + [a^2(r)]}}{[c^2]} = 1 - \frac{[a(r)c]^2}{(1 + f^2)[a^2(r)][c^2]}. \tag{7.43}$$

Since the covariance matrix (7 31) is a function of r and N, κ also depends on N. We evaluate (7.43) in the limit of $N \to \infty$. Table 7.5 contains the result.

It appears that considerable improvement is possible for the curvature, and hence momentum, measurement of a large drift chamber when one extra point outside can be measured with great precision. This is a consequence of the increase of the lever arm. As an example we consider the geometry of the ALEPH TPC. For a well measured track there are 21 measured points, equally spaced over $L = 130$ cm; each has $\varepsilon = 0.16$ mm, and therefore $\sqrt{[c^2]} = \sqrt{(720/25)} \times (0.16 \times 10^{-3})/1.30^2 = 5.1 \times 10^{-4}$ m^{-1}, according to (7.27). A silicon miniature vertex detector with $\varepsilon_v = 0.02$ mm, 35 cm in front of the first measuring point ($r = 0.77$), would result in an improvement factor of

Table 7.5. Factors $\kappa(f, r)$ of (7.43) by which the measurement of curvature becomes more accurate when a single measuring point with accuracy $\varepsilon_v = f\sqrt{[a^2(r)]}$ is added to a large number of regularly spaced wires at a distance rL from their centre

r	f^2: 0	0.1	0.2	0.3	0.5	1.0	2.0
0.5	0.67	0.70	0.73	0.76	0.79	0.85	0.90
0.6	0.53	0.59	0.63	0.67	0.72	0.80	0.87
0.7	0.43	0.51	0.57	0.61	0.68	0.77	0.85
0.8	0.37	0.46	0.53	0.58	0.65	0.75	0.84
1.0	0.28	0.40	0.48	0.54	0.62	0.73	0.83
1.5	0.18	0.35	0.44	0.51	0.60	0.72	0.82
2.0	0.13	0.33	0.43	0.49	0.59	0.71	0.82
infin.	0	0.30	0.41	0.48	0.58	0.71	0.82

$\kappa = 0.40$, according to Table 7.5. Here we have determined $\sqrt{[a^2(0.77)]}$ $= 7 \times 0.16 \text{ mm}/\sqrt{21} = 0.24 \text{ mm}$, using the first of (7.31) and Table 7.3; therefore $f = 0.02/0.24 = 0.08$.

Such enormous improvement could only be converted into fact if all the systematic errors were tightly controlled at a level better than these 20 µm. Also, for the moment we have left out any deterioration caused by multiple scattering, to which subject we will turn in Sect. 7.4.

7.3.2 Extrapolation to a Vertex

Since it is the first purpose of a vertex detector to determine track coordinates at a primary or secondary interaction vertex, we must carry our error calculation forward to the vertex. The situation is sketched in Fig. 7.4. The coordinate

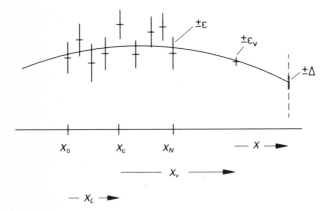

Fig. 7.4. Extrapolation of the quadratic fit to a vertex located a distance x in front of a vertex detector with accuracy $\pm \varepsilon_v$. The $N + 1$ wires of a chamber are uniformly spaced over the length L. The vertex detector is a distance x_v away from the centre of the wires. In the text, $r = x_v/L$

x represents the distance of the detector to the vertex, and the extrapolation error Δ of the vertex position is given by the covariance

$$\Delta^2 = [y^2] = [(a'(r) + b'(r)x + c'(r)x^2/2)^2]$$
$$= [a'^2] + 2x[a'b'] + x^2([b'^2] + [a'c'])$$
$$+ x^3[b'c'] + x^4[c'^2]/4 . \qquad (7.44)$$

Here the $[a'^2], [a'b'], \ldots$ represent the elements of the covariance matrix E (7.42) at the position r of the vertex detector.

The evaluation of (7.44) is too clumsy when done in full generality, and we will consider the special case of a vertex detector that is much more accurate than the drift chamber at r, as presented in Table 7.3:

$$f^2 \ll 1$$

or

$$\varepsilon_v^2 \ll [a^2(r)] . \qquad (7.45)$$

The covariance matrix at r for the ensemble of chamber and vertex detector, in first order of $\varepsilon_v^2/[a^2(r)]$, takes the form

$$E = \begin{pmatrix} \varepsilon_v^2 & \varepsilon_v^2[ab]/[a^2] & \varepsilon_v^2[ac]/[a^2] \\ \varepsilon_v^2[ab]/[a^2] & [b^2] - [ab]^2/[a^2] & [bc] - [ab][ac]/[a^2] \\ \varepsilon_v^2[ac]/[a^2] & [bc] - [ab][ac]/[a^2] & [c^2] - [ac]^2/[a^2] \end{pmatrix} .$$
$$\qquad (7.46)$$

The elements of E depend on N and r. We evaluate (7.46) in the limit of $N \to \infty$ and for some special values of r. For $r = 1/2$ (i.e. the vertex detector at the same place as the last wire), using (7.44), (7.45), (7.31), (7.25), one finds that

$$\Delta^2 = \frac{\varepsilon^2}{N+1}\left(48\frac{x^2}{L^2} + 120\frac{x^3}{L^3} + 80\frac{x^4}{L^4}\right) + \varepsilon_v^2\left(1 + 8.0\frac{x}{L} + 6.7\frac{x^2}{L^2}\right) .$$

Here we have counted x/L positive if x and x_v point in the same direction. We see that the vertex extrapolation accuracy deteriorates as x/L increases. There is one term proportional to the accuracy squared of the vertex detector alone, and one term proportional to the quantity $\varepsilon^2/(N + 1)$, which characterizes the accuracy of the drift chamber.

$$\Delta^2 = R^2\left(r, \frac{x}{L}\right)\frac{\varepsilon^2}{N+1} + S^2\left(r, \frac{x}{L}\right)\varepsilon_v^2 . \qquad (7.47)$$

Figure 7.5 contains graphs of $R(r, x/L)$ and $S(r, x/L)$ for four values of r in the range of typical applications.

Taking again the geometry of the ALEPH experiment as a numerical example, we read from Fig. 7.5 that for $r = 0.77$ and $x_v = 6.5$ cm the vertex extrapolation accuracy can reach $\Delta = \sqrt{(0.28^2\varepsilon^2/21 + 1.14^2\varepsilon_v^2)}$. Using $\varepsilon = 160\,\mu m$ and $\varepsilon_v = 20\,\mu m$ one obtains $\Delta = 25\,\mu m$ in this particular case.

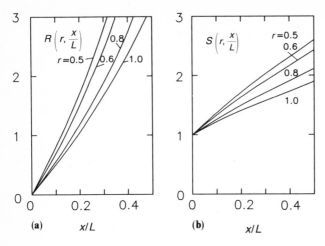

Fig. 7.5a, b. Graph of the two functions in (7.47) for four positions r of the vertex detector; (a) $R(r, x/L)$, (b) $S(r, x/L)$

7.4 Limitations Due to Multiple Scattering

The achievable accuracy of a chamber may be reduced by multiple scattering of the measured particle in some piece of the apparatus. There it suffers a displacement and a change in direction, and we want to know the influence on the accuracy of vertex reconstruction and of momentum measurement.

7.4.1 Basic Formulae

The theory of multiple scattering has been developed by Moliere, Snyder, Scott, Bethe and others. Reviews exist from Rossi [ROS 52], Bethe [BET 53] and Scott [SCO 63]. Our interest here must be limited to the basic notions.

After a particle has traversed a thickness x of some material, it has changed its direction and position, and we want to know the two-dimensional probability distribution $P(x, y, \theta_y)$ of the angular and spatial variables θ_y and y. Here we describe in a Cartesian coordinate system the lateral position at a depth x by the coordinates y and z. The corresponding angles are denoted by θ_y (in the x–y plane) and θ_z (in the x–z plane). Because of the symmetry of the situation, the two distributions $P(x, y, \theta_y)$ and $P(x, z, \theta_z)$ must be the same. In the simplest approximation they are given [ROS 52] by

$$P(x, y, \theta_y)\,dy\,d\theta_y = \frac{2\sqrt{3}}{\pi}\frac{1}{\Theta_s^2 x^2}\exp\left[-\frac{4}{\Theta_s^2}\left(\frac{\theta_y^2}{x} - \frac{3y\theta_y}{x^2} + \frac{3y^2}{x^3}\right)\right], \qquad (7.48)$$

where the only parameter Θ_s^2 is called the mean square scattering angle per unit length. The angular distribution, irrespective of the lateral position, is given by the integral

$$Q(x, \theta_y) = \int_{-\infty}^{\infty} P(x, y, \theta_y)\, dy = \frac{1}{\sqrt{\pi}}\frac{1}{\Theta_s\sqrt{x}}\exp\left[-\frac{\theta_y^2}{\Theta_s^2 x}\right]. \tag{7.49}$$

We have for this projection a Gaussian distribution with a mean square width of

$$\langle\theta_y^2\rangle = \Theta_s^2 x/2 . \tag{7.50}$$

Since the basic process is the scattering with a single nucleus, we understand that the variance of θ_y must be proportional to the number of scatterings, and hence x.

The corresponding spatial distribution, irrespective of the angle of deflection, is given by the integral over the other variable:

$$S(x, y) = \int_{-\infty}^{\infty} P(x, y, \theta_y)\, d\theta_y = \sqrt{\frac{3}{\pi}}\frac{1}{\Theta_s\sqrt{x^3}}\exp\left[-\frac{3y^2}{\Theta_s^2 x^3}\right]. \tag{7.51}$$

This Gaussian distribution has a mean square width of

$$\langle y^2\rangle = \Theta_s^2 x^3/6 . \tag{7.52}$$

For completeness we mention that the covariance is given by

$$\langle y\theta_y\rangle = \int y\theta_y P(x, y, \theta_y)\, dy\, d\theta_y = \Theta_s^2 x^2/4 . \tag{7.53}$$

The mean square scattering angle per unit length, Θ_s^2, depends on particle velocity β and momentum p as well as on the radiation length X_{rad} of the material:

$$\Theta_s^2 = \left(\frac{E_s}{\beta c p}\right)^2 \frac{1}{X_{\mathrm{rad}}} . \tag{7.54}$$

The constant E_s, which has the dimension of energy, is given by the fine-structure constant α and the electron rest energy mc^2:

$$E_s = \left(\frac{4\pi}{\alpha}\right)^{1/2} mc^2 = 21\,\mathrm{MeV} . \tag{7.55}$$

Using (7.55) and (7.54) in (7.53) and (7.52) we may express the projected r.m.s. deflection and displacement as

$$\sqrt{\langle\theta_y^2\rangle} = \frac{15\,\mathrm{MeV}}{\beta c p}\sqrt{\frac{x}{X_{\mathrm{rad}}}} , \tag{7.56}$$

$$\langle y^2\rangle = \frac{x}{3}\sqrt{\langle\theta_y^2\rangle} . \tag{7.57}$$

A short list of scattering lengths for some materials of interest in our context is given in Table 7.6. It is extracted from [TSA 74]; see also [PAR 90].

Table 7.6. Radiation lengths X_{rad} and densities ρ of some materials relevant for drift chambers

Material	ρ	X_{rad}	$X_{rad}\rho$
Gases (N.T.P.)	(g/l)	(m)	(g/cm^2)
H_2	0.090	6800	61.28
He	0.178	5300	94.32
Ne	0.90	321.6	28.94
Ar	1.78	109.8	19.55
Xe	5.89	14.4	8.48
Air	1.29	284.2	36.66
CO_2	1.977	183.1	36.2
CH_4	0.717	648.5	46.5
Iso-C_4H_{10}	2.67	169.3	45.2
Solids	(g/cm^3)	(cm)	(g/cm^2)
Be	1.848	35.3	65.19
C	2.265	18.8	42.70
Al	2.70	8.9	24.01
Si	2.33	9.36	21.82
Fe	7.87	1.76	13.84
Pb	11.35	0.56	6.37
Polyethylene CH_2	0.92–0.95	~ 47.9	44.8
Mylar $C_5H_4O_2$	1.39	28.7	39.95

In Moliere's theory [MOL 48], which represents the measurements very well (see BET 53), the Gaussian distribution of the scattering angle is only one term in a series expansion. The effect of the additional terms is that the Gauss curve acquires a tail for angles that are several times as large as the r.m.s. value. Although only at the level of a few per cent compared to the maximum, they are many times larger than the Gauss curve for these arguments. The tail is caused by the few rare events where one or two single scatters reach a deflection several times as large as the r.m.s. sum of the large majority of events, which are dominated by small-angle scattering alone. This behaviour is typical for Coulomb scattering as described by the Rutherford formula; for hadrons, nuclear elastic scattering also makes a contribution to the tail. All the theories that go beyond the simple approximation discussed above are analytically quite complex.

Highland [HIG 75] has taken a practical standpoint by fitting the width of the scattering-angle distribution of the Moliere theory to a form similar to (7.56), using the radiation lengths as listed in the Particle Data Tables (and partly in our Table 7.6).

He finds that the width of Moliere does not vary appreciably with the nuclear charge number Z; he introduces an x-dependent correction factor $(1 + \varepsilon(x))$ to (7.56), which redefines the energy constant. His result is

$$\theta_y^{1/e} = \frac{14.1 \text{ MeV}}{\beta c p} \sqrt{\frac{x}{X_{\text{rad}}}} \left(1 + \frac{1}{9} \log_{10} \frac{x}{X_{\text{rad}}} \right). \tag{7.58}$$

$\theta_y^{1/e}$ is the angle in the x–y plane at which the Moliere probability density is down by a factor of $1/e$ from the maximum. Formula (7.58) is supposed to be accurate to 5% in the range $10^{-3} < x/X_{\text{rad}} < 10$, except for the very light elements, where the accuracy is 10–20%.

7.4.2 Vertex Determination

At colliders the observation of primary vertices and of secondary vertices from particles with short lifetimes is through a vacuum tube; in addition there is the material of the vertex chamber itself. We would like to compute the vertex localization error due to multiple scattering in these materials. The situation is sketched in Fig. 7.6. We assume for simplicity that the track is infinitely well measured outside the scattering material. Upon extrapolation, it misses the vertex position by the distance d. The average over the square of d is equal to

$$\langle d^2 \rangle = \langle \vartheta_1^2 \rangle r_1^2 + \langle \vartheta_2^2 \rangle r_2^2 + \cdots, \tag{7.59}$$

where the mean squared projected scattering angle $\langle \vartheta_i^2 \rangle$ is given by the thickness t_i of the layer i in units of radiation lengths of the material i according to (7.56), where $\langle \vartheta^2 \rangle = \langle \theta_y^2 \rangle$. We express the r.m.s. value of d in terms of a 'Coulomb scattering limit' S_c of the vertex localization accuracy. By combining (7.59) and (7.56) we have

$$\sqrt{\langle d^2 \rangle} = \frac{S_c}{\beta p c}, \tag{7.60}$$

with

$$S_c = 15 \text{ MeV} \left(\sum t_i r_i^2 \right)^{1/2}. \tag{7.61}$$

Practical values of S_c range between 45 and 200 μm GeV for various experiments; compare Table 10.3.

The above calculation gives correct vertex errors in the limit that the Coulomb scattering is much larger than the measurement errors and that all the

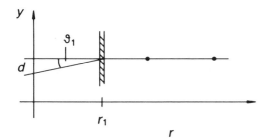

Fig. 7.6. Coulomb multiple scattering of an infinitely well measured track

scattering material is between the sense wires and the vertex. For the general case the estimation of vertex errors is more involved. Although the quadratic addition of measurement errors (7.8) and the multiple-scattering errors (7.60) gives the right order of magnitude for the total, more sophisticated estimation methods give better results. Lutz [LUT 88] has developed an optimal tracking procedure by fitting the individual contributions to multiple scattering between the measuring points.

7.4.3 Resolution of Curvature for Tracks Through a Scattering Medium

We have seen in (7.56) that the mean square scattering angle increases as the momentum of the particle becomes smaller. Let us imagine a drift chamber in a spectrometer where the momentum of a particle is low enough to make the multiple-scattering error the dominating one. We would like to know the accuracy for a measurement of the curvature in this situation; this implies we suppose that the accuracy of the apparatus itself is infinitely better than the multiple-scattering errors.

We resume the arguments of Sect. 7.2 at (7.2.1). In order to evaluate $[y_n y_m]$ in the case of multiple scattering, we observe that

$$[y_n y_m] = [y_n(y_n + (x_m - x_n)\theta_{yn})] = [y_n^2] + (x_m - x_n)[y_n\theta_{yn}] \tag{7.62}$$

for $x_m \geq x_n$. Now we make use of (7.52) and (7.53) and obtain

$$[y_n y_m] = \Theta_s^2 x_n^2 (3x_m - x_n)/12 . \tag{7.63}$$

Gluckstern [GLU 63] has evaluated the correlation coefficients at the first of $N + 1$ uniformly spaced wires with the result

$$[c^2]_{MS} = \frac{\Theta_s^2}{2L} C_N ,$$

$$\left[cb\left(\frac{1}{2}\right)\right]_{MS} = -\frac{\Theta_s^2}{2} D_N , \tag{7.64}$$

$$\left[b^2\left(\frac{1}{2}\right)\right]_{MS} = \frac{\Theta_s^2}{2} L E_N .$$

The N-dependent coefficients are listed in Table 7.7.

Let us note that the measurement accuracy $\sqrt{[c^2]}$ for the curvature in this multiple-scattering-limited case is inversely proportional to the square-root of the length. Also, the variations of b and c are much more independent than they were in the measurement-limited case (7.35), because the ratio $D_N/\sqrt{(C_N E_N)}$ is relatively small.

The problem of optimal spacing and optimal weighting has been treated by Gluckstern [GLU 63] as well as by some other authors cited in his paper.

Table 7.7. Values of the factors C_N, D_N and E_N in (7.64) for various values of N

N	C_N	D_N	E_N
2	1.33	0.167	0.167
3	1.25	0.125	0.154
4	1.25	0.124	0.160
5	1.26	0.132	0.167
9	1.31	0.156	0.187
infin.	1.43	0.214	0.229

7.5 Spectrometer Resolution

Much of the effort of constructing precise drift chambers was for the goal of high-resolution magnetic spectrometers. In this section we want to collect our results concerning the measurement errors of curvature and to express them in terms of momentum resolution. We are concerned with measurements inside a homogeneous magnetic field. There were two limiting cases, the one caused by measurement errors, in the absence of multiple scattering, and the other due to multiple scattering, with zero measurement errors.

7.5.1 Limit of Measurement Errors

We noted before that the variance $[c^2]$ of the curvature is a function of the track length L, the point-measuring accuracy ε, and the number $N + 1$ of wires, but it does not depend on the curvature itself. For uniform wire spacing we had

$$[c^2] = \frac{\varepsilon^2}{L^4} A_N \quad (A_N \text{ in Table 7.2}) \tag{7.27*}$$

and for wires with optimum spacing

$$[c^2] = \frac{\varepsilon^2}{L^4} \frac{256}{N + 1}. \tag{7.37*}$$

The track length is understood to be measured in a plane orthogonal to the magnetic field.

   In order to calculate the momentum resolution, we use (7.10) and (7.12) and express the r.m.s. error $\delta(1/R) = \sqrt{[c^2]}$ as

$$\sqrt{[c^2]} = \delta\frac{1}{R} = eB\,\delta\left(\frac{1}{p_T}\right) = -eB\frac{\delta p_T}{p_T^2} = \frac{-3}{10}\left(\frac{\text{GeV/c}}{\text{T m}}\right)B\frac{\delta p_T}{p_T^2}. \tag{7.65}$$

Introducing θ, the angle between the momentum vector and the magnetic field, the transverse component p_T is given by

$$p_T = p \sin \theta . \tag{7.66}$$

We drop the minus sign and write (7.65) as

$$\frac{\delta p}{p} = p \sin \theta \frac{10}{3} \left(\frac{\mathrm{T\,m}}{\mathrm{GeV}/c} \right) \frac{1}{B} \sqrt{[c^2]} . \tag{7.67}$$

This formula shows that the spectrometer resolving power, expressed as a percentage of the measured momentum, deteriorates proportionally to p. This follows from the fact that $[c^2]$ is independent of c; it is also visible directly in the sagitta s (Sect. 7.2.3), whose size can be measured with less relative accuracy $\delta s/s$ as it shrinks with increasing momentum.

7.5.2 Limit of Multiple Scattering

In this case the variance $[c^2]$ depends on the track length L, the mean square scattering angle per unit length, Θ_s^2, and very weakly on the number of measuring points. Θ_s^2 in turn is a function of the scattering material and of the βp of the particle. Using (7.54) and (7.64) we had for uniform wire spacing

$$[c^2]_{\mathrm{MS}} = \left(\frac{21\,\mathrm{MeV}}{\beta c p} \right)^2 \frac{1}{X_{\mathrm{rad}}} \frac{C_N}{2L} \quad (C_N \text{ in Table 7.7}) . \tag{7.68}$$

L is the length the track has travelled in the medium and $[c^2]$ the variance of the total curvature. Only a projection on a plane perpendicular to the magnetic field can be confused with a curvature variation caused by a variation in momentum,

$$[c^2_{\mathrm{proj}}] = [c^2]_{\mathrm{MS}}/\sin^4 \theta , \tag{7.69}$$

because the projected radius of curvature scales with the square of the projected length. It is $[c^2_{\mathrm{proj}}]_{\mathrm{MS}}$ to which (7.67) applies in the present case, and we find for the momentum resolution

$$\begin{aligned}
\frac{\delta p}{p} &= \frac{p}{\sin \theta} \frac{10}{3} \left(\frac{\mathrm{T\,m}}{\mathrm{GeV}/c} \right) \frac{1}{B} \sqrt{[c^2]_{\mathrm{MS}}} \\
&= \frac{21\,(\mathrm{MeV}/c)}{\beta \sin \theta} \frac{10}{3} \left(\frac{\mathrm{T\,m}}{\mathrm{GeV}/c} \right) \frac{1}{B} \sqrt{\frac{C_N}{2LX_{\mathrm{rad}}}} .
\end{aligned} \tag{7.70}$$

We see that the multiple-scattering limit of the momentum resolution is independent of momentum because, as p goes up, the decrease in bending power is compensated by the decrease in the scattering angle. According to (7.70), a muon traversing 1 m of magnetized iron (2 T) at right angles to the field cannot be measured better than to 22%, even if there were infinitely many measuring points with infinite accuracy inside the iron. Depending on the apparatus, L will often be a function of the angle θ.

References

[BET 53]   H.A. Bethe, Moliere's theory of multiple scattering, *Phys. Rev.* **89**, 1256 (1953)

[BOC 90]   R.K. Böck, H. Grote, D. Notz, M. Regler, Data analysis techniques for high-energy physics experiments (Cambridge University Press, Cambridge, UK 1990)

[GLU 63]   R.L. Gluckstern, Uncertainties in track momentum and direction due to multiple scattering and measurement errors, *Nucl. Instr. Meth.* **24**, 381 (1963)

[HIG 75]   V.L. Highland, Some practical remarks on multiple scattering, *Nucl. Insrtrum. Methods* **129**, 497 (1975)

[LUT 88]   G. Lutz, Optimum track fitting in the presence of multiple scattering, *Nucl. Instrum. Methods Phys. Res. A* **273**, 349 (1988)

[MOL 48]   G. Moliere, Theorie der Streuung schneller geladener Teilchen II, Mehrfach und Vielfachstreuung, *Z. Naturforsch.* **3a**, 78 (1948)

[PAR 90]   Particle Data Group, Review of Particle Properties, *Phys. Lett. B* **239** (1990)

[ROS 52]   B. Rossi, High-Energy Particles (Prentice-Hall, Englewood Cliffs, NJ, USA 1952)

[SCO 63]   W.T. Scott, The theory of small-angle multiple scattering of fast charged particles, *Rev. Mod. Phys.* **35**, 231 (1963)

[SEL 72]   S.M. Selby (ed.), *Standard Mathematical Tables* (The Chemical Rubber Co., Cleveland, Ohio, USA 1972)

[TSA 74]   Y.-S. Tsai, Pair production and bremsstrahlung of charged leptons, *Rev. Mod. Phys.* **46**, 815 (1974)

8. Ion Gates

Here we wish to take up the subject of ion shutters, which are sometimes placed in the drift space in order to block the passage of electrons or ions. In Chap. 3 we introduced such wire grids in the context of the electrostatics of drift chambers, stressing the relation between the various grids and electrodes that make up a drift chamber. So far we have dealt with the most straightforward type of gating grid, where all the wires are at the same potential.

For applications in large drift chambers there are more sophisticated forms of ion shutters which, in some modes of operation, have different transmission properties for electrons and for heavy ions.

We will begin this chapter by inspecting the conditions that make it necessary to involve gating grids (Sect. 8.1). Then we will survey various forms of gating grids that are in use in particle experiments (Sect. 8.2). Finally we compute values for the transparency under various conditions for which we make use of the results of Chap. 2 concerning ion drift velocities and the magnetic drift properties of electrons in gases.

8.1 Reasons for the Use of Ion Gates

8.1.1 Electric Charge in the Drift Region

Any free charges in the drift volume give rise to electric fields which superimpose themselves on the drift field and distort it. Chambers with long drift lengths L are particularly delicate, especially when they are operated with a low drift field. The displacement σ_x of an electron drift path at its arrival point is of the order of the drift length times the ratio of the disturbing field E_{dist} over the drift field E_{drift}:

$$\sigma_x = L \frac{E_{dist}}{E_{drift}} \, . \tag{8.1}$$

This holds if the disturbing field is orthogonal to the drift field, thus causing a displacement of the field lines to the side, but it also holds if the disturbing field acts in the same direction as the drift field, thus causing a change in the velocity and hence of the arrival time, provided the chamber works in the unsaturated mode of the drift velocity (cf. Chap. 11).

Free charges can originate as electrons or gas ions from the ionization in the drift volume, travelling in opposite directions according to their proper drift velocities, or they can be ions from the wire avalanches that have found their way into the drift region, where they move towards the negative high-voltage electrode. This last category contributes the largest part because for every incoming electron which causes G avalanche ions to be produced at the proportional wire, there will be $G\varepsilon$ in the drift space, where ε is the fraction that arrives in the drift space rather than on the cathodes opposite the proportional wire. If there is no gating grid, $G\varepsilon$ will be of the order of 10^3 in typical conditions, but could also be larger (see remarks below).

In order to estimate the size of disturbing fields we consider two examples. The first is a drift chamber in which a gas discharge burns continuously between some sense wires and their cathodes. Such a self-sustained process could be caused by some surface deposit on the cathode, or otherwise. Let the total current be 1 μA, and let $\varepsilon = 5\%$ of it penetrate into the drift volume. If the ion drift velocity there is 2 m/s, then the linear charge density in the column of travelling ions is $\lambda = 25 \times 10^{-9}$ A s/m. Using Gauss' theorem in this cylindrical geometry, the resulting radial field a distance r away amounts to

$$E_{\text{dist}} = \frac{\lambda}{2\pi r \varepsilon_0} = 450(V)\frac{1}{r} . \tag{8.2}$$

Here we have neglected the presence of any conductors.

The second example is the ALEPH TPC, irradiated by some ionizing radiation, say cosmic rays or background radiation from the e^+e^- collider. Let the rate density of electrons liberated in the sensitive volume be R $(s^{-1}m^{-3})$. For every electron, $G\varepsilon$ ions will appear in the drift space, where they travel with velocity v_D. Therefore the total charge density in the volume has the value

$$\rho = \frac{eRLG\varepsilon}{v_D} , \tag{8.3}$$

where e is the charge of the electron and L is the length of the drift volume. ρ could depend on the radius r, as one would expect from radiation originating along the beam. Let the TPC be approximated by the space between two infinite coaxial conducting cylinders with radii $r_1 = 0.3$ m and $r_2 = 2$ m, with the drift direction parallel to the common axis. The cylinders are grounded. The radial field E_r is calculated from Maxwell's equations,

$$\nabla E_r = \frac{1}{r}\frac{\partial}{\partial r} r E_r = \frac{\rho}{\varepsilon_0} , \tag{8.4}$$

$$\int_{r_1}^{r_2} E_r \, dr = 0 .$$

If $\rho(r) = \rho_0$ does not depend on r, as in the case of irradiation by cosmic rays,

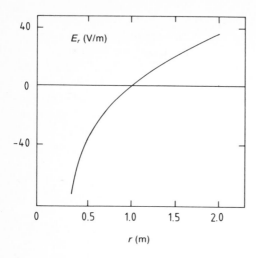

Fig. 8.1. Radial dependence of the electric field produced by a uniform charge distribution in a cylindrical TPC. The scales are from the example discussed in the text

the resulting radial disturbing field takes the form

$$E_r(r) = \frac{\rho_0}{2\varepsilon_0} \left(r - \frac{1}{2r} \frac{r_2^2 - r_1^2}{\ln(r_2/r_1)} \right). \tag{8.5}$$

If we insert into (8.3) values that are characteristic for the ALEPH TPC irradiated by cosmic rays at sea level ($R = 2 \times 10^6 \, \text{s}^{-1} \, \text{m}^{-3}$, $v_D = 1.5 \, \text{m/s}$, $G\varepsilon = 10^3$, $L = 2 \, \text{m}$), we find that $\rho_0 = 0.43 \times 10^{-9} \, \text{A s m}^{-3}$. Figure 8.1 shows the radial field that results from (8.5) in this particular example. We notice that it is negative close to the inner field cage and positive close to the outer one. Compared to the regular axial drift field of $10^4 \, \text{V/m}$, it amounts to a fraction of a per cent.

It should be mentioned that the coefficient ε is not very well known. A lower limit can be evaluated assuming that the amplification produced ions uniformly around the proportional wire. In this case ε is equal to the fraction of electric field lines that reach from the sense wire into the drift region. Using the formalism of Sect. 3.1 ε is equal to the ratio of the surface charge densities on the high-voltage plane and on the sense wire grid:

$$\varepsilon = \frac{|\sigma_\text{p}|}{\sigma_\text{s}}, \tag{8.6}$$

typically between 3 and 5%.

However, it is known that the amplification is not isotropic around the proportional wire. This has been discussed in Sect. 4.3. Large avalanches tend to go around the wire; small avalanches develop on the side where the electrons have arrived. The ions from the small avalanches may go into the drift space more efficiently. Nothing quantitative has been published concerning this question.

Some further uncertainty exists concerning the value of v_D. The effect of charge transfer between the travelling ions and any gas molecules of low ionization potential will substitute these molecules, which are often much slower, for the original ions; see Table 2.2. This effect creates a space charge that is not uniform in the drift direction.

The most important contribution to space charge is usually created by the background radiation at the accelerator. Background conditions can be very different from one experiment to the other. In the ALEPH TPC at LEP (sensitive volume $45\,m^3$) the sense-wire current is $\sim 0.5\,\mu A$ under standard operating conditions. The UA1 central detector at the CERN $Sp\bar{p}S$ (sensitive volume $25\,m^3$) had a current of $\sim 120\,\mu A$ at the peak luminosity of $3 \times 10^{29}\,cm^{-2}\,s^{-1}$ [BEI 88].

8.1.2 Ageing

There is experimental evidence of a deterioration in performance of drift and proportional chambers after they have been used for some time. Once the total charge collected on the anode wires during their lifetime exceeds some value between 10^{-4} and 1 C per cm of wire length, a loss of gain and excessive dark currents are observed in many chambers. Whereas a short account of this problem is given in Sect. 11.4, we mention it in the context of ion gates in order to underline their effect on the ageing process.

The dose mentioned above is the product of the incoming charge and the gain factor. The limit of observable deteriorations depend very much on the gas composition and on the electric field on the electrode surfaces; under carefully chosen conditions, a limiting dose of the order of 0.1 to 1 C/cm should be achievable (cf. Sect. 11.4). The ageing process can be stretched in time if only part of the radiation is admitted to the sense wires, using a gate that is triggered only on interesting events.

8.2 Survey of Field Configurations and Trigger Modes

In order to describe the various forms of gating grids employed in particle experiments we use two classifications, one according to drift paths on the basis of different (electric and magnetic) field configurations, the other according to the dynamical behaviour, i.e. how the gates are switched in connection with events in time.

8.2.1 Three Field Configurations

The simplest gate is the one which has all its wires at the same potential (*monopolar gating grid*). We have stated in Sect. 3.2 that it is closed to incoming electrons and outgoing ions if all the field lines terminate on it, provided the common potential is large enough and positive (Fig. 3.10b).

There is a technical difficulty connected with this configuration: any transition from the closed to the open state must occur in a time interval ΔT comparable to the electron travel time over a distance small compared to the sensitive drift length, say $\Delta T \approx 1$ μs or less. The charge brought onto the grid in this short time causes an enormous disturbance on the nearby sense wires, orders of magnitudes larger than a signal. One solution to this problem is the introduction of a 'shielding grid' between the gating grid and the anode [BRY 85].

Another way to circumvent this difficulty is to close the gating grid by ramping two opposite potentials $+ \Delta V_g$ and $- \Delta V_g$ on neighbouring wires [BRE 80, NEM 83]. In this case the net amount of charge on the gating grid does not change, and in a first approximation there is no charge induced in the other electrodes of the drift chamber. This type of grid is called a *bipolar gating grid*. In the closed state it has positive charge on every second wire and negative charge on the wires in between; the drift field lines terminate on the positive charges.

In the presence of a magnetic field B the drift paths of electrons are no longer the electric field lines, because their drift-velocity vector has components along B and along $[E \times B]$ according to (2.6). The exact behaviour is governed by the parameter $\omega\tau$, or the ion mobility multiplied by the magnitude of the B field. This parameter is always very small for ions in drift chambers but can be much larger than one for electrons in suitable gases (cf. Chap. 2). If this is the case and ΔV_g is increased from zero, the ions are stopped first, while many of the electrons are still able to penetrate. The reason is the reduced electron mobility towards the gating grid wires. As ΔV_g is further increased, the gate is finally closed also for the electrons.

This *bipolar gating grid in a magnetic field*, with a suitable value of ΔV_g, can therefore be operated as a diode ([AME 85-1] and later but independently [KEN 84]). The electrons from the drift region make their way through the grid on trajectories which are bent in the wire direction as well as in the direction orthogonal to both the wires and the main drift direction. See Fig. 8.2 for an illustration of the principle.

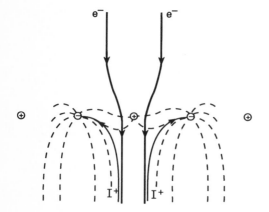

Fig. 8.2. Scheme of the bipolar grid immersed in a magnetic field, which makes it transparent to electrons while it remains opaque to ions

8.2.2 Three Trigger Modes

The simplest case is the bipolar gating grid operated in a magnetic field as a diode. It does not have to be triggered, and one avoids all switching circuitry and all pick-up problems. There is a disadvantage to this solution: although the positive ions from the wire avalanches are not admitted back into the drift space, all the electrons from the drift region are allowed to produce avalanches. Under conditions where chamber ageing is a problem, one gives away a factor of the lifetime of the chamber which depends on the background conditions of the experiment.

The tightest trigger is the one where some counters outside the drift chamber select the wanted event and open the gate by applying a (bipolar) voltage pulse to the grid in order to remove the closing potential(s). Depending on the time delay T_d between the moment of the event and the moment the gate has been opened, one loses a length L of sensitivity given by the electron drift velocity u:

$$L = T_d u \; .$$

This loss can sometimes be avoided at the expense of some background – if there is a regular time pattern when the events occur, even if they come with low probability. The ALEPH TPC at LEP is triggered 'open' a few µs before every bunch crossing of the collider. When there is no event, the gate is switched back to 'closed'. In this way it is 'open' for 6 µs out of every 22 µs, but it stays 'open' long enough to read an event. This mode has been termed a *synchronous trigger*, to be contrasted with the *asynchronous trigger* described in the previous paragraph. The synchronous trigger also avoids the disturbance on the signal lines which remains large even with a bipolar gating grid, owing to small accidental asymmetries between neighbouring gating-grid wires.

8.3 Transparency under Various Operating Conditions

The transmission properties of a gate can be defined with respect to the incoming electrons. The electron transparency T_e is the ratio between the number of electrons traversing the grid and the number travelling towards it. In the absence of magnetic field, T_e is given by the corresponding ratio of the numbers of field lines. As a first example, we have calculated T_e in Chap. 3 for the monopolar gate as a function of the grid potential.

Similarly the ion transparency T_i is the fraction of ions traversing the grid, compared to all the ions travelling towards it. For the synchronous trigger mode we must also know $\langle T_i \rangle$, the time-averaged ion transparency.

In this section, computed and measured transparencies are presented for the static and for the synchronously pulsed bipolar gate, and for the bipolar gate in a magnetic field, operated as a diode. For the graphical representation of results and for the comparison with measurement we use a standardized system of electrodes according to Fig. 3.3 with wire positions as in the ALEPH TPC.

8.3.1   Transparency of the Static Bipolar Gate

In order to describe the problem it is convenient to define two surface charge densities, separating the contribution of the wires ramped at positive and negative ΔV_g. The (electron and ion) transparency is zero when the surface charge density of positive charges is equal in absolute value to the surface charge density on the high-voltage plane (or larger). The electric field configurations for full and for zero transparency are shown for our standard case in Fig. 8.3.

To compute the potential difference ΔV_g needed to close the gating grid we first consider the situation in which both the high-voltage plane and the zero-grid plane are grounded. The general solution is then obtained by superimposing the solution calculated in Sect. 3.4.1 when a common voltage V_g is applied to the wires of the gating grid.

Using the formalism of Sect. 3.2 it can be shown that by applying ΔV_g to the wires of the gating grid we induce a positive charge on the wires at positive

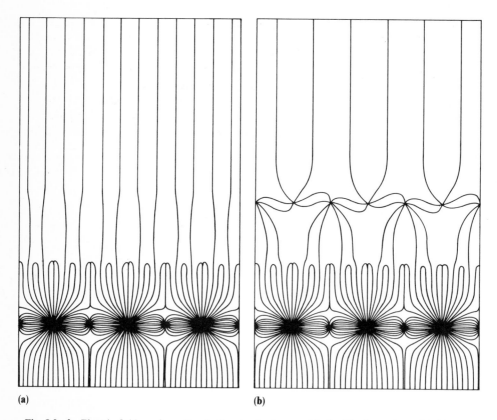

(a)   (b)

Fig. 8.3a, b. Electric field configuration in the standard case with the bipolar gating grid. (a) gate open, (b) gate closed

potential producing a positive surface charge density

$$\sigma_\Delta^+ = \frac{-\varepsilon_0}{\frac{s_3}{\pi} \ln \frac{\pi r_g}{2 s_3}} \Delta V_g \, . \tag{8.7}$$

The wires at negative potential contribute to a negative surface charge density

$$\sigma_\Delta^- = -\sigma_\Delta^+ \, , \tag{8.8}$$

and the total charge variation on the gating grid is 0.

This solution can be superimposed on the one calculated in Sect. 3.2.1 since it has the same boundary conditions. The total surface charge density of the wires at potential $V_g + \Delta V_g$ is

$$\sigma_\Delta^+ + \frac{\sigma_g}{2} \, , \tag{8.9}$$

where σ_g is computed using (3.2.2).

When $\sigma_\Delta^+ + \sigma_g/2$ is positive the transparency of the gating grid is given by

$$T = 1 - \frac{\sigma_\Delta^+ + \sigma_g/2}{|\sigma_p|} \, . \tag{8.10}$$

Only half of the wires of the gating grid contribute to the positive density. This explains the factor 2 of (8.10). Formula (8.10) has been tested experimentally using a chamber with wires arranged as shown in Fig. 3.3. We depict in Fig. 8.4 the calculated and measured electron transparencies as functions of ΔV_g. There is perfect agreement except for the large values of ΔV_g where the gate was found

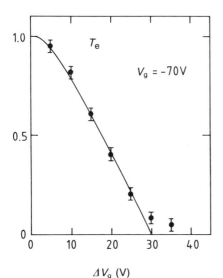

Fig. 8.4. Electron transparency T (here called T_e) of a gating grid with 2 mm pitch, as a function of the voltage difference ΔV_g applied to adjacent wires. The points are measurements by [AME 85-1]; the line is a calculation according to (8.10)

not to be as opaque as calculated, probably because of diffusion, which was omitted in the calculation.

The conditions for a closed grid are

$$\sigma_\Delta^+ + \frac{\sigma_g}{2} > |\sigma_p| = \varepsilon_0 E \, , \tag{8.11}$$

where E is the drift field.

When σ_g has the limiting value for the 'full transparency' condition (see 3.20) we obtain

$$\sigma_\Delta^+ > \varepsilon_0 E \left(1 + 2\pi \frac{r_g}{s_3} \right) . \tag{8.12}$$

In the general case, using (3.22), (8.7) and (8.11) one can calculate the minimum ΔV_g needed to close the grid:

$$\Delta V_g > -\frac{s_3}{\pi} \ln \left(\frac{\pi r_g}{2s_3} \right) \left(E - \frac{E(z_3 - z_2) + V_g - V_z}{2 \left(z_3 - z_2 - \frac{s_3}{2\pi} \ln \frac{2\pi r_g}{s_3} \right)} \right) . \tag{8.13}$$

Figure 8.5 shows how the electric field lines are organized around the bipolar gate as the differential voltage ΔV_g is increased. We notice that for small ΔV_g the grid is partly transparent and that for larger ΔV_g the field lines from the drift region end on the wires of the gating grid. The configuration of Fig. 8.5b is the case where the gate is just closed. It has field lines from the two regions in close proximity, so that diffusion will allow a small fraction of electrons to traverse the grid. This is not the case in Fig. 8.5c and d, where the gate is more firmly closed.

8.3.2 Average Transparency of the Regularly Pulsed Bipolar Gate

The passage of a drifting electron or ion through a gate requires some time. This time is much longer for ions than for electrons because of their smaller drift velocity u. Since the field region across a bipolar grid extends over some fraction of the pitch s (see Fig. 8.5) we may estimate the time to cross the gate to be roughly

$$\tau = s/u \, , \tag{8.14}$$

say between 10 and 1000 µs for ions and three orders of magnitude less for electrons.

A bipolar grid can be operated at a frequency f by switching the closing voltage $\pm \Delta V_g$ during some part of every cycle so that it is 'open' for the remaining time Δt ($\Delta t < 1/f$); the switching is schematized in Fig. 8.6.

If $\tau < \Delta t$, the grid can be traversed in one cycle, and the average transparency is

$$\langle T \rangle \sim f \Delta t \, . \tag{8.15}$$

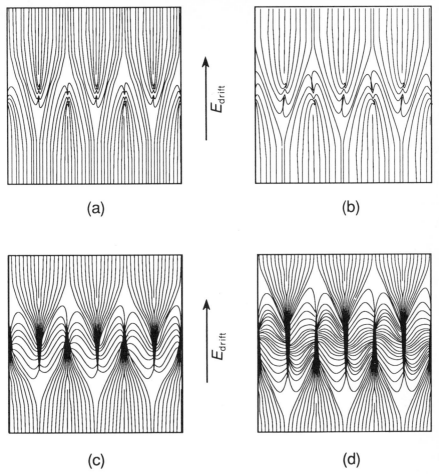

Fig. 8.5a–d. Electric field lines near the gating grid for different values of ΔV_g. $E_{drift} = 110$ V/cm and $V_g = -70$ V throughout. (a) $\Delta V_g = 30$ V, (b) $\Delta V_g = 45$ V, (c) $\Delta V_g = 100$ V, (d) $\Delta V_g = 200$ V

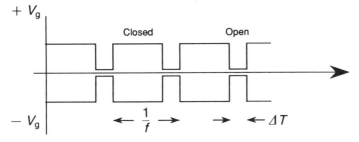

Fig. 8.6. Regular switching of the bipolar grid at a frequency f. ΔT is the 'open' time

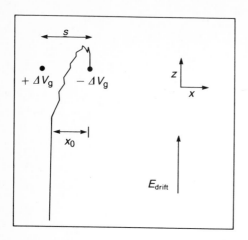

Fig. 8.7. Sketch of the ion path near the gating grid pulsed at some frequency

When $\tau \gg \Delta t$, many cycles are needed to traverse the grid. The electric field changes many times while the ions are near the grid, and they follow a path that alters its direction with the variation of the electric field; see Fig. 8.7 for an illustration. Each cycle, during the 'closed' condition, the ion makes a step towards the gate wire at lower potential, and under favourable conditions it reaches the wire and is absorbed.

For an accurate computation of the transparency we would have to follow the relevant field lines at every step. But we are satisfied with an order-of-magnitude calculation, which will exhibit the main variables and their influence on the average ion transparency.

We call the effective depth of the grid S – it will be some fraction of the pitch s – and we assume that the drift field E_d acts all the time whereas the transverse field E_x between the wires of the gating grid acts only during the time the gate is closed.

Assuming that an ion needs n cycles for the traversal of the grid,

$$n = \frac{Sf}{v_x} , \tag{8.16}$$

it is absorbed if

$$x_0 < n\left(\frac{1}{f} - \Delta t\right)v_x . \tag{8.17}$$

Here v_z and v_x are the two components of the drift velocity and x_0 is the initial transverse distance of the ion from the closest gating wire at lower potential (see Fig. 8.7).

The transparency is zero if

$$s < \frac{S}{v_z}f\left(\frac{1}{f} - \Delta t\right)v_x , \tag{8.18}$$

where s is the pitch of the gating grid.

We can calculate the transparency in the general case assuming that the ions are uniformly distributed along x when they reach the region of the gating grid. Using (8.16) and (8.17) we obtain

$$\langle T \rangle = 1 - \frac{S}{s}\frac{v_x}{v_z}(1 - f\Delta t) . \qquad (8.19)$$

Since the drift velocity of the ions is proportional to the electric field, the ratio between its components can be replaced by the ratio between the components of the electric field. The z component of the electric field is the drift field E_d and the x component can be approximated by $2\Delta V_g/s$. We obtain for the time-averaged ion transparency

$$\langle T \rangle = 1 - \frac{S}{s}\frac{2\Delta V_g}{sE_d}(1 - f\Delta t) . \qquad (8.20)$$

Equation (8.20) shows that within the limits of our very coarse approximations, the average transparency is a linear function of ΔV_g and the frequency f. This is also borne out by experiment: in Figs. 8.8 and 8.9 we see the result of measurements [AME 85-2] made with a gating grid operating under conditions similar to those of the ALEPH TPC. Formula (8.20) describes both curves quite well with values of S/s between 0.4 and 0.5.

8.3.3 Transparency of the Static Bipolar Gate in a Transverse Magnetic Field

A magnetic field along the main drift direction z changes the path of the electrons according to the parameter $\omega\tau$; see (2.6). The component u_x of the drift velocity perpendicular to the magnetic field and perpendicular to the grid wires

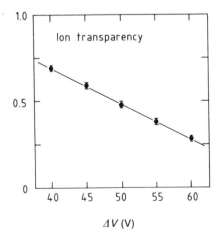

Fig. **8.8.** Average transparency $\langle T \rangle$ for ions through a bipolar grid with a pitch of 2 mm, which was pulsed at 100 kHz with an 'open' time of 6 µs and a 'closed' time of 4 µs, as a function of the voltage difference ΔV applied to adjacent wires. The points are measured by [AME 85-2]; the straight line was drawn to connect the points

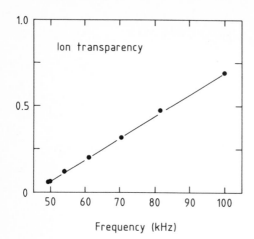

Ion transparency

Frequency (kHz)

Fig. 8.9. Average transparency $\langle T \rangle$ for ions through a bipolar grid (pitch 2 mm), which was pulsed with $\Delta V_g = \pm 45$ V for a constant 'open' time of 6 μs, as a function of the frequency. The points are measured by [AME 85-2]; the straight line was drawn to connect the points

is reduced by the factor $1 + \omega^2\tau^2$, whereas the drift velocity u_z in the main direction stays the same:

$$u_x = \frac{1}{1 + \omega^2\tau^2} \mu E_x ,$$

$$u_z = \mu E_z .$$

(8.21)

Here μ is the electron mobility. We see that there is a strong influence of the magnetic field on the operation of the bipolar gate – it is closed at much higher values of ΔV_g, since the relevant drift-velocity component u_x, which moves the electrons towards the wires of the grid, is reduced.

We can construct the drift trajectories when we know the behaviour of $\omega\tau$ with the electric field strength E. This could be calculated according to the principles discussed in Sect. 2.2.3. But for our example we simply use

$$\omega\tau = 6(T^{-1})B \quad \text{for } E < 100 \text{ V/cm} ,$$

$$\omega\tau = 6(T^{-1})B100(\text{V/cm})/E \quad \text{for } E > 100 \text{ V/cm} .$$

Figure 8.10 shows the electron drift lines in the x–z plane for a number of conditions. The electron transparency is computed counting the fraction of drift velocity lines that cross the grid. Comparing Figs. 8.10a and b, one observes that the magnetic field of 1.5 T changed the transparency from zero to about 80%.

Figure 8.11 displays the measured electron transparency of a gating grid with a pitch of 2 mm in a drift field of 100 V/cm for different values of the magnetic field [AME 85-1]. One observes that the closing voltage goes up roughly linearly with the magnetic field.

Displacement of the Electrons along the Wire Direction. The electrons cross the gating grid in the presence of a component (x) of the electric field perpendicular to the magnetic field. Owing to the $E \times B$ term of the drift-velocity equation (2.6)

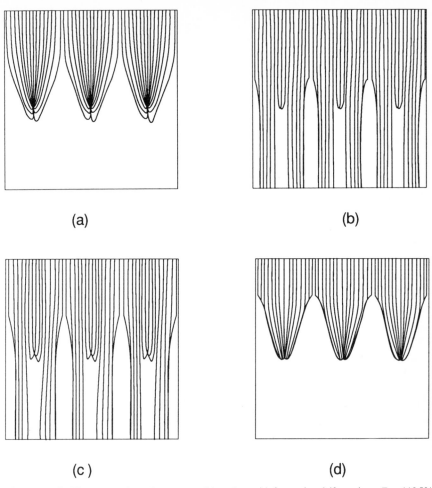

Fig. 8.10a–d. Electron trajectories approaching the grid from the drift region. $E = 110 \, \text{V/cm}$. (a) $\Delta V_g = 45 \, \text{V}$, $B = 0$, (b) $\Delta V_g = 45 \, \text{V}$, $B = 1.5 \, \text{T}$, (c) $\Delta V_g = 70 \, \text{V}$, $B = 1.5 \, \text{T}$, (d) $\Delta V_g = 200 \, \text{V}$, $B = 1.5 \, \text{T}$

they are also displaced along the direction (y) parallel to the wires of the grid. The component of the drift velocity along the wire direction is

$$u_y = \frac{\omega \tau}{1 + \omega^2 \tau^2} \mu E_x \; .$$

This effect is similar to the displacement of the electrons along the sense-wire direction when they approach the sense wire from the zero-grid region (see Sect. 6.3.1).

Figure 8.5b shows the electric field lines in the region of the gating grid when the grid is closed for the ions (but not for the electrons if a strong magnetic field

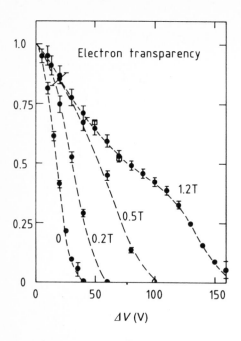

Fig. 8.11. Electron transparency vs. ΔV_g (here called ΔV) for various magnetic fields, as measured by [AME 85-1]. The lines are drawn to guide the eye. The two squares show the result of calculations according to Sect. 8.3.3

is present). The electric field is roughly constant between two sides of the grid, and its x component changes sign on the two sides of the same wire. All the electrons that cross the grid on the same side of a given wire are displaced along y in the same direction and roughly by the same amount; all the others that pass on the other side are displaced by the same amount but in the opposite direction. This displacement can be of the order of fractions of a millimetre and depends on ΔV_g, on the magnetic field and on the gas mixture.

The distribution of the arrival point of the electrons on the sense wire is modified. This effect influences the spatial resolution in a similar way as the $E \times B$ effect at the sense wire.

If the pitch of the gating grid and the sense-field grid are the same, it is possible to choose the sign of ΔV_g in such a way that the y displacement at the gating grid has sign opposite to the y displacement at the sense wire [AME 86]. In this case there is a compensation of the overall $E \times B$ effect with a possible benefit to the spatial resolution; nothing quantitative has appeared in the literature.

References

[AME 85-1]   S.R. Amendolia et al., Influence of the magnetic field on the gating of a time projection chamber, *Nucl. Instrum. Methods Phys. Res. A* **234**, 47 (1985)
[AME 85-2]   S.R. Amendolia et al., Ion trapping properties of a synchronously gated time projection chamber, *Nucl. Instrum. Methods Phys. Res. A* **239**, 192 (1985)

[AME 86]   S.R. Amendolia et al., Gating in the ALEPH time projection chamber, *Nucl. Instrum. Methods Phys. Res. A* **252**, 403 (1986) and CERN-EF preprint 84-11 (1984) (unpublished)

[BEI 88]   S.P. Beingessner, T.C. Meyer, M. Yvert, Influence of chemical trace additives on the future ageing of the UA1 Central detector. *Nucl. Instrum. Methods Phys. Res. A* **272**, 669 (1988)

[BRE 80]   A. Breskin, G. Charpak, S. Majewski, G. Melchart, A. Peisert, F. Sauli, F. Mathy, G. Petersen, High flux operation of the gated multistep avalanche chamber, *Nucl. Instrum. Methods* **178**, 11 (1980)

[BRY 85]   D.A. Bryman et al., Gated grid system used with a time projection chamber, *Nucl. Instrum. Methods Phys. Res. A* **234**, 42 (1985)

[KEN 84]   Joel Kent, Positive ion suppression with untriggered TPC; College de France Preprint 84-17 (1984) (unpublished)

[NEM 83]   P. Nemethy, P.J. Oddone, N. Toge, A. Ishibashi, Gated time projection chamber, *Nucl. Instrum. Methods Phys. Res.* **212**, 273 (1983)

9. Particle Identification by Measurement of Ionization

Among the track parameters, the ionization plays a special role because it is a function of the particle velocity, which can therefore be indirectly determined through a measurement of the amount of ionization along a track. For relativistic particles this dependence is not very strong, and therefore the amount of ionization must be measured accurately in order to be useful. The ionization has a very broad distribution, and a track has to be measured on many segments (several tens to a few hundred) in order to reach the accuracy required. Fortunately only relative values of the ionization need to be known between tracks of different velocities.

Whereas in Chap. 1 we dealt with the ionization phenomenon in general terms, the present chapter is devoted to a discussion of those aspects of ionization that are important for particle identification. After an explanation of the principle we elaborate on the factors that determine the shape of the ionization curve, and the achievable accuracy. Calculated and measured particle separation powers are discussed next. One section is devoted to cluster counting. Finally, the problems encountered in practical devices are reviewed.

9.1 Principles

In a magnetic spectrometer one measures the particle momentum through the curvature of its track in the magnetic field. If a measurement of ionization is performed on this track, there is the possibility of determining the particle mass, thus identifying the particle. The relation between momentum p and velocity v involves the mass m:

$$v = c^2 p / E = c^2 p / \sqrt{(p^2 c^2 + m^2 c^4)} \,, \tag{9.1}$$

where c is the velocity of light.

Any quantification of ionization involves the peculiar statistics of the ionization process. Let us recapitulate from Chap. 1 that the distribution function of the number of electrons produced on a given track length has such a tail towards large numbers that neither a proper mean nor a proper variance exists for the number of electrons. In order to characterize the distribution one may quote its most probable value and its width at half-maximum. The most probable number of electrons may be taken to represent the 'strength' of the ionization. It is *not* proportional to the length of the track segment. (We have seen in Figs. 1.10 and 1.11 how the most probable value of the ionization per unit length goes up with the gas sample length.) From a number of pulse-height measurements on one track one derives some measure I of the strength of the ionization; details are given in Sect. 9.3. Since I depends on the velocity of the particle and is proportional to the square of the charge Ze of the particle, we write

$$I = Z^2 F_{g,m}(v) \ . \tag{9.2}$$

$F_{g,m}(v)$ depends, firstly, on the nature of the gas mixture and its density (index g). But it depends also on the length and number of samples and on the exact way in which the ionization of the track is calculated from the ionization of the different samples (index m); in addition, there is the small dependence of the relativistic rise on the sample size. If we divide $F_{g,m}(v)$ by its minimal value (at $v_{\min}$), we obtain a normalized curve

$$F_{g,m}^{(\text{norm})}(v) = F_{g,m}(v)/F_{g,m}(v_{\min}) \ . \tag{9.3}$$

It turns out that this curve is approximately independent of the exact manner of the ionization measurement:

$$F_{g,m}^{(\text{norm})}(v) \approx F_g(v) \ . \tag{9.4}$$

(Some qualifications are mentioned in the end of Sect. 9.8.1.) Most analyses of the ionization loss have been done with this approximation, i.e. one neglects any variation of the relativistic rise on the sample size and works with one universal normalized curve.

Figure 1.19 contained theoretical values for the most probable ionization in argon, as well as experimental values of the truncated mean in argon–methane mixtures. The abscissa is the logarithm of $\beta\gamma$ ($\beta = v/c$, $\gamma^2 = 1/(1 - \beta^2)$), and the vertical scale is normalized to the value in the minimum.

One cannot resolve (9.1), (9.2) and (9.4) into a form $m = m(I, p)$ because $F_g(v)$ is not monotonic. Therefore we discuss particle identification by plotting $F_g(v)$ for various known charged particles as a function of p. Using the argon curve of Fig. 1.19 we obtain the family of curves shown in Fig. 9.1. We see five curves of identical shape, for electrons, muons, pions, kaons, and protons; they are displaced with respect to each other on the logarithmic scale by the logarithms of the particle mass ratios.

The simultaneous measurement of the momentum and the ionization of a particle results in a point in the diagram of Fig. 9.1. Complete particle identification takes place if, inside the measurement errors, this point can be associated with only one curve. In practice the measurement errors of the ionization are very often so large that several particles could have caused the measured ionization; if this is the case it may be possible to exclude one or more particles.

For good particle identification the measurement accuracy δI is obviously as essential as the vertical distance between the curves in Fig. 9.1. It is therefore our first task to review the gas conditions responsible for the shape of the curve and

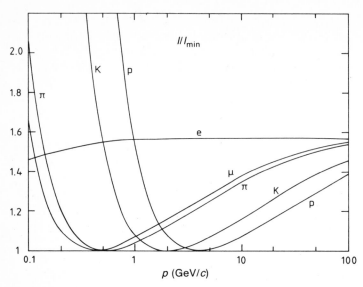

Fig. 9.1. Most probable values of the ionization (normalized to the minimum value) in argon at ordinary density as function of the momenta of the known stable charged particles

in particular for the amount of the relativistic rise. Then we have to understand the circumstances that have an influence on δI, and which value of δI can be reached in a given apparatus. The ratio of the accuracy over the relativistic rise characterizes the particle separation power.

But before going into these details let us make an estimate of what we can expect from this method of particle identification. Imagine some drift chamber with a relative accuracy of $\delta I/I = 12\%$ FWHM (5% r.m.s.). If we assume that a respectable degree of identification is already achievable on a curve that is only twice this r.m.s. value away from any other one, if we remove the muon curve (because it is hopelessly close to the pion curve and because muons can be well identified with other methods), and if, finally, we assume that the error of the momentum measurement is negligible, then we can redraw those branches of the curves in Fig. 9.1 that would allow a respectable degree of particle identification in argon. The result is seen in Fig. 9.2.

Under the assumed circumstances, these particles could be identified in the ranges shown in Table 9.1. It is characteristic of the complicated shape of the ionization curve that there are certain bands of overlap in which unique identification is impossible. A useful separation between kaons and protons at high momenta would require the measurement accuracy for the ionization to be $\delta I/I = 8.5\%$ FWHM or better (in argon at normal density). Such values have actually been attained by specialized particle identifiers, see Sect. 9.8.

After having described the principle of particle identification based on the *amount of ionization* along a given length of track we should also mention the

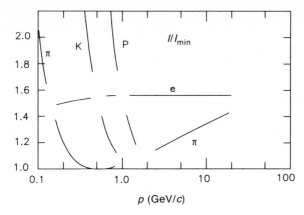

Fig. 9.2. The same as Fig. 9.1 with all lines left out that either belong to a muon or that have a neighbouring line closer than 10% in ionization (corresponding to a separation of 2 standard deviations when the r.m.s. accuracy is 5%)

Table 9.1. Momentum ranges above 0.1 GeV/c for a 2 σ identification of particles in argon (normal density), for two measuring accuracies of the ionization (μ's removed)

Particle	Momentum ranges (GeV/c)		
$\delta I/I = 5\%$ r.m.s.			
e	0.17–0.4	0.6 –0.8	1.3–19
π	(0.1) –0.12	0.16–0.8	2.7–19
K	(0.1) –0.4	0.6 –0.8	
p	(0.1) –0.8	1.1 –1.5	
$\delta I/I = 3.5\%$ r.m.s.			
e	0.16–0.5	0.6 –0.9	1.1–30
π	(0.1) –0.13	0.16–0.9	2.2–30
K	(0.1) –0.5	0.6 –0.9	6 –46
p	(0.1) –0.9	1.1 –1.5	6 –46

possibility based on the *frequency of ionization* along the track. If we were to count the number of ionization clusters rather than the amount of ionization on some track length, there would be a profit in the statistical accuracy. A brief description of this interesting method is given in Sect. 9.6.

Much easier than the identification of the known particles would be the recognition of a stable quark with charge $e/3$ or $2e/3$, because it would create only 1/9 or 4/9 of the normal ionization density. Quarks aside, one recognizes that relatively small differences in $\delta I/I$ may be decisive for particle identification. Therefore we must study in some detail how the shape of the ionization curve and the achievable accuracy depend on the parameters of the drift chamber.

9.2 Shape of the Ionization Curve

We recall from Chap. 1 that the curve reaches its minimum near $\beta\gamma = 4$. The shape of a curve that has been normalized to this minimum is characterized by the height of the plateau ('amount of relativistic rise', R, above the minimum) and by the value of the relativistic velocity factor, γ^*, at which the plateau is reached. Both numbers depend on the gas and its density as well as the sample length. In Table 9.2 (Sect. 9.5) we find relativistic rise values calculated by Allison [ALL 82] for various gases at normal density and 1.5-cm-long samples. One observes that R is highest for the noble gases and much lower for some hydrocarbons.

The density dependence is contained in the logarithmic term (1.59) and was demonstrated in Fig. 1.20. The curve of Fig. 9.3 shows how the relativistic rise of argon goes down as the density increases. Measured ionization ratios for particles with 15 GeV/c momentum are contained in Fig. 9.4. All these data show that the relativistic rise becomes better (higher) as the gas density is

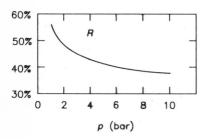

Fig. 9.3. Calculated variation of the relativistic rise R of the most probable ionization with the gas density (pressure p at constant temperature) [ALL 82]

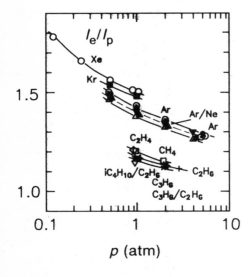

Fig. 9.4. Measured ratio of the ionization of electrons to that of pions with momentum of 15 GeV/c as a function of the gas pressure for various gases. The ionization is the average over the lowest 40% of 64 samples, each 4 cm long [LEH 82b]

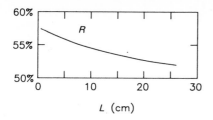

Fig. 9.5. Calculated variation of the relativistic rise as a function of the gas sample length for argon at normal density [ALL 82]

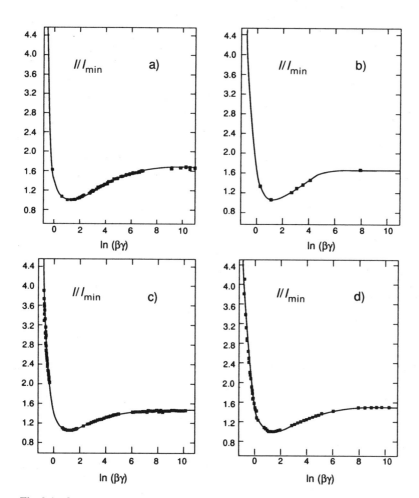

Fig. 9.6a–d. Four different measurements of the normalized ionization strength I/I_{min} fitted to the five-parameter form (9.5) [ASS 91]. (**a**) Data in Ar + CH$_4$ (various small concentrations up to 10%) at 1 bar, combined from [LEH 78], [LEH 82] and ALEPH data (unpublished). (**b**) Data in Ar(90%) + CH$_4$(10%) at 1 bar [WAL 79a]. (**c**) Data in Ar(80%) + CH$_4$(20%) at 8.5 bar [COW 88]. (**d**) Data in Ar(91%) + CH$_4$(9%) at 1 bar [ALE 91]. A truncated mean was used in all cases to measure I

decreased and as one passes from smaller to larger atomic numbers of the elements that make up the gas.

The small influence of the sample length on the relativistic rise of argon is displayed in Fig. 9.5. The relativistic rise goes down from 57 to 53% as the sample length increases from 1.5 to 25 cm at one bar.

In a particle experiment, a determination of the ionization as a function of the particle velocity is done with identified particles over a range of momenta. Once the function is established it can be used for the identification of other particles. For this purpose it would be useful to have a mathematical description of the ionization curve. Neither the theory of Bethe, Bloch, and Sternheimer nor the model of Allison and Cobb offer a closed mathematical form. Fits have sometimes been based on (1.66) and a piecewise parametrization of the quantity δ, using five or more parameters for $\delta(\beta)$. Hauschild et al. have found it possible to work with only two free parameters for $\delta(\beta)$ and with a total of four for a description of their data [HAU 91]. – We have sometimes used the following form for a description of measured ionization curves:

$$F_g(v) = \frac{p_1}{\beta^{p_4}} \left\{ p_2 - \beta^{p_4} - \ln\left[p_3 + \left(\frac{1}{\beta\gamma}\right)^{p_5} \right] \right\}, \tag{9.5}$$

where $\beta = v/c$, $\gamma^2 = 1/(1 - \beta^2)$, and the p_i are five free parameters. Equation (9.5) is a generalization of (1.56), valid in the model of Allison and Cobb for each transferred energy. Figure 9.6 shows four examples of fits to this form.

9.3   Statistical Treatment of the n Ionization Samples of One Track

The ionization measurements in the various cells, which we imagine to be of equal size, are usually entirely independent – the transport of ionization from one cell to its neighbour caused by the rare forward-going delta rays can be neglected, and we exclude the effect of diffusion or electric cross-talk between neighbouring cells. With each well-measured track we obtain n electronic signals which, up to a common constant, represent n sample values of the ionization distribution in a cell. The task is to derive from them a suitable estimator for the strength of the ionization in the cell.

The most straightforward, the average of all n values, is a bad estimator which fluctuates a lot from track to track, because the underlying mathematical ionization distribution has no finite average and no finite variance (see (1.34)). A good estimator is either derived from a fit to the shape of the measured distribution or from a subsample excluding the very high measured values.

In many cases one knows the shape of the signal distribution, $f(S)dS$, up to a scale parameter λ that characterizes the ionization strength. The n signals

$S_1, \ldots, S_n$ are to be used for a determination of λ. Let λ be fixed to 1 for the normalized reference distribution $f_1(S) \, dS$. Then there is a family of normalized distributions

$$f_\lambda(S) \, dS = (1/\lambda) \, f_1(S/\lambda) \, dS \, , \tag{9.6}$$

and one must find out which curve fits best the n signal values. The most efficient method to achieve this is the maximum-likelihood method. It requires one to maximize with respect to λ the likelihood function

$$\mathscr{L} = \prod_i (1/\lambda) \, f_1(S_i/\lambda) \, , \tag{9.7}$$

where the product runs over all n samples. The result is a λ_0 and its error $\delta\lambda_0$. One may take λ to represent the most probable signal of the distribution; then $\lambda_0 \pm \delta\lambda_0$ is the most probable signal measured for the track at hand.

So far, it has been assumed that the shape of the ionization curve is the same (at a given gas length) for all ionization strengths, i.e. particle velocities. For very small gas lengths this is not really true (see Fig. 1.21). In this case one needs a table of different curves rather than the simple family (9.6). The rest goes as before.

For a general treatment of parameter estimation, the likelihood method, or the concept of efficiency, the reader is referred to a textbook on statistics, e.g. Cramér [CRA 51] or Fisz [FIS 58] or Eadie et al. [EAD 71].

If one does not know the shape of the signal distribution beforehand, one may use a simplified method, the method of "the truncated mean". It is characterized by a cut-off parameter η between 0 and 1. The estimator $\langle S \rangle_\eta$ for the signal strength is the average of the ηn lowest values among the n signals S_i:

$$\langle S \rangle_\eta = \frac{1}{m} \sum_1^m S_j \, ,$$

where m is the integer closest to ηn, and the sum extends over the lowest m elements of the ordered full sample ($S_j \le S_{j+1}$ for all $j = 1, \ldots, n-1$). The quantity that matters is the fluctuation of $\langle S \rangle_\eta$ divided by its value; if for a typical ionization distribution one simulates with the Monte Carlo method the measurement of many tracks in order to determine the best η, one finds a shallow minimum of this quantity as a function of η between 0.35 and 0.75. In this range of η it is an empirical fact that the values of $\langle S \rangle_\eta$ are distributed almost like a Gaussian. In many practical experiments the method of the truncated mean was the method of choice, because of its simplicity and because the likelihood method had not given recognizably better results. From a theoretical point of view, the likelihood method should be the best as it makes use of all the available information.

Whatever the exact form of the estimator, we assume that it is a measure of the amount of ionization, which obeys (9.2), and that the approximation (9.4) is sufficiently well fulfilled for a practical analysis.

9.4 Accuracy of the Ionization Measurement

We have to understand how the accuracy of the ionization measurement of one track varies with its length L and the number n of samples, with the nature of the gas and with its pressure p. Since the distribution in a given gap with length $x = L/n$ depends only on the cluster-size distribution and on the number of primary interactions in the gap, the ionization distribution varies with the density (or pressure at constant temperature) in the same way as it does with x, and therefore the width W of the distribution depends on x and p through their product xp. The relative accuracy will therefore be of the form

$$\frac{\delta I}{I} = f(xp, n, \text{gas}) \quad \text{or} \quad g(Lp, n, \text{gas}) \, .$$

9.4.1 Variation with n and x

If a track of length L is sampled on n small pieces of length x, we may expect that the accuracy δI_{mp} in the determination of the most probable-value improves when we increase n and L, keeping x fixed. Different values of n may arise for tracks in one event, depending on the number of samples lost because of overlapping tracks. If the basic distribution of the ionization gathered in one piece x were well behaved in the sense of having a mean and a variance – a prerequisite for an application of the central-limit theorem of statistics – then the accuracy would become better with increasing n and L according to the law $\delta I_{mp} \sim 1/\sqrt{n}$ in the limit of large n. It turns out that with the very special cluster-size distribution of Chap. 1 the gain is more like $n^{-0.46}$ for practical values of n when dealing with a maximum-likelihood fit for the most probable value [ALL 80]. Walenta and co-workers obtained $n^{-0.43}$ when dealing with the average $\langle I \rangle_{40}$ of the lowest 40% [WAL 79]. The differences in the exponent seem to be small but they count for large n! In any event one deals here with empirical relations that have not been justified on mathematical grounds.

If we vary L and x simultaneously, keeping n fixed, we must consider how the width of the ionization distribution changes with the size x of the individual sample. The variation of L and x at constant n could be produced by increasing the gas pressure in a suitable multiwire drift chamber.

As the gap is made larger, W decreases. This has been treated in Chap. 1 (see the discussion around (1.38) and Fig. 1.10). The experiments suggest a power behaviour of the relative width W on a single gap, which varies like

$$(W/I_{mp})_1 : (W/I_{mp})_2 = [(px)_1 : (px)_2]^k = [(pL)_1 : (pL)_2]^k \tag{9.8}$$

(I_{mp} = most probable ionization). Based on the PAI model and supported by experimental data one finds that $k = -0.32$ [ALL 80].

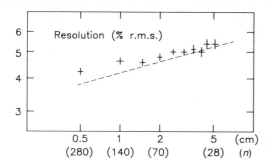

Fig. 9.7. Ionization resolution at constant track length as a function of the sample length. *Crosses*: measured in the ALEPH-TPC (average track length 140 cm, argon (91%) + methane (9%) at 1 bar, diffusion < 1.4 mm r.m.s.). *Line*: prediction of Allison and Cobb, our Eq. (9.9)

Next we keep the length $L = nx$ fixed and increase n, reducing x. Now there is a net gain in the accuracy δI proportional to

$$n^{-0.14} \quad \text{or} \quad n^{-0.11}$$

because the better accuracy from the increase of n is partly offset by the worsening from the decrease in x.

Although there is agreement between the different authors about these facts, they disagree about the question down to what subdivision this rule holds. Whereas Allison and Cobb, on the basis of their theory and in accordance with the above, recommend sampling the ionization as finely as possible [ALL 80], Walenta suggests that below 5 cm bar in argon the improvement is negligible [WAL 79]. Lehraus et al., from a more practical point of view, argue similarly [LEH 82a]. It must be said that, at the time, such fine samples had not been systematically explored. More recently the ALEPH and the DELPHI TPCs, working at normal gas density and with 4 mm sense-wire spacing, sample remarkably thin gas layers. Figure 9.7 shows an ALEPH study [ASS 90] in which lepton tracks from e^+e^- interactions were analyzed by adding the signals from neighbouring sense wires, thus varying n and x at fixed L. Though not quite reaching the theoretical prediction (see below), the measured accuracy as a function of n lends some support to the idea that fine subdivisions down to half a cm bar may still increase the accuracy.

9.4.2 Variation with the Particle Velocity

We have seen how the single-gap width W is reduced by multiple measurements. The value of W itself results from the summation over the cluster-size distribution (see Sect. 1.2.4). Does W depend on the particle velocity? This must be suspected, because the number of primary interactions does, and so does the cluster-size distribution, although only very weakly. In the PAI model also this detail has been computed. We see in Fig. 9.8 the dependence of the relative width W/I_{mp} on the relativistic velocity factor $\beta\gamma$, which reflects the characteristic shape of the ionization curve. In the example at hand, which is 1.6 cm of argon, W/I_{mp} oscillates inside a narrow band of $(80 \pm 7)\%$ as $\beta\gamma$ increases from 1 to the

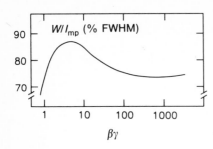

plateau. One concludes that the velocity dependence of the relative width can usually be neglected in this range.

9.4.3   Variation with the Gas

We finally have to involve the properties of the gas itself; they will determine the absolute scale of the accuracy $\delta I/I$ when px and n are specified. The pulse-height distributions of many gases have been measured in single and multiple gaps. From the work of the Lehraus group [LEH 82b] we present in Fig. 9.9 a compilation of measured single-gap ionisation widths of various gases at different densities. It turns out that the noble gases have wider distributions than the organic molecular gases.

With the PAI model, Allison and Cobb calculate for argon ([ALL 80], equation 4.3) a value which represents some average over the particle velocities:

$$\delta I_{mp}/I_{mp} = 0.96 n^{-0.46} (px)^{-0.32} \quad \text{(FWHM)} . \tag{9.9}$$

A systematic comparison between different gases is possible using the dimensionless scaling variable

$$\xi/I = 2\pi N r_e^2 mc^2 x/(\beta^2 I) . \tag{9.10}$$

We have encountered the variable ξ with the dimension of an energy, as the main factor in the Bethe–Bloch formula (1.65) determining the energy lost in the gas thickness x, to be multiplied only by a β-dependent factor of the order of unity. (N = electron density, r_e = classical electron radius = 2.82 fm, mc^2 = rest energy of the electron, β = particle velocity in units of the velocity of light.) We have also met the variable ξ as the scaling factor (1.49) in the Landau distribution.

The mean ionization potential I is the only constant in the Bethe–Bloch formula that characterizes the gas; see Table 1.3. The ratio ξ/I will be roughly proportional to the number of electrons liberated over the distance x, and therefore it is plausible that it might serve in a comparison of accuracies $\delta I/I$. For $\beta = 1$ and gas layers of 1 cm, ξ/I takes on the values 0.32, 0.50, 0.62, 0.65, and 0.70 for the noble gases He, Ne, Ar, Kr, and Xe at N.T.P., respectively.

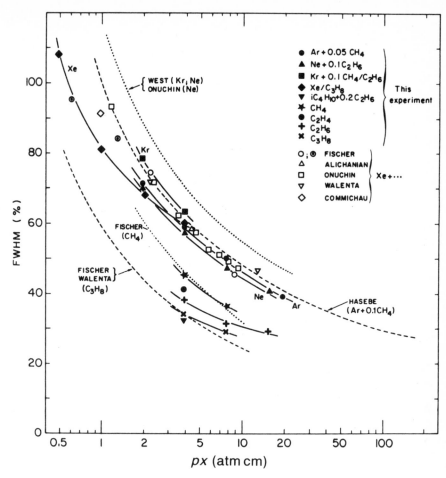

Fig. 9.9. Single-gap resolution measured for various gases and pressures [LEH 82b] as a function of the product of pressure p and sample length x

From the work of Ermilova, Kotenko and Merzon [ERM 77], Fig. 9.10 contains calculated values of the width of the ionization distribution in a single gap, for various noble gases, plotted as a function of the ratio ξ/I. The full width at half maximum is expressed in units of ξ. We notice that the points of the different gases and sample lengths x fall roughly on one universal curve, which only depends on ξ/I. We also notice that for ξ/I near 1 the theoretical predictions for the width of Landau and Blunck–Leisegang (see Sect. 1.2.5) differ by half an order of magnitude; on the other hand, the Monte Carlo calculation of Ermilova et al., which is very similar to the subsequent PAI model, accurately predicts the measured widths. The three calculations and the experimental points fall generally together only for ratios ξ/I larger than about 50.

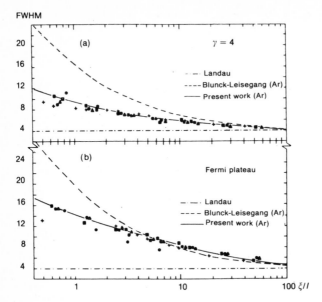

Fig. 9.10. The widths of the ionization distributions in units of ξ for the noble gases, calculated for various sample lengths by Ermilova et al., plotted as a function of ξ/I (9.10) [ERM 77]. The points calculated by a Monte Carlo method follow a universal dependence of ξ/I

Using the universal variable ξ/I Allison and Cobb generalized relation (9.9) for all gases:

$$\delta I_{mp}/I_{mp} = 0.81 n^{-0.46}(\xi/I)^{-0.32} \quad \text{(FWHM)} . \tag{9.11}$$

It should be emphasized that this formula has not yet been tested experimentally in a systematic comparison between different gases. The value of the constant is equal to the value of $0.96(\xi/I)^{0.32}$ for 1 cm of argon.

In (9.11) the length of the device is contained in ξ and n, the gas pressure in ξ. If we want to make their influence more visible we may calculate from (9.9) the relative accuracy obtainable with a device 1 m long containing argon gas at N.T.P., and 100 samplings; the result is $\delta I_{mp}/I_{mp} = 0.115$. The value of ξ varies between different gases in the ratio of their atomic charge number Z. Therefore we have for the general case of a gas, characterized by Z and I,

$$\frac{\delta I_{mp}}{I_{mp}} = 0.115 \left(\frac{1 \text{ m bar}}{Lp}\right)^{0.32} \left(\frac{Z_{Ar} I}{Z I_{Ar}}\right)^{0.32} \left(\frac{100}{n}\right)^{0.14} \quad \text{(FWHM)} , \tag{9.12}$$

where L is the length, p the gas pressure and n the number of samplings; I indicates the average ionization potential as used in the Bethe–Bloch formula.

9.5   Particle Separation

In order to judge the power of particle separation for a certain gas and wire arrangement we must combine the results of Sects. 9.2 and 9.4. Generally speaking, the figure of interest is the ratio $\delta I/(IR)$. The noble gases have a wider pulse-height distribution but also a larger relativistic rise, compared to the hydrocarbons investigated. The same compensation is at work with the gas density: as the accuracy $\delta I/I$ becomes better with higher density, the relativistic rise R gets worse; in addition the saturation of the plateau starts at a lower γ^*. Finer sampling is an advantage even in the approximation (9.4). In addition, there is the tiny increase, if any, of R with finer samples.

The performance of various gases, according to calculations with the PAI model, is compared in Table 9.2 [ALL 82]. These results refer to normal density and to tracks 1 m long, which are sampled in 1.5 cm intervals. The relativistic rise R is indicated for the most probable value of the ionization distribution and for the lower and upper points at half maximum. The full width W_{66} of the ionization determination derivable from a maximum-likelihood fit to all 66 measurements is also shown. The last 6 columns contain particle-pair separation powers S, i.e. for a given pair of particles at some momentum the difference of the peak positions, divided by W_{66}. Remarkably, the separation powers of different gases are very similar.

Measurements of the separation power for various gases and pressures have been performed in a beam of identified particles by Lehraus et al. [LEH 82b]. In their detector with 64 cells, each 4 cm long, they determined simultaneously the difference between the ionization of electrons, pions and protons and the

Table 9.2. Performance of different gases according to calculations with the PAI model [ALL 82]. A track of 1 m length, sampled in 66 layers of 1.5 cm is assumed. See text for explanation

Gas	R (% of min. value) Peak lower upper)			W_{66} (% FWHM)	S at 4 GeV/c			S at 20 GeV/c		
	Peak	lower	upper	(% FWHM)	e–π	π–K	K–p	e–π	π–K	K–p
He	58	86	45	15	1.5	0.9	0.3	0.3	0.9	0.5
Ne	57	72	57	13	1.8	1.1	0.4	0.6	1.1	0.6
Ar	57	66	49	12	1.6	1.1	0.4	0.6	0.9	0.6
Kr	63	65	61	11	2.2	1.3	0.5	0.9	1.1	0.7
Xe	67	76	64	13	2.2	1.2	0.4	1.0	1.0	0.6
N_2	56	59	48	11	1.8	1.3	0.4	0.7	1.0	0.7
O_2	54	54	47	9	2.0	1.5	0.5	0.7	1.1	0.8
CO_2	48	52	41	8	2.0	1.6	0.5	0.6	1.1	0.9
CH_4	43	45	39	9	1.6	1.5	0.5	0.5	1.0	0.8
C_2H_4	42	46	38	8	1.7	1.5	0.5	0.6	1.0	0.8
C_2H_6	38	41	34	8	1.6	1.6	0.5	—	—	—
C_4H_{10}	23	24	21	6	1.8	1.3	0.3	0.6	1.0	0.8
80% Ar^+ 20% CO_2	55	62	48	12	1.9	1.2	0.4	—	—	—

accuracy δI (r.m.s. width of the distribution). The gas and its pressure were varied. Considering the separation of these particles at 15 GeV/c, the results are expressed as the ratios D/σ, equal to

$$\frac{\langle I^{\pi}\rangle_{40} - \langle I^{p}\rangle_{40}}{\delta\langle I^{\pi}\rangle_{40}} \text{ or } \frac{\langle I^{e}\rangle_{40} - \langle I^{\pi}\rangle_{40}}{\delta\langle I^{\pi}\rangle_{40}}, \tag{9.13}$$

plotted against pressure in Fig. 9.11. This diagram contains a summary of much careful and systematic study. We observe a general decrease below 1 bar and often a decrease above 2 bar, although there are several gas mixtures, where the pion–proton separation increases up to 4 bar. In any event, the values of the ratios stay the same within a few tenths between 1 and 4 bar.

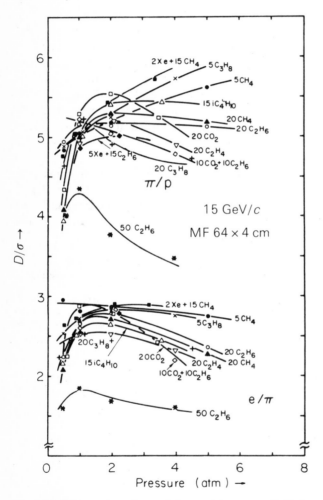

Fig. 9.11. Pressure dependence of the resolving power (9.13) for various argon gas mixtures, as measured by [LEH 82a] at 15 GeV/c. The numbers close to the chemical symbols of the quench gases refer to the concentrations; for example, 20CH$_4$ means 80% Ar + 20% CH$_4$

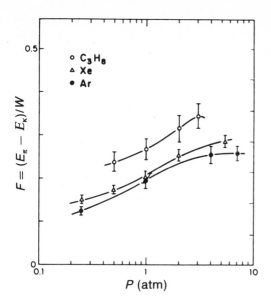

Fig. 9.12. Single-gap measurements for three different gases by [WAL 79a]: ratio of the peak displacement between pions and kaons at 3.5 GeV/c to the width of the distribution, plotted as a function of the gas pressure

Single-gap measurements in three gases were reported by Walenta et al. [WAL 79] at 3.5 GeV/c. The measured ratio of peak displacement over the full width is plotted against pressure in Fig. 9.12. An increase in separation power can be observed at pressures above 1 bar. The overall picture is that with some gases a profit can be made by going above 1 bar.

9.6 Cluster Counting

The statistical distribution of the number of ionization electrons produced on a given track length has a width W (measured at one half of the maximal probability), which was plotted relative to the most probable ionization I_{mp} in Fig. 1.12 as a function of the track length. This width is determined by the fluctuation in the number of primary clusters that occur over this length and by the fluctuation in the cluster size. It is the latter that dominates; for example, there are 28 primary clusters on 1 cm bar of argon whose Poissonian fluctuation is only 44% FWHM, whereas W/I_{mp} is approximately equal to 1, and for 10 cm bar of argon, the corresponding numbers are 14% compared to $W/I_{mp} \approx 0.5$. The cluster-size distribution $P(k)$ (see Sect. 1.2.4) dominates because of its $1/k^2$ behaviour at large k.

Since the number of primary clusters is given with much more relative accuracy than the corresponding amount of ionization, one would like to make use of it for particle identification. The primary ionization cross-section, as a function of velocity, has a similar behaviour to the ionization curve, the slope of the relativistic rise being essentially the same. The plateau, however, is reached at a lower saturation point; this makes the total relativistic rise R_p from the minimum to the plateau somewhat smaller than it is for the amount of ionization. For details see Allison [ALL 80].

Measurements of the relativistic rise of the primary ionization cross-section were reported by Davidenko et al., who used an optical streamer chamber filled with helium at 0.6 bar [DAV 69]; they measured $R_p = 0.38$ (difference between plateau and minimum divided by the value at the minimum). Blum et al. found $R_p = 0.6$ in Ne(70) + He(30) at 1 bar using spark chambers [BLU 74]. Although a systematic evaluation of primary ionization curves is lacking, the effect is obviously strong enough to make the counting of individual ionization clusters a method which is interesting because of its promise of accuracy – if only a way could be found to separate clusters in drift chambers.

One proposal was made by Walenta [WAL 79b]. In his 'Time Expansion Chamber', described in more detail in Chap. 10, the drift region is electrically separated from the amplification region; by using a very low drift velocity, it is possible to separate individual clusters in time and to register individual electrons with some efficiency. The basic problem in argon at atmospheric pressure is that the clusters are so close to each other (28 per cm in the minimum) that even a small amount of diffusion will wash out the primary ionization pattern. For larger distances between clusters one would have to employ lighter gases and/or lower gas pressures. We are not aware of any particle experiment which has successfully used cluster counting for the identification of particles.

9.7 Ionization Measurement in Practice

The large multiwire drift chambers provide many coordinate measurements on each track, and more often than not this includes recording the wire pulse heights. These are then a measure of the ionization density of each track segment, provided the chamber is operated in the proportional regime. In the universal detectors the particles go in all directions, and their ionization is sampled in different ways, depending on how the track is oriented with respect to the wire array that records it.

There are also more specialized drift chambers, which were built with the express purpose of particle identification behind fixed-target experiments. There, the particles travel generally into one main direction, which makes the ionization sampling much more uniform than it is in the universal detectors.

For good ionization measurements a balance has to be found between two conflicting requirements. On the one hand, the wire gain must be kept small in order to have a signal accurately proportional to the incoming ionization charge. When the gain is too large, the amplification will drop when the incoming charges are concentrated along a short piece of wire, owing to space charge near the wire (see Sect. 4.5), resulting in a gain variation with the amount of diffusion and the track-wire angle. Furthermore, the positive ions created in the wire avalanches penetrate to a certain extent into the drift space, where they cause field and hence track and gain distortions. On the other hand, large signals are required for the coordinate measurement in order to overcome the electronic amplifier noise; this is particularly important for coordinate measurements along the wire, using charge division.

The pulse integration must work in such a way that the result is exactly proportional to the collected charge, and especially that it is independent of the length of the pulse. Otherwise it has to be corrected by a function of the track angle and the drift length, which both have an influence on the pulse length. In order to have a clean pulse integration, overlapping tracks must be eliminated with some safety margin. The number of signals on a track in the middle of a jet of other particles can therefore be considerably reduced. It is not uncommon that in a large drift chamber the average number of signals is only little more than one half of the number of wires that could in principle measure a track.

In order to derive a meaningful ionization of a track, truly independent of all the external circumstances of the measurement, a number of calibrations and corrections are required at the percentage level of accuracy. These may be grouped in two categories, according to whether or not they depend on the track parameters.

9.7.1   Track-Independent Corrections

Without going into the details, we just mention the following effects which change the apparent ionization and therefore have to be kept under control: gas density (pressure and temperature), concentration ratios of the components, electron attachment, base line and pedestal shift of the electronic channels.

The response of the individual wires must be the same throughout the drift chamber. At first sight one may argue that the wires measuring the same track are allowed to have inequalities among each other up to a value small compared only to the fluctuations of a wire signal (say small compared to 70%). On a closer look, the requirement of uniformity is more stringent, because the situation changes from one track to the next. A group of wires may not contribute to the next track because the track is displaced or because it overlaps with other tracks. So in practice the wire response must be equalized to a narrower measure. A quantitative evaluation would obviously have to take into account the geometrical details of the situation. Although such equalization can be achieved by determining a correction factor for each wire, matters are simpler when the hardware is already made quite uniform by respecting tight tolerances on the gain and time constants of the charge-sensitive amplifiers.

9.7.2   Track-Dependent Corrections

The amount of track ionization collected on one wire of a grid plane depends on the track orientation. Let α be the angle between the track and the plane, and let

β be the angle in the plane that the projected track has with respect to the normal to the wire direction (a track parallel to the wire has $\alpha = 0$ and $\beta = 90°$). The length x of track contributing to a signal is equal to $x_0/(\cos\alpha\cos\beta)$, where x_0 is the distance between sense wires. We know from Chap. 1 (Fig. 1.10 and the discussion after (1.50)) that the most probable value I_{mp} of the ionization is not a linear function of x but can better be approximated by

$$I_{mp}(x)/I_{mp}(x_0) = (x/x_0)\,(1 + C_1\ln(x/x_0))\,, \tag{9.12}$$

where the value of C_1 can be obtained from Fig. 1.11 or from a Monte Carlo simulation, or experimentally by grouping wires to make larger x; C_1 is typically between 0.1 and 0.3. The signals from every track segment have to be divided by the factor (9.12) in order to be normalized to the standard length x_0.

The *effect of saturation* is described by a length λ on the wire over which the incoming electrons are distributed; λ is equal to $x_0\tan\beta$ and has to be convoluted with the transverse diffusion, which in turn is a function of the drift length z. A first-order correction could be the following expression, which depends on a constant C_2:

$$I(\beta)/I(90°) = 1 - C_2/\lambda(\beta, z)\,. \tag{9.13}$$

This correction is complicated because it is a function of the drift length z as well as of the angle β; it is also not quite independent of the amount of ionization itself and of the angle α.

In a geometry where the wires run orthogonal to a magnetic field, the $E \times B$ effect causes the drifting electrons to arrive on the wires at an angle ψ (Fig. 6.5); in this case λ is equal to $x_0\tan(\beta - \psi)$.

The signal attenuation due to *electron attachment* requires a correction which depends on the drift length z through a constant C_3:

$$I(z) = I(0)\,e^{-C_3 z} \tag{9.14}$$

Electronic cross-talk between wires can give rise to an α-dependent correction because the small 'illegal' pulses induced on neighbouring wires are prompt, whereas the normal signal develops according to the arrival time of the electrons. The time differences between the normal and the small induced pulses depend on α and will give rise to a variation in the pulse integration.

The corrections discussed above are all interconnected, and it is difficult to isolate the individual effects. In fact, we do not know a particle experiment in which such a programme has been successfully completed in all details. Nonetheless, with the corrections applied, in practice it has usually been possible for the determination of track ionization to reach a level of accuracy that is not far from that predicted by (9.9). The reader who is interested in more details of the correction procedure is referred to the articles by the groups of ISIS2 [ALL 82, ALL 84], OPAL [BRE 87, HAU 91], CRISIS [TOO 88] and ALEPH [ATW 91].

9.8   Performance Achieved in Existing Detectors

It is appropriate now to describe a small number of wire chambers used in particle experiments and to compare their ionization-measuring capability with the theory. Although a systematic discussion of existing drift chambers begins in Chap. 10, we want to include a few devices here as examples in the context of ionization measurement. We distinguish specialized chambers and universal detectors. The emphasis is on measurements in the region of the relativistic rise.

9.8.1   Wire Chambers Specialized to Measure Track Ionization

As the progress in accelerator technology made fixed-target experiments possible in particle beams with momenta up to one hundred GeV/c or more, the relativistic rise of the ionization in gases was put to the service of identifying the secondary particles. The first and largest such undertaking was the External Particle Identifier ('EPI') behind the Big European Bubble Chamber ('BEBC'). It consisted of a stack of 64 rectangular chambers, each 1 m high, 2 m wide and 12 cm deep. Vertical proportional wires – surrounded by field wires – were strung to make sensitive regions in the gas, shaped like rectangular boxes, 1 m high, and with a 6×6 cm^2 ground surface, two layers to a frame, each layer 32 boxes wide; see Fig. 9.13. This arrangement was used as a multiwire proportional chamber hodoscope – the coordinate information came in steps of 6 cm

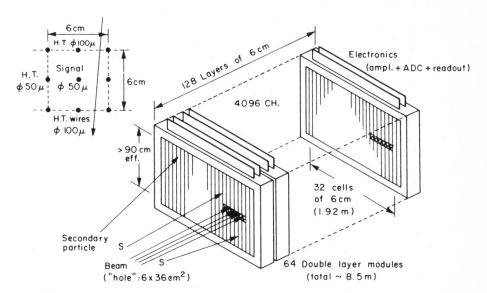

Fig. 9.13. Layout of the External Particle Identifier and cell structure. The position of the insensitive beam hole is indicated

horizontally, but vertically there was no subdivision. The ensemble of the 64 frames was 8.5 m long (deep) and weighed 16 ton. Mounted four metres above the ground, it still looked small compared to the big magnet of BEBC, which was nearly 16 m away (centre to centre), so the solid-angle coverage of secondary particles was only 8 m sr. The whole apparatus had to be movable sideways in order to be able to accept at least a small fraction of the secondary particles that had the same charge sign as the incoming beam. The limiting aperture was a hole in the iron mantle of BEBC through which the secondary particles had to leave the zone of interaction.

If the spatial resolution was modest, the same can be said for the time resolution. The device had to register all the events in one burst, which lasted for only 3 μs. These details are important in the context of pattern recognition and an unambiguous association of track segments.

For the ionization measurement, enough length was available to reach a very fine precision. For beam tracks, Lehraus and collaborators reached a resolution of 6 to 7% FWHM, varying slightly with particle species and momentum. This is almost as good as the theoretical limit of 5.8% derived from (9.9) based on the PAI model.

In Fig. 9.14 we see the ionization of beam tracks consisting of protons and pions with a momentum of 50 GeV/c. The resolution is so good that a clean identification of almost every individual particle is possible.

When analysing tracks that emanate from interactions in the bubble chamber, one loses track segments overlapping with others in the same region (box) of sensitivity. This loss in the number of ionization samples N is an important cause for the observed deterioration of resolution. The resolution in this overlap region seemed to deteriorate more quickly than according to the rule $N^{-0.46}$ or $N^{-0.43}$ discussed in Sect. 9.4.1 for clean measurements. In EPI the average number of samples $\langle N \rangle$ that remained above a minimal requirement N_{min}, for example, was between 72 and 91 (momentum dependent), resulting in resolutions between 10.6 and 8.7%; see Table 9.3.

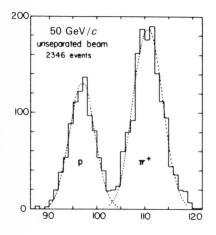

Fig. 9.14. Distribution of ionization measured with EPI in an unseparated positive beam line, 900 m downstream of the target, at 50 GeV/c momentum [LEH 78]

Table 9.3. Wire chambers specialized in ionization measurement in fixed-target experiments

Chamber	EPI		ISIS2[a]	CRISIS
Experiment	Big European Bubble Chamber		European Hybrid Spectrometer with HOLEBC	FNAL Hybrid Spectrometer with 30' BC
Reference	[LEH 78] [BAR 83]		[ALL 84]	[TOO 88] [GOL 85]
Chamber type[b]	MWPC Stack		type 3 drift chamber	type 3 drift chamber
Max. no. of samples per track	128		320	192
Sample length (cm)	6		1.6	1.6
Total length of sensitive volume (m)	7.7		5.1	3.1
Size of entrance window hor. × vert. (cm)	192 × 90		200 × 400	98 × 102
Max. drift length (cm)	6		200	25
Gas mixture	$Ar(95) + CH_4(5)$		$Ar(80) + CO_2(20)$	$Ar(80) + CO_2(20)$
Pressure (bar)	1		1	1
Gas amplification factor	4000		10000	—
Estimation method employed[c]	$\langle S \rangle_{40}$		M–L	$\langle S \rangle_{75}$
Analysis of beam tracks:				
Observed resolution (FWHM, per cent)	6.0–7.0		—	7.6–8.0
Theoretical limit (our equation 9.9)	5.8		5.8	7.4
Analysis of secondary tracks:				
Min. no. of useful samples required	50	80	100	50
Average no. of useful samples obtained	$(72–91)^d$	$(91–109)^d$	250	140[e]
Observed resolution (FWHM, per cent)	(10.6–8.7)	(8.7–7.8)	8.3	10
Theoretical limit (our Eq. (9.9))	(7.6–7.8)	(6.8–6.4)	6.5	8.7
Relativistic rise (per cent over the minimum)		58	56	≃ 45

[a] Chamber also described in Sect. 10.7.5.
[b] Drift chamber definitions, see Sect. 10.1.
[c] Notation, see Sect. 9.3; M–L is maximum likelihood.
[d] Depending on momenta of secondaries.
[e] Our estimate.

Comparable in size was the large drift chamber ISIS2 (which stands for the 'Identification of Secondaries by Ionization Sampling'); it had a length of 5 m and an aperture of 2 m horizontally times 4 m vertically. A description is found in Chap. 10. Here we summarize its capabilities as an ionization-measuring device. With more frequent and finer samplings than EPI, but with reduced length, it should have the same resolving power, according to (9.9). Although the drift-time measurement allowed the vertical track coordinates in ISIS to be known to a few millimetres, the detailed track filtering and subsequent hit association to tracks proved to be critical for achieving a good resolution. For instance, in a typical study of secondary tracks in an experiment 10 m behind the small bubble chamber HOLEBC, the average number of samples retainable per track was $\langle N \rangle = 250$, if the required minimum was set to $N_{min} = 100$. The achieved accuracy was 8.3% FWHM instead of the theoretical 6.5% expected for this value of $\langle N \rangle$. Whereas such details depend entirely on the track density in the chamber, they are quoted here to illustrate how the ionization resolving power goes down with an increasing overlap of tracks.

With this apparatus the most probable ionization density of each track could be measured with an accuracy of 3.5% r.m.s. and the relativistic rise was 56%. Using the dependence of the ionization on the particle velocity, as well as the momenta determined in the spectrometer, useful particle identification was possible up to 60 GeV. In order to give an impression of the achieved identification reliability, we show in Fig. 9.15 the uniqueness f of the electron–pion separation as a function of particle momentum. The parameter f is defined to be the percentage of correctly identified pions (probability $> 1\%$) whose ionization is incompatible with an electron at the 1% level, and this is equal to the percentage of correctly identified electrons incompatible with a pion. (These ratios could be measured because there was independent kinematical particle identification for a sample of events.) Figure 9.15b contains the same for the uniqueness of the pion–kaon separation.

The achieved degree of particle identification is also demonstrated in the scatter plot of momentum vs. ionization of Fig. 9.16. Working with secondary

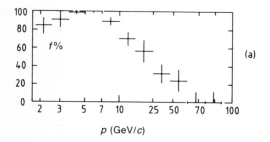

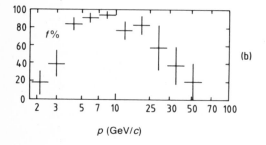

Fig. 9.15a, b. ISIS2 performance: the uniqueness f of particle separation as a function of particle momentum: (a) electron–pion separation; (b) pion–kaon separation. (For the definition of f, refer to the text.)

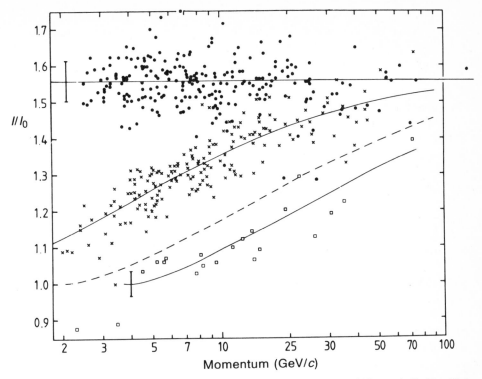

Fig. 9.16. ISIS2 scatter plot of the measured ionization vs. momentum of kinematically identified tracks with more than 100 samplings. *Full circle*: electron; *cross*: pion; *open square*: proton. The lines are the expected curves for electrons, pions, kaons (*dashed*) and protons. The error bars represent one standard deviation for tracks with 250 samplings [ALL 84]

particles of known masses and momenta, Allison and colleagues were able to compare the ionization measurement with expectations based on the PAI model. In the plot we see electrons, pions and protons, but no muons or kaons, because these were not in the sample of kinematically identified tracks; this simplified the situation. One can easily make out the regions in the drawing where the measured tracks are unambiguously associated to one line, i.e. identified. Some confusion between pions and electrons sets in above 10 GeV/c; protons from pions can be separated to higher momenta, in the absence of kaons.

The idea of secondary particle identification in a fixed-target experiment was also the purpose of the construction of CRISIS, which served in the Hybrid Spectrometer at FNAL together with the 30' bubble chamber. It derived its name as well as its design principle from ISIS, of which it was a 'Considerably Reduced' version: the length was 3.1 m, the aperture 1 m by 1 m. The drift length was subdivided into 4 separate shorter regions, thus reducing the problem of track overlap. In the analysis of 100 GeV/c beam tracks, a resolution of 7.6 to 7.9% FWHM was reported, compared to 7.4 from (9.9).

The measurement of secondaries from beam–nucleus interactions at 100 GeV/c momentum resulted in a width $\delta I/I = 10\%$ (FWHM) after requiring more than 50 charge samples for a track. For a comparison with ISIS2 we form the ratio $(\delta I/I)/R$, which is 0.22 for CRISIS, compared to 0.15 for ISIS2. The group determined the ratio of the numbers of pions to the numbers of protons/antiprotons as a function of momentum on a statistical basis; the presence of kaons could be inferred.

The fact that their plateau was observed near $R = 45$ per cent above the minimum is very interesting. In columns 1, 2 and 3 of Table 9.3 we find 58, 56 and ≈ 45; these numbers should be equal to within a few per cent. This shows that ionization measurements of the kind described here depend very much on the individual experimental procedures (pulse integration, pedestal and other corrections, statistical treatment), therefore, a universal curve for all experiments with the same gas and pressure is not applicable. The resolving powers of measurements in the relativistic-rise region achieved in different experiments are reasonably compared using the ratio $(\delta I/I)/R$.

With this experience, some qualifications are in order concerning our statement (9.4). The approximate independence of $F_{g,m}^{(\mathrm{norm})}(v)$ from the index m is with respect to small variations in sample size (as they occur in one experiment owing to different track directions) and with respect to the statistical estimator. Between different experiments, $F_g(v)$ can evidently come out different, as a function of the experimental procedures.

9.8.2 Ionization Measurement in Universal Detectors

From the large detectors described in Chap. 10 we have selected three examples for a discussion of their capability to determine the ionization of tracks. They are the ones that show the greatest promise of accuracy using (9.9). All operating essentially with argon gas, they span a range of gas pressures between 1 and 8.5 bar.

The OPAL chamber looks at a radial track with 159 sense wires, spaced 1 cm; with the gas pressure at 4 bar, (9.9) predicts a resolution of 6.0% FWHM. When selecting isolated e- and μ-tracks that can be well measured, the group reaches 7.3%.

In the analysis of hadron jets, losses occurred in the number of useful samples, owing to the overlap of tracks; the hit numbers of the remaining ones followed a distribution between 0 and 159, peaking at 110 with 89% above 40. On average, well-measured tracks had 94 samples, and a resolution of 8.9% FWHM was obtained. Equation (9.9) predicts 7.6% for these. Again, the deterioration is worse than would be expected from the reduced number of samples alone.

The OPAL group has made a special study of this effect by plotting in Fig. 9.17 the accuracy reached within jets for minimally ionizing pions as a function of the number of useful samples. This is compared with data for

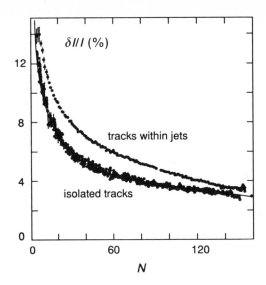

Fig. 9.17. OPAL study of the dependence of the resolution $\delta I/I$ (r.m.s.) on the number of ionization samples N used per track. *Upper curve:* Minimally ionizing pions within jets. *Lower curve:* Single μ and e tracks (number of good ionization samples artificially reduced) [HAU 91]

isolated particles. These data contain usually the (almost) full number of wires, and these are then artificially reduced by randomly removing good signals; the remaining number N of samples was found to determine the accuracy proportional to $N^{-0.43}$. The curve for particles in jets is worse than (i.e. lies above) the one for isolated particles, for every N. We may suspect that problems of space charge near the wires and in the drift region as well as residual problems of pattern recognition are the cause of this behaviour.

The PEP-4 TPC can measure a good track 183 times. With the wires spaced at $x = 0.4$ cm and the gas pressure 8.5 bar, the individual sample length px is very similar to that of the OPAL chamber. On the basis of (9.9), one expects a resolution of 5.9% FWHM, and 6.9% is reached for isolated tracks.

The tracks in particle jets are measured with a resolution of 8.3% FWHM. The number of lost samples is our own estimate; as in the case of OPAL, it does not fully account for the reduction in accuracy, compared to the isolated tracks.

The ALEPH TPC is almost twice as large in radius as the PEP-4 TPC and can measure radial tracks on 340 wires. Operating with atmospheric gas pressure, its individual sample length is only $px = 0.4$ cm bar, smaller than in any other drift chamber. The signals are therefore extremely feeble. In the minimum of ionization, only eleven clusters contribute on average to the signal of a wire (cf. Table 1.1). According to (9.9), a resolution of 8.8% FWHM should be expected, and 10.3% has been reached for isolated tracks.

When studying tracks inside jets and of lower momenta, the large magnetic field (1.5 T; see Table 10.4) poses a special problem: particle tracks below 0.5 GeV/c coil up inside the chamber and require angular corrections (cf. Sect. 9.7.2) over a large range of angles. The resolution for hadron tracks is 13.3% FWHM at the average value of $N = 250$ samples; this can be compared to the 10.2% expected from (9.9).

In a comparison of these three detectors one must consider the different gas pressures. The relativistic rise R, expressed as the percentage difference of the plateau over the minimum, is better for the smaller pressures. On the other hand, the resolution $\delta I/I$ is better for the higher pressures. A useful over-all criterion is the measured ratio $\delta I/(IR)$, which we have computed in the last line of

Table 9.4. Ionization-measuring capability of some universal drift chambers in collider experiments[a]

Drift chamber	OPAL Jet Chamber	PEP 4 TPC	ALEPH TPC
Reference	[BRE 87] [HAU 91]	[COW 88]	[ATW 91]
Drift chamber type[b]	2	3	3
Max. no. of samples per track	159	183	340
Sample length (cm bar)	4	3.4	0.4
Max. drift length (cm)	3–25	100	220
Gas mixture	Ar(88) + CH_4(10) + i-C_4H_{10}(2)	Ar(80) + CH_4(20)	Ar(91) + CH_4(9)
Pressure (bar)	4	8.5	1
Gas amplification factor	10000	—	5000
Estimation method employed[c]	$\langle S \rangle_{70}$	$\langle S \rangle_{65}$	$\langle S \rangle_{60}$
Analysis of isolated tracks:			
Min. polar angle required (deg):	45	45	45
Observed resolution (FWHM, per cent)	7.3	6.9	10.3
Theoretical limit[d] (our Eq. (9.9))	6.0	5.9	8.8
Analysis of tracks inside jets:			
Min. polar angle required (deg):	45	45	45
Min. no. of useful samples required	40	80	150
Average no. of useful samples obtained	94	140[f]	250
Observed resolution[e] (FWHM, per cent)	8.9	8.3 ± 0.7	13.3
Theoretical limit[g] (our Eq. (9.9))	7.6	6.7	10.2
Relativistic rise observed (per cent over the minimum)	47	37	57
Measured resolution for isolated tracks (FWHM), divided by the observed relativistic rise	0.155	0.186	0.181

[a] More details on these chambers can be found in Tables 10.2 and 10.4
[b] Definitions Sect. 10.1
[c] Definitions Sect. 9.3
[d] Using lines 4 and 5 above (the effects of slightly larger average sample size and any loss of samples are neglected)
[e] Pions in the minimum of ionization
[f] Our estimate
[g] Using lines 5 and 16 above

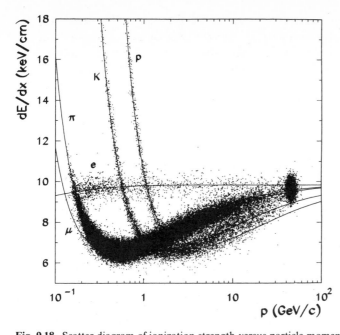

Fig. 9.18. Scatter diagram of ionization strength versus particle momentum for multihadronic and dimuon events in Z^0 decays, as measured with the OPAL Jet Chamber [HAU 91]

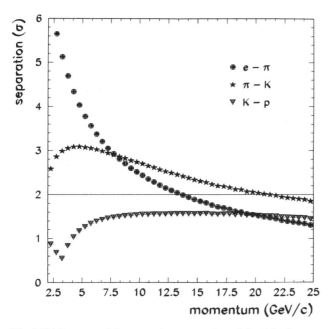

Fig. 9.19. R.m.s. particle-separation power D as defined in the text, achieved with the OPAL Jet Chamber [HAU 91]

Table 9.4. The best figure of merit belongs to the OPAL Jet Chamber; for it we show the scatter diagram of ionization strength versus particle momentum in Fig. 9.18. The OPAL r.m.s. particle-separation power D for various pairs of particles is plotted in Fig. 9.19 as a function of momentum. The quantity D_{12} is defined as the ratio of the difference of the ionization strengths of the two particles 1 and 2 to the average resolution

$$D_{12} = \frac{|I_1 - I_2|}{(\delta I_1/I_1 + \delta I_2/I_2)/2} .$$

Experimentally, the two terms in the denominator are equal.

References

[ALL 80]   W.W.M. Allison, Relativistic charged particle identification by energy-loss, *Ann. Rev. Nucl. Sc.* **30**, 253 (1980)

[ALL 82]   W.W.M. Allison, Relativistic particle identification by dE/dx: the fruits of experience with ISIS, in *Int. Conf. for Colliding Beam Physics, SLAC, 61* (1982)

[ALL 84]   W.W.M. Allison, C.B. Brooks, P.D. Shield, M. Aguilar Benitez, C. Willmot, J. Dumarchez and M. Schouten, Relativistic charged particle identification with ISIS2, *Nucl. Instrum. Methods Phys. Res.* **224**, 396 (1984)

[ASS 90]   R. Assmann, Ionisationsmessung und Teilchenidentifizierung in der zentralen Spurendriftkammer (TPC) von ALEPH, Diploma Thesis University of Munich 1990, unpublished

[ASS 91]   We thank R. Assmann (Munich) for the computation of the fits

[ATW 91]   W.B. Atwood et al., Performance of the ALEPH time projection chamber, *Nucl. Instrun. Methods Phys. Res. A* **306**, 446 (1991)

[BAR 83]   V. Baruzzi et al., Use of a large multicell ionization detector – the External Particle Identifier – in experiments with the BEBC hydrogen bubble chamber, *Nucl. Instrum. Methods* **207**, 339 (1983)

[BLU 74]   W. Blum, K. Söchting, U. Stierlin, Gas Phenomena in spark chambers, *Phys. Rev. A* **10**, 491 (1974)

[BRE 87]   H. Breuker et al., Particle identification with the OPAL jet chamber in the region of the relativistic rise, *Nucl. Instrum. Methods Phys. Res. A* **260**, 329 (1987)

[COW 88]   G.D. Cowan, Inclusive π, K and p, p$^-$ production in e$^+$e$^-$ annihilation at $\sqrt{s} = 29$ GeV, Dissertation, University of California, Berkeley (1988), also Berkeley preprint LBL-24715

[CRA 51]   H. Cramér, *Mathematical Methods of Statistics* (Princeton University Press 1951)

[DAV 69]   V.A. Davidenko, B.A. Dolgoshein, V.K. Semenov, S.V. Somov, Measurements of the relativistic increase of the specific primary ionization in a streamer chamber, *Nucl. Instrum. Methods* **67**, 325 (1969)

[EAD 71]   W.T. Eadie, D. Dryard, F.E. James, M. Roos, B. Sadoulet, *Statistical Methods in Experimental Physics* (North-Holland, Amsterdam 1971)

[ERM 77]   V.C. Ermilova, L.P. Kotenko and G.I. Merzon, Fluctuations and the most probable values of relativistic charged particle energy loss in thin gas layers, *Nucl. Instrum. Methods* **145**, 555 (1977)

[FIS 58]   M. Fisz, *Wahrscheinlichkeitsrechnung und mathematische Statistik* (translated from Polish) (VEB Verlag der Wissenschaften, Berlin [11]1989)

[GOL 85]   D. Goloskie, V. Kistiakowsky, S. Oh, I.A. Pless, T. Stroughton, V. Suchorebrow, B.

Wadsworth, O. Murphy, R. Steiner and H.D. Taft, The performance of CRISIS and its calibration, *Nucl. Instrum. Methods Phys. Res. A* **238**, 61 (1985)

[HAU 91]   M. Hauschild et al., Particle identification with the OPAL Jet Chamber, CERN preprint PPE/91-130 (August 1991), to appear in *Nucl. Instrum. Methods Phys. Res.*

[LEH 78]   I. Lehraus, R. Matthewson, W. Tejessy and M. Aderholz, Performance of a largescale multilayer ionization detector and its use for measurements of the relativistic rise in the momentum range of 20–110 GeV/c, *Nucl. Instrum. Methods* **153**, 347 (1978)

[LEH 82a]   I. Lehraus, R. Mathewson and W. Tejessi, Particle identification by dE/dx sampling in high pressure drift detectors, *Nucl. Instrum. Methods* **196**, 361 (1982)

[LEH 82b]   I. Lehraus, R. Mathewson and W. Tejessi, dE/dx measurements in Ne, Ar, Kr, Xe and pure hydrocarbons, *Nucl. Instrum. Methods* **200**, 199 (1982)

[TOO 88]   W.S. Toothackier et al., Secondary particle identification using the relativistic rise in ionization, *Nucl. Instrum. Methods Phys. Res. A* **273**, 97 (1988)

[WAL 79a]   A.H. Walenta, J. Fischer, H. Okuno and C.L. Wang, Measurement of the ionization loss in the region of relativistic rise for noble and molecular gases, *Nucl. Instrum. Methods* **161**, 45 (1979)

[WAL 79b]   A.H. Walenta, The time expansion chamber and single ionization cluster measurement, *IEEE Trans. Nucl. Sc.* **NS-26**, 73 (1979)

10. Existing Drift Chambers – An Overview

Drift chambers have in common that the drift of the ionization electrons in the gas is used for a coordinate determination by measurement of the drift time. In this chapter we want to take a look at the large variety of forms in which drift chambers have been built for particle physics. As a complete coverage of all existing drift chambers does not correspond to our plan we subdivide the material according to a geometrical criterion into three basic types, and then we discuss typical forms within each type. The selection of chambers is intended to represent the different choices that have been made in this field, but a certain arbitrariness in choosing the examples was unavoidable. Of the chambers we know, we have preferred those that are already working to similar ones that are still in a state of preparation; and those that are well documented we have preferred to similar ones with less detailed descriptions. We have considered drift chambers for the detection of charged particles, but not those for single photons, so the image chambers useful in biology and medicine have been left out as well as those used to detect Cerenkov radiation.

A description of drift chambers in the context of particle-physics experiments is given in the review article by Williams [WIL 86].

Let us begin by defining the three basic types.

10.1 Definition of Three Geometrical Types of Drift Chambers

Type 1: The simplest drift chamber is ideally a *sensitive area* placed across the path of a particle in order to measure one, or perhaps both, coordinates of the point of penetration. It was developed out of the multiwire proportional chamber, or MWPC – a hodoscope of parallel proportional wires – by supplementing the electronics of the sense wires with equipment to measure the time of the avalanche pulse and hence the drift time of the ionization electrons. The electric field in the drift region is shaped, usually with conductors on defined potentials like field wires, but sometimes also by charges deposited on insulators.

In practice, the ionization electrons of a short piece of track are collected on the nearest sense wire. For a complete determination of track parameters, such chambers are often used in stacks of several parallel chambers. The first working drift-chamber system, which was built by Walenta, Heintze and Schürlein [WAL 71], was of type 1.

Types 2 and 3: The particle traverses a *sensitive volume*. The ionization electrons of a number of track pieces are collected on an equal number of sense wires, and several coordinates are measured along the track. Such an arrangement allows the measurement of track directions and, if a magnetic field is present, of track curvature. If the sampling of ionization along the track is done often enough, there is the possibility of determining the ionization density to a degree useful for particle identification.

One type of volume-sensitive drift chamber consists of an arrangement of a large number of parallel or almost parallel sense wires that span the volume, usually interleaved with equally parallel field-shaping wires: this we call type 2.

Chambers of type 2 may be thought of as consisting of many chambers of type 1 in the same volume. Obviously there is no sharp dividing line between types 1 and 2. We will call a drift chamber that presents a sensitive area to the penetrating particle, but measures the particle track with several sense wires, a '*type 1 multisampling chamber*', its depth in the particle direction being considerably smaller than the linear dimensions that span the area.

A quite different type of volume-sensitive drift chamber is characterized by a sensitive volume which is free of wires; the wires are located on one or two surfaces that delimit the drift region. This we call type 3. Well-known examples are the universal track detectors that carry the name 'time projection chambers' (TPCs). In their original form they have segmented cathodes (pads) behind their sensitive wire planes, and the electric field is parallel to the magnetic field; later on, TPCs were also built without magnetic field. Other type 3 chambers include the large drift volumes built for particle identification where all tracks are drifted onto one central wire plane for ionization measurement, or the spiral drift chamber where the electrodes are situated on a cylinder mantle.

For a type 3 drift chamber, the use of the name 'time projection chamber' is as vague as the name is fanciful (time cannot be projected). There seems to be a tendency now for many constructors of new type 3 drift chambers to call their device a 'TPC'.

10.2 Historical Drift Chambers

Throughout this book, we are usually not concerned about the originators of each of the many ideas which together represent our knowledge about drift chambers. In this section we want to make an exception and follow the roots of the drift chamber itself as well as of its two later configurations, the types 2 and 3.

The possibility of using the time of the signal for a coordinate determination was already recognized by the authors of the paper that introduced the multiwire proportional chamber, or 'Charpak chamber', as it was then called [CHA 68].

The first studies exploiting the drift time for a coordinate measurement were done by Bressani, Charpak, Rahm and Zupančič at CERN in 1969 [BRE 69]. It is here that the word 'drift chamber' makes its appearance in the literature. A 3-cm-long drift space was added to a conventional multiwire proportional chamber with dimensions 12×12 cm^2; it was separated from the drift region by a wire mesh with 90% transparency. The ionization electrons of a beam particle drifted orthogonally through it onto the sense-wire plane, where they were measured on 15 proportional wires that were spaced at 2 mm. We reproduce in Fig. 10.1 the historic picture of this first drift chamber. Drift-velocity measurements and resolution studies were done with it. The crucial role of diffusion for the accuracy was recognized.

The documentation of this work is not easy to find because it was published in the proceedings of an International Seminar on Filmless Spark and Streamer Chambers in Dubna, with contributions mainly written in Russian. The early investigations were later reported in a wider context by Charpak et al. [CHA 70].

The first operational drift-chamber system including electronic circuitry and digital readout was built by Walenta, Heintze and Schürlein [WAL 71]. The chamber was organized in a novel way, consisting of a multiwire proportional chamber with large wire spacings as shown in Fig. 10.2, where the drift-time measurement gave the coordinate between the wires, thus improving accuracy and using fewer electronic channels compared to the MPWC. The ionization of a measured particle was amplified on one proportional wire. This geometry, which in our classification is of type 1, was later to be developed into a form where the drift field was extended and made more homogeneous by introducing field wires on graded potentials. An example is the type 1 chamber by Breskin et al. schematically shown in Fig. 10.3 [BRE 75]. Further important developments

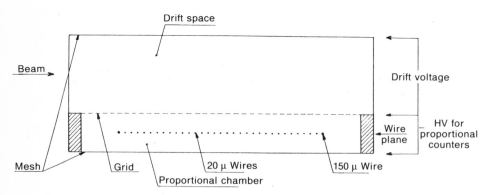

Fig. 10.1. Disposition of the electrodes of the first drift chamber. Original drawing by Bressani et al. [BRE 69]

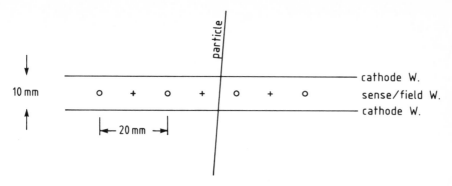

Fig. 10.2. Disposition of the electrodes in the chamber built by Walenta et al.

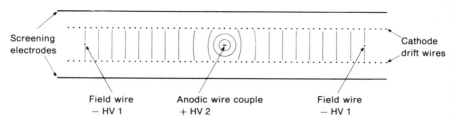

Field wire
– HV 1

Anodic wire couple
+ HV 2

Field wire
– HV 1

Fig. 10.3. Disposition of the electrodes in the chamber built by Breskin et al. [BRE 75], cathode wires at uniformly decreasing potentials produce a long and homogeneous drift field

of the drift-chamber technique had been reported by the CERN group [CHA 73].

If the creation of the drift chamber out of an array of proportional wires was an important invention which produced many branches of useful particle detectors, the appearance of one of its principal configurations, the type 2 drift chamber, was part of a rapid development towards larger drift chambers and represents a much smaller 'quantum jump'. When the new e^+e^- storage rings like DORIS and SPEAR required particle tracking in all directions around the interaction point, the most obvious design was the one where several cylindrical type 1 chambers with increasing diameters were inserted into one another, each one an entity that could be pulled out for maintenance. The MARK I drift chambers and the PLUTO spark chambers were examples of this geometry.

The first genuine type 2 drift chamber was that of the old MARK II detector at SPEAR [DAV 79]. Its designers had taken the essential step of filling the entire sensitive volume with wires strung between two opposite end-plates, having done away with all intermediate ring-shaped structures and gas foils, thus paving the way for a new generation of larger, volume-sensitive drift chambers of very high accuracy. A description of the state of the art in 1981 may be found in [WAG 81] and [FLÜ 81].

The origin of the type 3 drift chambers is in Oxford. The ISIS project for the Identification of Secondaries by Ionization Sampling was first described in 1973

[MUL 73, ALL 74], and a prototype with an 85-cm-long drift was tested the same year [see ALL 82]. The basic idea was a large sensitive volume that contained only gas and an electric field, the track ionization being collected and measured on a single wire plane. Multihit electronics recorded the drift times of all the tracks of one event, thus producing a direct image of the event in one projection. The purpose of the apparatus was not primarily a precise determination of coordinates (for this the diffusion was too large as a consequence of the long drift) but the measurement of ionization density for particle identification. A later version of ISIS is described in Sects. 10.7 and 9.7. ISIS can be considered a predecessor of the TPC.

The first time the TPC concept appeared on paper was in a laboratory report by D. Nygren [NYG 74]. The chamber was subsequently built at Berkeley and SLAC for the PEP-4 experiment. Again, it is not easy to find access to a documentation of this work. A very detailed description of the PEP-4 TPC is contained in the experiment proposal of the collaborating institutes [CLA 76]. It was presented to the appropriate bodies for approval and later widely circulated among workers in the field. This proposal and its numerous appendices reported on many aspects of the feasibility of a TPC. Some of this important material has never been published.

Conscious [NYG 83] of the latest work at CERN and at Oxford, Nygren invented the TPC by performing a synthesis into one instrument of all the known elements: long drift, diffusion suppression through parallel E and B fields, 3-dimensional coordinate measurement by pick-up electrodes for x and y, and drift-time measurement for z. This was made possible by a novel storage of the pulse train of every electrode over the full length of the drift time. The PEP-4-TPC is described in more detail in Sect. 10.7.

10.3 Drift Chambers for Fixed-Target and Collider Experiments

A typical fixed-target experiment would have an arrangement of type 1 chambers that record (together with other detectors) the reaction products behind a target. If in addition to their directions one also measures the momenta of these particles with the help of a magnet, then we have a magnetic spectrometer, a typical form of which is schematically drawn in Fig. 10.4. Stacks of planar type 1 chambers have been in wide use for such spectrometers.

As for the momentum-measuring accuracy $\delta p/p$ at a given bending power p_B of the magnet, the contribution of each arm is, in a first approximation, proportional to $\delta x/L$, the ratio of the chamber point-measuring accuracy to the length of the lever arm (or thickness of the stack):

$$\frac{\delta p}{p} = \text{const}\, \frac{p}{p_B} \frac{\delta x}{L}.$$

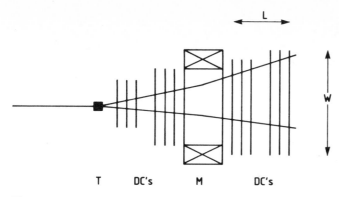

Fig. 10.4. A typical magnet spectrometer behind a fixed target T, using stacks of type 1 drift chambers (DCs) in front of and behind the magnet M

Often there is freedom to increase L; then the value of δx is less critical, but the dimensions W of the chambers must be increased in proportion.

For the study of very-high-multiplicity events in heavy-ion fixed-target experiments, even type 3 chambers have become popular. The direct three-dimensional measurement of track elements is an advantage for such events (see Sect. 10.7.4).

An important group of drift chambers has been built around the interaction regions of particle colliders where they are – together with the installed calorimeters – the principal instruments for the investigation of the particles from the collisions. Close to the interaction point we find the *vertex detectors* around the beam pipe whose purpose it is to determine the interaction vertex with high precision. Further out in radius the *main drift chambers* measure the direction and momenta of the charged particles and, in many cases, their ionization. In the design of these tracking devices the choice of the magnetic field and its orientation is a primary concern.

10.3.1 General Considerations Concerning the Directions of Wires and Magnetic Fields

For symmetric electron–positron machines where one studies point-like interaction – these cause essentially isotropic particle distributions – the magnetic field B was most naturally created by a solenoid on axis with the particle beams. A good measurement of particle momenta requires the curvature of tracks to be determined with the greatest accuracy in the *azimuthal* direction. In drift chambers of type 2 it is the drift-time measurement that provides the accuracy; hence the wires are essentially parallel to the magnetic field ('axial wire chambers'). In drift chambers of type 3 the azimuthal accuracy is provided by cathodic pick-up electrodes so that the electric drift field can be made parallel to

B. Therefore, in e^+e^- machines we find axial wire chambers and TPC's along the direction of the beams. Along **B**, particle curvature is not measurable.

For the study of hadron interactions in the $\bar{p}p$ and pp colliders, the magnetic field **B** has sometimes been chosen to be at right angles to the beam direction, on account of the important flux of high-momentum particles in the direction of the beams – the UAI-experiment at the CERN $\bar{p}p$ collider and the Split Field Magnet at the CERN Intersecting Storage Rings being well-known examples. The orientation of the wires of type 1 or type 2 drift chambers again has to be parallel to **B**, which is also the flight direction of the particles with vanishing curvature.

At the e^-p collider HERA at DESY there is an inherent asymmetry in the energies of the colliding particles, because 30 GeV electrons collide with 800 GeV protons. The solution adopted for each of the two experiments is a solenoid, coaxial with the beams, instrumented with a coaxial type 2 chamber, and complemented in the forward direction of high momenta with several planar type 1 stacks for particles with polar angles θ approximately between 10 and 30°. In these type 1 chambers we find the sense wires orthogonal to the magnetic field, thus giving an accurate measurement of θ and a good double track resolution in this angular range.

10.3.2 The Dilemma of the Lorentz Angle

In drift chambers of type 2, for an accurate determination of the curvature of an ionization track one compares the times of arrival of the electrons from different track segments on different wires (Fig. 10.5a). The best accuracy is obtained

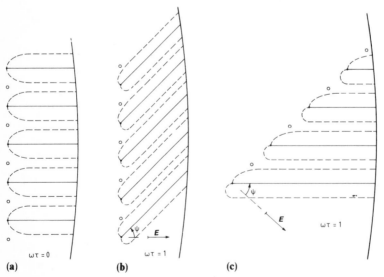

Fig. 10.5a–c. Drift paths of ionization electrons from the particle track to the sense wires. (**a**) $\omega\tau = 0$ (no magnetic field); (**b**) $\omega\tau = 1$, sense wire plane parallel to the tack; (**c**) $\omega\tau = 1$, sense-wire plane inclined

when the electron drift is at right angles to the track; for then the time differences measure the curvature directly, and also the electrons of every track segment that go to the same sense wire are most concentrated in arrival times and hence give the smallest variance in the measurement (cf. Sect. 6.3).

However, the magnetic field B orthogonal to the electric drift field E forces the electrons towards the direction given by the vector product $-[E \times B]$. The *Lorentz angle* ψ between the drift direction u and E, in the approximation of Sects. 2.1 and 2.2 (2.11) is given by

$$\tan \psi = \omega \tau .$$

Figure 10.5b shows the resulting disadvantage: not only does one have to measure the sagitta under an angle of projection, but also inside every track segment the arrival times are spread out more than before. This gives rise to the 'angular wire term' in (6.27), with $\alpha = \psi$, even if the track is perfectly radial. Therefore, in this geometry one is obliged to keep $\omega\tau$ small, although – in the interest of a good momentum measurement – one wants the largest possible B field. This is the dilemma of the Lorentz angle.

Since τ is related to the electron mobility μ by the approximate relation $\mu = (e/m)\tau$ (for more details see Chap. 2), one way out of the dilemma is to choose a gas with a low mobility (small drift velocity at high E field) and/or a high gas density N to which τ is inversely proportional by the relation $\tau = 1/N\sigma c$; cf. Sects. 2.2 and 11.4.

Another way out, but only of part of the dilemma, is a rearrangement of the wires according to Fig. 10.5c: the plane of the sense wires is made *inclined* to the radial direction by the Lorentz angle ψ. This solution has sometimes been adopted for high magnetic fields. Although this solution brings the drift direction back to the normal to the stiff radial tracks, the angular wire effect in (6.27) does not vanish for them. The importance of this effect depends on the radial sense-wire spacing.

10.3.3 Left–Right Ambiguity

The relation between drift time and space coordinate is not always unique. If ionization electrons can reach a proportional wire from two opposite sides, then there are two space coordinates to correspond to a measured signal time, one of them correct, the other a 'ghost'. The situation arises in drift chambers of types 1 and 2, but usually not in the type 3 chambers. (An exception is treated in Sect. 10.7.5). This *left–right ambiguity* can sometimes be resolved with pattern-recognition methods which take into account that a track produces correct coordinates that are continuous and extrapolate to realistic points of origins whereas the ghost coordinates do not make sense – provided the wires are arranged accordingly. Another possibility is the '*staggering*' of the sense wires, i.e. alternate sense wires are slightly displaced from their original positions in opposite directions; the correct coordinates are the ones that combine to form a continuous track. Still another method makes use of the development of avalanches, which do not go around the whole wire when small (Sect. 4.3).

Suitable pick-up electrodes can identify the side of approach of the drifting electrons [WAL 78, BRE 78].

10.4 Planar Drift Chambers of Type 1

10.4.1 Coordinate Measurement in the Wire Direction

Whereas the principle coordinate measurement in the wire chambers is through the drift time and therefore orthogonal to the wire direction, for a coordinate measurement along the wire other methods are needed. As we discussed in Sect. 6.1, there are three such methods: 'charge division' and 'time difference', based on a pulse measurement at the two wire ends, and 'cathode strips', based on a measurement of pulses in sections of the cathode.

With these methods it is possible to record the positions of every track segment (the ionization electrons collected on the wire) in all three dimensions, the position of the wire and the drift time giving the additional two. For a true three-dimensional measurement of track segments with cathode strips these must be *short* so that every cathode pulse can be associated from the beginning with its proper wire signal. With *long* cathode strips it is only after track reconstruction that this association can be made. The use of cathode strips in drift chambers has so far been limited to type 1 and type 3 chambers, and short strips ('pads') are so far only found in type 3 chambers.

Many spectrometers were built in which a particle had to successively traverse several parallel drift chambers, which measured only one drift coordinate for the track; but by arranging the sense wires in different directions, the track was measured in different stereo directions, thus allowing the full track reconstruction in three dimensions to be made. For example, a group of 9 parallel chambers could have chambers 1, 4 and 7 with vertical wires (angle of inclination $\alpha = 0$), chambers 2, 5 and 8 with wires inclined by $\alpha = 45°$, and chambers 3, 6 and 9 with wires at $\alpha = -45°$.

Different from a true coordinate measurement along the wires, this stereo measurement of tracks does not immediately yield three-dimensional coordinate information for every track segment. In the pattern of the drift times one would first find the tracks in the projection of the first group of chambers, then in the second, and so on; later, the various projected tracks would be combined in space.

10.4.2 Five Representative Chambers

In Table 10.1 we have listed five chambers of type 1 that are or were part of important experiments in high-energy physics. In the first two chambers, particle tracks are sampled once, whereas the last three belong to the multisampling kind.

The CDHS chambers were located between iron plates that were at the same time the target and the calorimeter plates of this neutrino experiment, so they had to be large and thin but were not particularly accurate. A relatively simple field-wire configuration was found to produce a sufficiently uniform drift field if the argon–isobutane gas mixture was used near saturation. A drift cell and its equipotential lines are shown in Fig. 10.6. Three identical units ('modules') of these hexagonally shaped chambers, with wire orientations at 0, +60 and −60°, were stacked between the iron plates.

The muon identifiers of the DELPHI experiment are the first example of a drift chamber that has the wire amplification in the limited streamer mode.

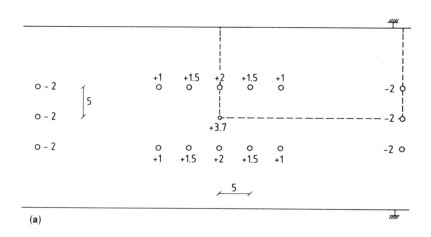

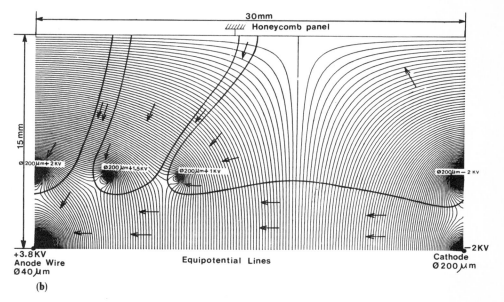

Fig. 10.6a, b. A drift cell of the CDHS chambers. (**a**) sense wires (*small circles*) and field wires (*large circles*) with their potentials in kV; some distances are also indicated in mm; (**b**) equipotential lines, delineating sensitive areas for one quarter of the drift cell, as indicated in (a)

Table 10.1. Some planar drift chambers of type 1

Name of experiment	CDHS	DELPHI1	NA34/HELiOS	L3	H1
Name of chamber	DC	Muon Identifier	DC 1	Middle Muon Chamber	FTD (radial wires)
Amplification mode	Proportional	Limited-streamer	Proportional	Proportional	Proportional
Method of z-measurement	Stereo (successive modules)	Time difference (delay line)	Stereo (successive modules)	Special chamber	Charge division (along r)
Reference	[MAR 77]	[DAU 86]	[BET 86] [FAB 90]	[ADE 90]	[BEC 89] [GRA 89]
Geometry					
Width (cm) × length (cm) (one chamber)	400 × 400 (hexagonal)	20 × 435	24 (diam)	163 × 556	30 (inner diam) 152 (outer diam)
Total area of these chambers in experiment (m²)	720	600	0.4	280	5
No. of sense wires per cell	1	1	8	24	12
No. of cells per chamber	130	1	8	16	48
Sense-wire spacing (mm) along the track	—	—	6 (2)	9	10
Sense-wire diameter (μm)	40	100	20	30	50
Max. drift length (cm)	3	10	1.5	5	1–5 (proportional to r)
Gap (mm)	30	20	48	360	120

Gas and fields

Gas (percentage concentration)	Ar(70) + i-C$_4$H$_{10}$(30)	Ar(14) + CO$_2$(70) + i-C$_4$H$_{10}$(14) + iso-propyl-alcohol (2)	CO$_2$(90) + C$_2$H$_6$(10)	Ar(62) + C$_2$H$_6$(38)	Xe(30) + C$_2$H$_6$(40) + He(30)
Electric drift field (kV/cm)	≥ 1	0.7	1.2	1.1	1.2
Magnetic field (T)	0	0	0	0.51	1.2
Drift velocity (cm/μs)	5.0	0.75	1.0	4.9	3
Performance (1)					
Point-measuring accuracy per wire					
In drift direction σ_x (mm)	< 0.7 (c)	1 (b)	0.12 (c) 0.06 (b)	0.17 (c) (3)	0.11–0.15 (b)
In wire direction σ_z (mm)	—	3 (b)	—	—	20 (b)
Double track resolution Δ (mm)	30 (a)	10	0.6 (b) 1.0 (c)	Not known	Approx. 1.5 (b)

1. The performance figures are based on (a) calculation and laboratory tests; (b) prototype and test-beam measurements; (c) measurements in the running experiment.
2. Charge collection per sense wire: 2.5–3 mm in track direction.
3. This led to a momentum resolution measured in the experiment equal to $\delta p_t / p_t^2 = 5.3 \times 10^{-4}$ (GeV/c)$^{-1}$ for high-momentum muons.

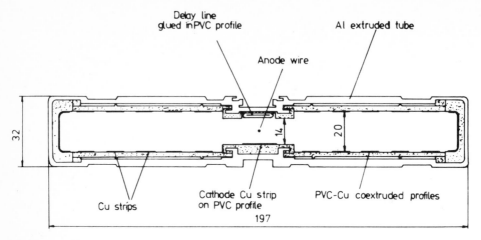

Fig. 10.7. Cross section of a drift-chamber module of the DELPHI Muon Identifier; the delay line is incorporated in the cathode

Shielded behind iron as they are, the increased dead time that is caused by the streamer regime was easily tolerable. The large pulses made it possible to use a slow delay line (0.6 µs/m) for a coordinate determination along the wire. Measuring the time difference the authors obtained an accuracy of better than 10^{-3} of the length. The delay line is incorporated in the cathode, as depicted in Fig. 10.7. The drift field is defined by copper strips in the plastic material of the body.

Of the drift chambers of the HELIOS experiment we discuss the small ones immediately behind the target. Their critical requirements were accuracy and double-track separation. The solution adopted involves a slow gas which in turn needed small drift cells in the high-multiplicity, high-rate environment. We see the drift cell in Fig. 10.8. It must provide a very homogeneous field because the gas is not saturated; therefore in each drift cell the eight sense wires are surrounded by a large number of field wires. In an attempt to reduce the influence of the drift-time variations, some of the field wires were given such potentials that only a fraction of the ionization along the track was collected on the sense wires. The whole stack consists of six chambers with three different wire orientations. The performance figures reveal that the outstanding measuring accuracy of 60 µm per wire, reached with the prototype, worsened and became 120 µm in the real experiment where the presence of many tracks deteriorated the critical field conditions.

The muon chambers of the L3 experiment are the result of a colossal effort to measure the momenta of fast muons with an accuracy of at least $\delta p_T/p_T^2 = 4 \times 10^{-4}\,\mathrm{GeV}/c^{-1}$, in cylindrical geometry. In the experiment, the radial space between 2.5 and 5.4 m was allocated to this task, inside a magnetic field of 0.5 T. The principle of the solution is sketched in Fig. 10.9; one observes that for the optimal measurement of momenta in the magnetic field, the

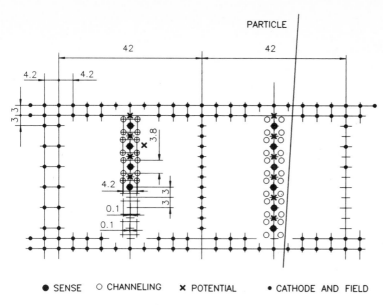

● SENSE   ○ CHANNELING   ✕ POTENTIAL   • CATHODE AND FIELD

Fig. 10.8. A drift cell of the DC1 chambers in the HELIOS experiment; dimensions in mm

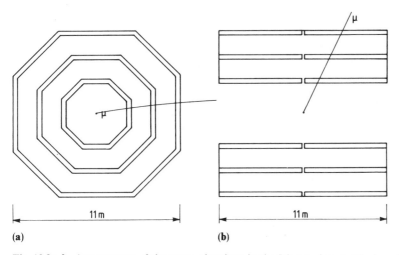

Fig. 10.9a, b. Arrangement of the muon chambers in the L3 experiment. (**a**) view along the axis; (**b**) side view

chambers are concentrated at the inner and outer edges, and in the middle of the radial space (cf. the discussion in Sect. 7.2.6). Their local measuring accuracy is increased by sampling each track N times, thus gaining a factor of $1/\sqrt{N}$ over the accuracy of a single wire – provided, the relative wire positions are sufficiently well known. For this purpose, the sense wires have been pulled over optically

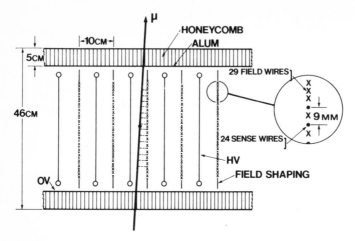

Fig. 10.10. One drift cell of a middle muon chamber of the L3 experiment

flat edges and are supported in the middle in a controlled fashion. We show in Fig. 10.10 the drift cell of the chamber type in the middle; it has $N = 24$ sense wires, 0.9 cm apart and 5.5 m long. Under operating conditions the Lorentz angle is $19°$.

In order to make full use of the local accuracy, the three chambers traversed by a fast muon must be in well-known and constant relative positions. This has been achieved to an accuracy of 30 μm, or 10^{-5} of the size of the mechanical structure. This combination of size and positioning accuracy in unprecedented in high-energy physics experiments.

The disk-shaped forward tracking chambers for the H1 experiment are in our collection on account of their circular symmetry and their *radial* sense wires. But they are also interesting because, in addition to measuring tracks, they serve for the detection of single photons created by transition radiators immediately upstream. For this double function the gas was chosen to contain a sizeable fraction of xenon with its short absorption length for soft X-rays.

The three-dimensional schematic drawing of Fig. 10.11 makes the 48 wedge-shaped drift cells visible. Each one has at its centre the radial plane of the eight radial sense wires, which are interspersed with field wires, and on both sides the cathode planes. So at the outer radius the drift space is larger than at the inner one in the ratio of these radii; and in order to maintain a uniform drift field in the cell, the cathode is divided into axial strips on such radially increasing potentials that the electric field is everywhere the same and orthogonal to the sense-wire plane. Suitable field degraders close the volume of every cell. The drift-time coordinate is in the azimuthal direction, thus appropriate for a measurement of the particle momentum in the axial magnet. (The influence of the magnetic field on the drift directions has been neglected here, cf. Sect. 10.3.2.) The radial track coordinate is obtained by charge division on pairs of wires. Similar chambers had previously been constructed for the CDF experiment at Fermilab.

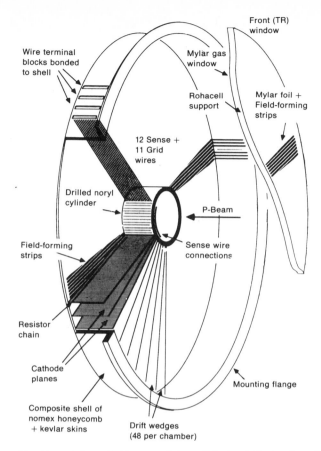

Front (TR) window

Wire terminal blocks bonded to shell

Mylar gas window

Rohacell support

Mylar foil + Field-forming strips

12 Sense + 11 Grid wires

Drilled noryl cylinder

P-Beam

Field-forming strips

Sense wire connections

Resistor chain

Cathode planes

Mounting flange

Composite shell of nomex honeycomb + kevlar skins

Drift wedges (48 per chamber)

Fig. 10.11. Schematic blow-up of the radial-wire drift chamber of the H1 experiment

10.4.3   Type 1 Chambers without Field-Shaping Electrodes

The drift field of a conventional drift chamber is established by conducting surfaces (field wires, metal strips etc.) at suitable potentials. But according to an idea of Allison et al. [ALL 82a], the drift field can also be established by deposition of electrostatic charge on the insulating surfaces that delimit the gas volume. They showed that in a large flat plastic box one can get a very homogeneous field just with one proportional wire in the middle and two small cathode strips along the parallel edges (Fig. 10.12). The deposition of charge is straightforward if the insulating surfaces are backed (on the outside) by grounded plates parallel to the surface. Looking at Fig. 10.13a, the field lines are as indicated at the moment that HV is applied to the sense wire. Any positive charge from the wire amplification will travel along the field lines and will fall onto the insulator and stay there to subsequently change the field configuration.

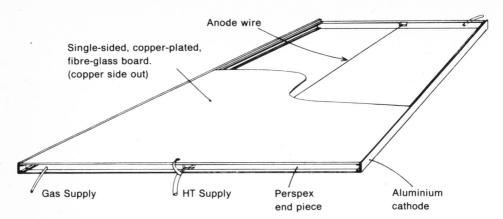

Fig. 10.12. A chamber without field-shaping electrodes

(a)

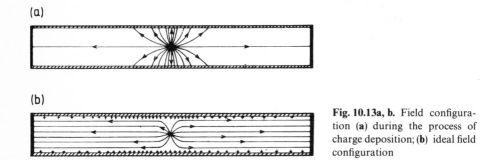

(b)

Fig. 10.13a, b. Field configuration (a) during the process of charge deposition; (b) ideal field configuration

This process continues until there are no more field lines pointing at the insulator, i.e. the ideal field configuration is reached (Fig. 10.13b). As long as there is any charge amplification going on at the wire, the ideal field configuration will automatically be reached. It should be noted that for a thin chamber, the charge on the wire must finally be equal to the charge on the insulating surface. This implies that the gas amplification factor, which is determined by the amount of charge on the wire, cannot be altered without changing the drift field in the volume. The problematic side of this method is in the stability of the field. Although the positive charges are easily brought into place, they are not so easily removed following small fluctuations – these may be due to local discharges, bursts of particles, deposits on the wire, etc., and are probably unavoidable. Allison et al. consider insulators with a non-zero conductivity; these would create a dynamic equilibrium, which would become a function of the rate of charge deposition.

On the other hand the new principle makes it easy to construct very simple drift chambers. Franz and Grupen [FRA 82] have described the properties of a flat, round chamber (1.4 cm × 50 cm diam) with only one short (1.4 cm) wire in the centre. Becker et al. [BEC 82] have reported measurements with long and thin (20 cm × 1.5 cm) plastic tubes, which could be bent. They have also discussed other applications of the new method. Dörr et al. [DOE 85] analyzed the behaviour of a circular chamber operating in the limited streamer-mode.

10.5   Large Cylindrical Drift Chambers of Type 2

In order to span a large cylindrical volume with wires parallel (or approximately parallel) to the axis, there must be strong end plates to hold the wires, which carry the combined pulling force of the wires and keep them in accurate positions. Usually the end plates, together with the outer and inner cylinder mantles, also form the envelope for the gas. The wire feedthroughs that transmit the high voltage and the signals are essential elements of the construction.

10.5.1   Coordinate Measurement along the Axis – Stereo Chambers

Where charge division and time-difference methods were found insufficient, the idea of varying the wire directions throughout the volume has been widely used. The equivalent of a stack of type 1 chambers with differing wire orientations is a (type 2) 'axial stereo chamber'. In a cylindrical chamber with many concentric layers of drift cells, a stereo layer can be formed by rotating the end points on one side of the wire that belong to this layer by the small 'stereo angle' α and by keeping the end points fixed on the other side. The cylindrical surface of this wire layer turns from a cylinder mantle into a hyperboloidal surface. The azimuthal wire position becomes a linear function, the radial wire position a quadratic function of z, to lowest order of αz, counting from the middle of the chamber. This inhomogeneity distorts the electric field to some extent. The achievable precision δz is equal to the measurement accuracy $dr\varphi$ achieved in the azimuthal direction divided by the stereo angle. Hence there is interest in making α as large as the increasing inhomogeneities permit. In practice, stereo angles are built to plus and minus a few degrees, alternating with zero degrees in successive layers.

Obviously a z measurement can also be provided outside the cylindrical wire chamber with a type 1 chamber. Special 'z chambers' have been built for this purpose in some experiments.

10.5.2   Five Representative Chambers with (Approximately) Axial Wires

Table 10.2 contains five chambers selected in the spirit explained at the beginning of this chapter. In quoting performance figures we have distinguished (see footnote 1 of the table) whether they are from early or from later stages of the development of the detector. It is not unusual that the prototype performed better than the full chamber eventually did in the experiment.

When the momentum-measurement accuracy actually achieved in an experiment is compared to that expected from the point-measuring accuracy, the number of points and the magnetic field strength (Sect. 7.5.1), it is often not as good as expected. The overall mechanical accuracy, local and global field distortions, imperfect calibrations all make their contribution. to the momentum-measuring accuracy, which sometimes improves only slowly as the apparatus is better and better understood.

10.5.3   Drift Cells

The minimal wire configuration needed to establish the drift and amplification field around the proportional wire (which is at a positive potential) is a group of field wires at a negative potential surrounding it. For a drift chamber of type 2, this elementary 'cell' is repeated to fill the whole volume. Figure 10.14 shows, as an example, drift cells of the ARGUS chamber; the full chamber consists of 5940 cells of this type. Advantages of this arrangement are the high degree of homogeneity over the full volume, the fine granularity, which allows single-hit electronics to be used, and the full track ionization that is employed. On the other hand, the local field around the wire in one cell is not very homogeneous and therefore requires a very careful calibration of the critical relation between drift time and wire–track distance. In Fig. 10.15 we see the electron drift paths calculated for such a cell type [HAR 84]. They have the form of spirals, created by the *B* field parallel to the wires. The contours of equal arrival times are circles near the sense wire, but they are shaped after the field-wire geometry far from the sense wire. This means in practice that the time–distance relation for tracks is a function of the orientation of the track with respect to the drift cell.

The design of the OPAL drift cell goes in the opposite direction (Fig. 10.16). The basic unit is a wedge-shaped sector that fills 15° of azimuth over the entire radial space so that the chamber consists of 24 of them. Each sector contains one radial plane of sense wires which are interspersed by field wires. The sense wires

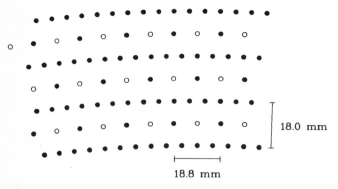

Fig. 10.14. Drift cells of the ARGUS chamber. The sense wire is, typically, at 2.9 kV and the cathode wires are grounded

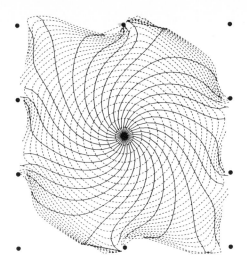

Fig. 10.15. Drift paths and lines of equal arrival times for a drift cell of the ARGUS chamber. The time lines are 15 ns apart

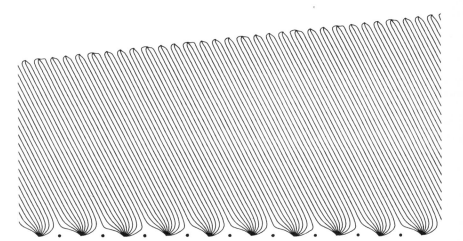

Fig. 10.16. A section of the OPAL drift cell with electron drift lines. Upper plane: cathode wires; lower plane: sense/field wires

are actually staggered to resolve the left–right ambiguity. Each sector is limited and separated from its two neighbours by two common radial planes of cathode wires at such radially increasing potentials as to create a homogeneous drift field orthogonal to the central sense/field-wire plane. The inner surface of the enclosing cylinder carries suitable field degraders. This geometry has also been called '*jet chamber*' geometry. In this design the homogeneity is constructed into each sector, which is relatively large. Close tracks are resolved electronically by using short pulses and digital pulse-shape analysis. The dilemma of the Lorentz angle is diminished by a high gas pressure and a low magnetic field.

Table 10.2. Some large cylindrical drift chambers of type 2

	Mark II	OPAL	ARGUS	SLD	CDF
Name of experiment	Mark II	OPAL	ARGUS	SLD	CDF
Name of chamber	New CDC	Jet-Chamber	Drift-Chamber	CDC	CTC
Method of internal z-measurement	Stereo	Charge division	Stereo	Charge division	Stereo
Reference	[ABR 89]	[FIS 89] [OPA 91]	[ALB 89]	[YOU 86] [SLD 85]	[BED 88] [BER 91]
Geometry					
Outer/inner radius (cm)	152/19	183/25	172/30	96/24	132/31
Length (cm)	230	320–400	200	180	321
No. of wires	36936	3826	30528	37760	36504
No. of sense wires	5832	159	5940	5120	6156
No. of sense wires per cell	6	159	1	8	12; 6
Sense wire diameter (μm)	30	50	30	25	40
Max. no. of measured points per track	72	159	36	80	84
Radial distance between sense wires (mm)	8.3	10	18	5	7
Max. drift distance (cm)	3	3–25	1	3	4
Stereo angle ($^\circ$)	4	0	2–5	3	3
Radial tilt of sense-wire plane ($^\circ$)	0	0	—	0	45

Gas and fields

Gas (percentage concentration)	Ar(89) + CO$_2$(10) + CH$_4$(1)	Ar(88) + CH$_4$(10) + i-C$_4$H$_{10}$(2)	C$_3$H$_8$(97) + CH$_3$OH(3) + H$_2$O(0.2)	CO$_2$(92) + i-C$_4$H$_{10}$(8)	Ar(49.6) + C$_2$H$_6$(49.6) + C$_2$H$_5$OH(0.8)
Gas pressure (bar)	1	4	1	1	1
Electric drift field (kV/cm)	0.7	0.9	1–2 (70%)	1.0	1.35
Magnetic field (T)	0.47	0.4	0.8	0.6	1.5
Drift velocity (cm/µs)	5	0.3	4–5	0.9	1
$\omega\tau$ (approx.)	0.4	0.35	< 1	0.05	
Performance (1)					
Point-measuring accuracy $\sigma_{r\varphi}$ (mm)	0.17 (c)	0.12 (b) 0.135 (c)	0.19 (c)	0.055 (b)	0.2 (c)
Charge division σ_z (mm)	—	30–40 (b) 60 (c)	—	9 (b)	—
Double-track resolution: $\Delta_{r\varphi}$ (mm)	3.8 (c)	2.5 (c)	9 (a)	1 (a)	5 (c)
Accuracy of ionization measurement (%)	7.2 (c)	3.5 (b) 3.8 (c)	4.5–5.5 (c)	—	—
Accuracy of momentum measurement (2) $\delta p_t/p_t^2$ (GeV/c)$^{-1}$	0.0046 (c)	0.0012 (b) 0.0022 (c)	0.009 (c)	0.0015 (a)	0.001 (c) (3) 0.002 (c) (2)

1. The performance figures are based on (a) calculation and laboratory tests; (b) prototype and test beam measurement; (c) measurements in the running experiment.
2. High p_t, full track length, chamber alone.
3. Using the constraint of the primary vertex.

The cell design of the new MARK II chamber follows a similar line. Radial sense/field-wire planes with radial cathode planes in between, but organized in 132 layers, 7.5 cm in radius. This is done in order to be able to create stereo angles in consecutive layers. There are 972 cells in the chamber. Figure 10.17 displays a cell with its electron drift trajectories.

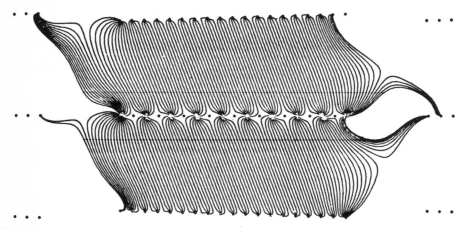

Fig. 10.17. Mark II drift cell with electron trajectories

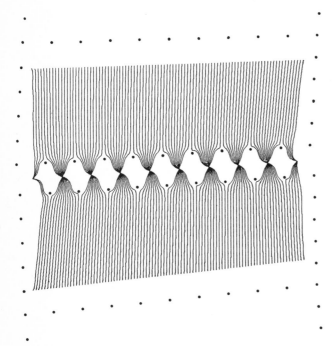

Fig. 10.18. SLD drift cell with electron drift lines

The small cell of the SLD drift chamber (Fig. 10.18) is designed with the purpose of better separating the amplification region from the drift region by displacing the field wires from their position between the sense wires to a position surrounding them. This is done to create a uniform drift region for the slow gas employed. Here the good diffusion properties of carbon dioxide have been combined with its low drift velocity at high electric field. Thus the Lorentz angle is kept near 6° at the field of 0.6 T. On the other side of the coin, this sets severe

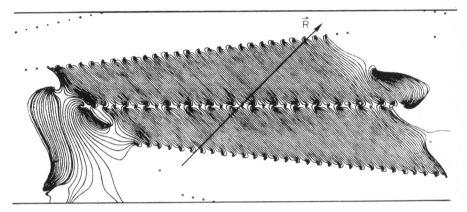

Fig. 10.19. A drift cell of the CDF Central Tracking Chamber, with electron drift lines

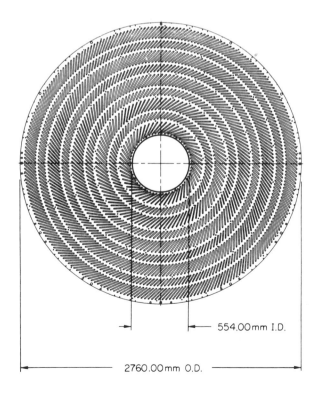

554.00 mm I.D.

2760.00 mm O.D.

Fig. 10.20. The end plate of the CDF Central Tracking Chamber with its drift cells, which are inclined with respect to the radial direction

requirements on the electric field uniformity, since the drift velocity under these conditions is proportional to the electric field ('unsaturated gas'). There are 640 drift cells in the detector. A very favourable point-measuring accuracy of 55 μm was obtained with the prototype. This requires that the wire displacements caused by mechanical inaccuracy, gravity and electrostatic forces as well as by the stereo arrangement are fully controlled to this precision.

The drift cells of the central tracking chamber of the CDF detector (Fig. 10.19) are inclined by 45° with respect to the radial direction in order to take into account the Lorentz angle at the employed magnetic field of 1.5 T. This solution also offers advantages for triggering and for pattern recognition. The arrangement of the drift cells, of which there are 660, in the end plate can be seen in Fig. 10.20. The cells are divided radially into nine superlayers which alternate between wires in the axial direction and wires at a $\pm 3°$ stereo angle. Tilted drift cells are also employed for the drift chambers of the H1 and ZEUS experiments at the HERA e⁻p collider.

10.5.4 The UA1 Central Drift Chamber

The detector fills a cylinder, 6 m long and 2.2 m in diameter, which contains the collider-beam pipe on its axis. Equally horizontal, but at right angles to the beam, is the magnetic field of 0.7 T; see Fig. 10.21. Along its length, the cylinder is divided into three cylindrical sections, each 2 m long: the central one and the two forward ones. The wires are all parallel to the magnetic field and organized in planes according to the following scheme: the anode planes have alternating sense (35 μm diam) and field (100 μm diam) wires every 5 mm; the cathode planes with 120 μm wires, also every 5 mm, are parallel to the anode planes, leaving drift spaces of 18 cm between them. A peculiarity of the UA1 detector is the orientation of these planes – they are vertical in the central cylinder part and horizontal in the forward parts. This arrangement is a consequence of the horizontal magnetic field; the solution optimizes the momentum-measuring accuracy, the track curvature being measured by the drift time. The charge-division technique is used for the coordinate along the wire direction. Filled with a gas mixture of 40% Ar and 60% C_4H_{10} at ambient pressure, the drifting electrons reach a velocity of 5.3 cm/μs under a Lorentz angle of 23° in their drift field of 1500 V/cm. The total number of wires is 22 800, of which 6110 are sense wires. A stiff vertical track is measured on 100 points, a stiff track near the forward directions on up to 180 points.

Performance figures were measured in the experiment as follows [NOR 90]: The average point measuring accuracy in the direction of the drift amounted to 350 μm. The relative accuracy of momentum measurement, which depends very much on the track orientation, was typically $\delta p/p = (0.01 \text{ GeV}/c^{-1})p$ for isolated tracks in a favourable direction (tracks that emanate from the interaction region under a polar angle of 90° and parallel to the magnetic field can

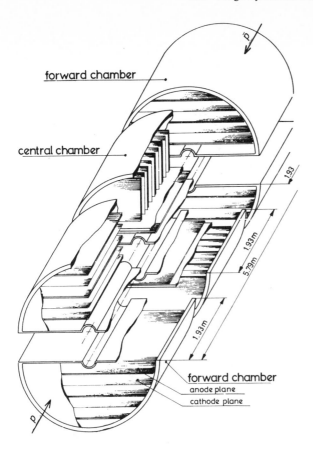

Fig. 10.21. Geometry of the UA1 central detector

obviously not get their curvature measured). The electronic two-track resolution limit on a wire was about 5 mm. The coordinate measurement accuracy along the wires was 3.5% r.m.s. of the wire length after the gain reduction by a factor of three that was necessary when the detector had to work in the high-luminosity environment; before this change, 2% had been reached. These numbers illustrate the conflicting requirements of having at the same time the smallest possible space charge in the drift region (small wire amplification) and the best possible charge-division accuracy (large amplification).

Ionization measurements of tracks in this complex detector were limited in accuracy and were not employed for particle identification.

In Fig. 10.22 we reproduce an event as reconstructed by the computer; the high number of measured points along each track make a clear picture even for the complicated events created in high-energy hadron collisions – in fact this picture is famous because it contains the first Z^0 particle ever observed. The two arrows mark the electron and the positron into which it decays.

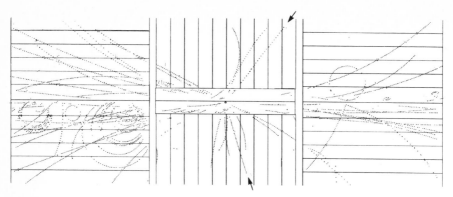

Fig. 10.22. Computer-reconstructed event in the UA1 central detector – the first Z_0 particle ever observed is seen to decay into an e^+ and an e^-. Only one in two wire signals is displayed

10.6 Small Cylindrical Drift Chambers of Type 2 for Colliders (Vertex Chambers)

Vertex detectors are needed for the measurement of the (primary) interaction vertex or any secondary vertices that may occur in an event from the decay of short-lived particles. The charm, bottom and τ particles have lifetimes of the order of 10^{-13} to 10^{-12} s, so the tracks from the secondary vertices will miss the pimary vertex by correspondingly small distances, which are of the order of 30 to 300 μm. Ideally, vertex chambers should have even better measurement accuracies.

This is the field where the fine-grained solid-state particle detectors with their superb spatial resolution are often a better choice than gas drift chambers. These have in their favour the simplicity of a proven technique and a somewhat better resistance against high-radiation background. Also they are more easily extended over large sensitive areas. Being closer to the interaction point they can often be built smaller and with shorter wires than the large axial chambers, so that the wire position can be extremely well defined. This is a prerequisite for the high measurement accuracy achieved in the vertex chambers. When measuring the position of a vertex inside the vacuum tube, multiple scattering in the wall of the tube is often a limiting factor, which depends on the momentum of the extrapolated particle as well as on the tube radius and the thickness of the wall.

In order to reach the best possible point-measuring accuracy and double-track resolution, the sense wires have to be as close as possible to each other (cf. the discussion in Sect. 10.8). The limit imposed by electrostatic stability depends on the wire length and the detailed electrostatic pattern, and it becomes more severe at higher gas pressures because of the corresponding higher voltages.

To find the correct gas, one has to balance the consequences of a slow 'cool' gas (see Sects. 2.2.4, 11.1) – slower electronics, high drift field with low diffusion, sensitivity to field inhomogeneity through unsaturated drift velocity – with the consequences of a fast 'hot' gas – faster electronics, lower sensitivity to field, temperature and pressure variations.

These questions have been studied in detail for the more recent vertex detectors. The chambers listed in Table 10.3 are again intended to represent the different choices in the field. They are all axial drift chambers of type 2. The ones that reach the highest precision are operated with slow cool gas at an elevated pressure. Hayes has compared vertex drift chambers for the LEP and SLC detectors [HAY 88]. The proceeding of a workshop dedicated mostly to advanced vertex detectors [VIL 86] offers some more material that describes the state of the art in 1986. A short review on straw chambers was given by Toki [TOK 90].

Among the more recent vertex detectors we find the drift chambers with the best measuring accuracy that are integrated into particle experiments. Higher accuracy has only been achieved with more specialized devices. These will be briefly discussed in Sect. 10.8.

10.6.1 Six Representative Chambers

The JDV of the UA2 experiment had to work in severe background conditions at the p̄p collider. The drift cell, shown in Fig. 10.23, has a geometry similar to that of the OPAL chamber; it operates in a fast gas at atmospheric pressure, and at the saturated drift velocity. The point-measuring accuracy achieved in the test beam (150 µm) deteriorated a lot (300 µm) in the large-multiplicity events of the real experiment.

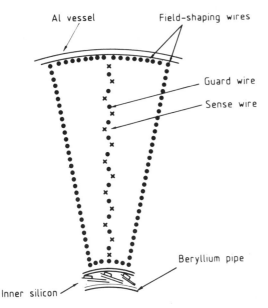

Fig. 10.23. Drift cell of the UA2 JVD chamber

Table 10.3. Some small drift chambers of type 2 (vertex chambers)

Name of experiment	UA2	MARK J	MARK II	OPAL	MAC	ALEPH
Name of chamber	JVD	TEC	DCVD	VCH	straw-VC	ITC
Method of z measurement	Charge division	—	—	Time difference + stereo	—	Time difference
Reference	[BOS 89]	[AND 86] [AND 88a] [AND 88b]	[ALE 89] [DUR 90]	[CAR 90] [OPA 91]	[ASH 87] [NEL 87]	[DEL 90] [BAR 89]
Geometry						
Outer/inner radius (cm)	13/4	12/5	17/5	22/10	9/3.5	28/13
Length (cm)	100	58	55	100	43	200
No. of wires	1232	2750	1800		324	5136
No. of sense wires	208	168	190	648	324	960
No. of sense wires per cell (6)	13	14	19	12 (6)	1	1
Sense-wire diameter (µm)	25	20	20	20	30	30
No. of measured points per track	13	14	19	18	6	8
Radial distance between sense wires (mm) (6)	6.4	2.4	2.9	5.0 (5.3)	6–10	11–19
Max. drift distance (cm)	2.5	1.3–2.6	3	0.8–1.9	0.35	0.47–0.65
Stereo angle (°)	0	0	0	4	0	0
Radial tilt of sense wire plane (°)	0	±4	15	0	—	—

Gas and fields

Gas (percentage concentrations)	Ar(60) + C₂H₆(40)	CO₂(80) + i-C₄H₁₀(20)	CO₂(92) + C₂H₆(8)	Ar(88) + CH₄(10) + i-C₄H₁₀(2)	Ar(49.5) + CO₂(49.5) + CH₄(1)	Ar(50) + C₂H₆(50)
Gas pressure (bar)	1	1.9	2	4	4	1
Electric drift field (kV/cm)	1	1.6	1.5	2.7	2–7 (70%)	3–5
Magnetic field (T)	0	0	0.47	0.4	0.6	1.5
Drift velocity (cm/µs)	5.2	0.7	0.6	4.0	1–3 (70%)	5.0
Approx. value of $\omega\tau$	0	0	0.02	0.1	small	small
Performance (1)						
Point measuring accuracy $\sigma_{r\varphi}$ (µm)	150 (b) / 180 (c) (3) / 300 (c) (5)	35 (b) / 40 (c)	30 (b) / 50 (c)	50 (b) / 55 (c)	70 (c) (4) / 45 (c) (3)	100 (b) / 150 (c)
Charge division or time difference accuracy σ_z (mm)	10 (b) / 25 (c) (3) / 35 (c) (5)	—	—	45 (c)	—	30 (b)
Double-track resolution $\Delta_{r\varphi}$ (mm)	2 (b)	0.3 (b) (7)	0.5 (c) (8)	2.0 (c)	6.9	11 (a)
Impact-parameter accuracy (2) σ_t (µm)	400 (c)	80 (b)	30 (c)	30 (c)	87 (c) (3) / 200 (c) (4)	130 (c)

1. The performance figures are based on (a) calculation and laboratory tests; (b) prototype, cosmic-ray or test-beam measurements; (c) measurements in the running experiment.
2. Vertex chamber plus main tracking chamber – high-momentum tracks.
3. Isolated tracks.
4. Multihadron events, including multiple scattering.
5. Multihadron events, average multiplicity 31 to 35.
6. Numbers in brackets refer to the OPAL stereo cell.
7. Track separation at which 50% of the tracks are correctly resolved.
8. Track separation at which approx. 100% of the tracks are resolved.

The vertex detector of the MARK-J experiment is built according to the principle of the 'time expansion chamber', or 'TEC'. Introduced by Walenta [WAL 79], it is characterized by a wire amplification region strictly separated from the drift region by a dense wire grid or wire mesh. The drift cell of the MARK-J TEC is shown in Fig. 10.24 and represents one twelfth of the entire chamber. One recognizes the two wedge-shaped drift spaces on either side of the 4 mm wide amplification region; they are not symmetrical because the sense/field wire planes are inclined alternatively by $\pm 4°$ in the service of pattern recognition. The gas is a slow and cool mixture of 80% CO_2 and 20% i-C_4H_{10}. The sense wires are read out by 100 MHz analogue-to-digital converters (flash-ADCs). An average point-measuring accuracy of 40 µm was determined with Bhabha events.

The MARK II vertex chamber was designed for the SLAC Linear Collider with its small beam pipe, which has a radius of 2.5 cm. The inner wall of the chamber and the beam pipe represent a total of 0.6% of a radiation length, thus contributing only 45 µm GeV/c to the Coulomb scattering limit S_c discussed in Sect. 7.4.2.

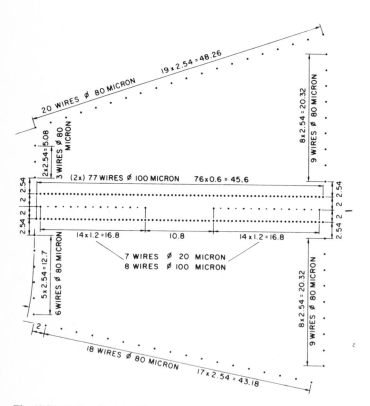

Fig. 10.24. Drift cell of the TEC of the MARK J experiment

The wires run parallel to the axis, and they are organized in cells that each span the full radial range between 5 and 17 cm (see Fig. 10.25). In the middle of each cell is the sense/field-wire plane, inclined by an azimuthal tilt angle of 15°; it contains 41 field wires, spaced out by 2.9 mm, and 40 sense wires between them. Two planes of grid wires, 1.8 mm on either side, separate the amplification region from the drift region and provide a certain amount of focussing for the electron drift paths, similar to what happens in the 'TEC'. The cathode planes with 59 wires, spaced out by 2 mm, bisect the angle between the sense/field-wire planes of two adjacent cells. Each drift cell is thus limited by the two cathode planes at radially increasing potentials and additional potential wires at the cylindrical surfaces. They complete the field cage which provides a uniform field despite the wedge of drift space it encloses.

The OPAL vertex chamber is constructed for LEP whose beam pipe is much larger than the one at the SLC. With its radius of 7.8 cm and its thickness of 0.86% radiation lengths the OPAL pipe contributes a Coulomb scattering limit $S_c = 110 \text{ GeV}/c$ to the impact-parameter resolution.

The wire pattern is shown in Fig. 10.26. The chamber is 100 cm long. It has one axial and one stereo layer at 20°. The axial layer between radii of 9.4 and 17.1 cm is of the 'jet chamber' type and has 36 radial planes of sense/field wires, each containing 12 sense wires spaced out by 5 mm, and the field wires between them. The left–right ambiguity, which does not occur when the sense-wire plane is tilted with respect to the radial direction, is resolved by staggering alternative sense wires by $\pm 41\ \mu\text{m}$ which increases to $\pm 80\ \mu\text{m}$ in the middle of the wires, owing to their electrostatic forces. The 10° in azimuth between two neighbouring sense/field-wire planes are bisected by the radial-wire-cathode planes, so that

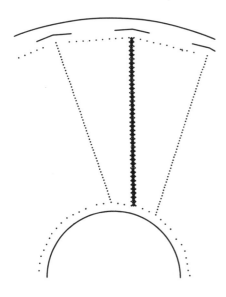

Fig. 10.25. Drift cell of the VCDV of the MARK II experiment

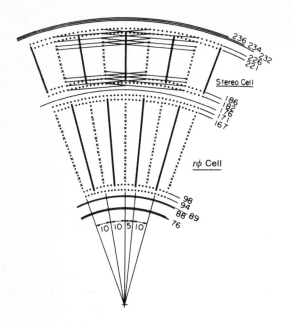

Fig. 10.26. Wire pattern of the OPAL vertex chamber

the maximal drift distance varies with radius from 9 to 15 mm. There is a double layer of potential wires to close the field cage of each cell. This produces a field that is uniform and at right angles to the sense/field-wire plane by grading the cathode voltages appropriately.

The stereo layer occupies the radial space between 18.2 and 22.6 cm; it is organized in the same fashion, but there are only half as many sense wires in this layer. The maximal drift distances are now between 1.6 and 1.9 cm.

Each of the 648 sense wires is read out at both ends and is timed with a constant-fraction discriminator and a time-to-digital converter (TDC). Thus, there is a coordinate measurement along the wire direction from the time difference, in addition to the azimuthal coordinate determined from the average of the two times.

The OPAL vertex detector is operated with the slow mixture CO_2 (92%) + i-C_4H_{10} (8%) at 2.5 bar. The space resolution achieved with the drift-time measurement is depicted in Fig. 10.27 as as function of the drift distance, both for the OPAL and the MARK II vertex detector prototypes. The resolution has reached the remarkable range between 20 and 40 μm for drift distances up to 3 cm. With fast gas and the same electronics, the resolutions were two or three times worse [HAY 88].

The drift chambers of the last two examples in our list are again constructed from cells that are built around the individual sense wires. The vertex chamber of the MAC detector consists of many thin tubes of aluminized mylar foil with the proportional wire in the centre. These 'straws' provide exact cylindrical symmetry and a high degree of electrical shielding between sense wires. Their

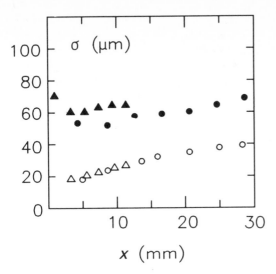

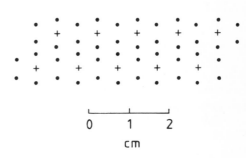

Fig. 10.27. Measured prototype vertex chamber resolutions σ as a function of the drift distance x after [HAY 86]. *Triangles*: OPAL; *circles*: MARK II. Open symbols: slow gas (CO_2 (80%) + i-C_4H_{10}(20%) at 2.5 bar for OPAL, CO_2 (92%) + i-C_4H_{10} (8%) at 3 bar for MARK II); *full symbols*: fast gas Ar (50%) + C_2H_4 (50%) at 3 bar for both chambers

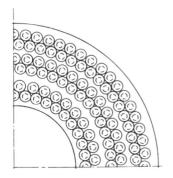

Fig. 10.28. One quarter of the end plate of the MAC straw vertex chamber (for dimensions, consult Table 10.3)

Fig. 10.29. Drift cells of the ALEPH Inner Track Chamber

inner diameter is 6.9 mm, and the wall thickness 50 μm. In Fig. 10.28 we see the layout of the vertex chamber end plate.

The cylindrical geometry produces a drift field that increases with the inverse distance from the wire. The drift velocity therefore also changes, depending on the gas used. In Table 10.3 we have indicated ranges for the field and the velocity that are valid in the outer 70% of the drift region. In a similar design for the MARK III experiment [ADL 89], the gas mixture was argon–ethane with a more uniform, and higher, velocity; the achieved point-measuring accuracy was comparable. Toki has recently reviewed straw chambers [TOK 90].

The purpose of the ALEPH-ITC is the production of an early trigger signal as well as precise tracking (a solid-state microstrip detector is mounted in addition). The drift cell is a group of six field wires around each proportional wire, in a flat hexagonal arrangement, which is depicted in Fig. 10.29. Each hexagon is characterized by two shape parameters, the diameters in the radial and in the circumferential directions; both vary with the wire layer. A time-difference measurement is the basis for a determination of the axial coordinate. It takes 0.5 μs for a trigger decision in r–ϕ and 2.5 μs in r–φ–z.

10.7   Drift Chambers of Type 3

An ionization track created in the sensitive gas volume of a type 3 chamber drifts to one of the end faces that delimit the volume. It is there that the amplification and signal generation take place. Large chambers have drift lengths of one or several metres. If a magnetic field is present in a TPC, it is parallel to the drift field, and in this orientation it can greatly improve the transverse diffusion of the travelling electrons, on account of the magnetic anisotropy of diffusion; see Sect. 2.2.6.

With the length of the drift path we have a correspondingly high drift time – often many tens of microseconds for electrons and several seconds for ions. This makes long type 3 chambers unsuited for measuring events at very high rates because the space charge from previous events in the drift space will distort the tracks. Ion shutters ('gating grids' – see Chap. 8) are suitable tools for reducing the amount of ion current that flows back into the drift volume from the amplification region.

The length of the drift path also sets severe requirements on the uniformity of the electric and magnetic fields: a non-uniformity of one per cent may cause transverse displacements of the order of one per cent of the drift length.

Finally we mention gas purity. Electrons may be attached to gas components that are present even in extremely low concentrations, the result being a signal loss that increases with the drift path. We have described electron attachment in Sect. 2.2.7.

Still a TPC can be a very powerful universal track detector. If all the above-mentioned problems have been overcome, one profits from its advantages: with its wires strung on the inner surface of the end-plates rather than through the volume the construction can be modular and technically very convenient. For every track segment one achieves an intrinsically three-dimensional coordinate determination – two coordinates being given by the position in the end plate of the pick-up electrodes that collect the segment, the third being measured through the drift time.

Furthermore, with a TPC one can totally avoid the dilemma of the Lorentz angle: if one puts $\boldsymbol{B}$ and $\boldsymbol{E}$ parallel in the drift volume, there is no shift of the drift direction, and the magnetic field may be increased as far as technically feasible, the gas pressure can be kept low, and the diffusion may be dramatically reduced; cf. the discussion in Chap. 11.

10.7.1 Double-Track Resolution in TPCs

The measurement of a track produces a dead region around this track where the chamber is insensitive to the measurement of other tracks. In a TPC with cathode pads, this dead region is a tube around the track, whereas in axial wire chambers and in TPCs with wire measurement it is a flat region that extends along the full length of the wire.

Concerning track measurements with pads, overlap occurs on the pad when it is still occupied with the pulse of a first track at a time when the pulse of a second track begins. The pulse length T is given by the intrinsic electronic pulse length t_e plus the drift-time difference t_d of the electrons from one track that reach the pad at different times, because of a track inclination θ and the size h of the pad. With reference to Fig. 10.30 we have an overlap region Δz in the drift direction, equal to

$$\Delta z = ut_e + ut_d = ut_e + h/\tan\theta \,, \tag{10.1}$$

where u is the drift velocity and θ the angle between the track and the normal to the pad (typically the polar angle). In Table 10.4 double-track resolutions are quoted for two polar angles.

In the azimuthal direction, two tracks will have to be 2 or 3 pad widths apart in order to be separable; a practical average may be 2.5 pad widths.

10.7.2 Five Representative TPCs

There are four large cylindrical TPCs that have been built around the interaction points of e^+e^- colliders (PEP-4, TOPAZ, ALEPH, DELPHI); see Table

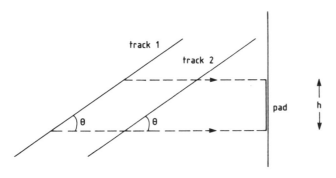

Fig. 10.30. Two inclined tracks; the first electrons from track 1 arrive on the pad at the same time as the last electrons from track 2

Table 10.4. Some TPCs built around colliders

Name of experiment Reference	PEP 4 [LYN 87] [AVE 89]	TOPAZ [KAM 86] [SHI 88]	ALEPH [DEC 90] [ATW 91]	DELPHI [BRA 89] [AAR 90] [SIE 90]	CDF (1) [SNI 88]
Geometry					
Outer/inner radius (cm)	100/20	127/30	180/31	116/32	21/7
Drift length (cm)	2×100	2×122	2×220	2×134	2×15
Number of sense wires	2196	2800	6336	1152	384
Sense-wire spacing (mm)	4	4	4	4	6.3
Number of pads	13824	8192	41004	20160	384
Pad dimensions ($r \times r\varphi$) (mm^2)	7×7.5	12×10 (4)	30×6	8×7	41×14
Max. no. of measured points per outgoing track					
On pads	15	10	21	16	3
On wires	183	175	338	192	24

Gas and fields	Ar(80) + CH$_4$(20)	Ar(90) + CH$_4$(10)	Ar(90) + CH$_4$(10)	Ar (80) + CH$_4$(20)	Ar(50) + C$_2$H$_6$(50)
Gas (percentage fractions)					
Gas pressure (bar)	8.5	3.5	1	1	1
Electric drift field (kV/cm)	0.55	0.35	0.12	0.15	0.25
Magnetic field (T)	1.32	1.0	1.5	1.2	1.5
Drift velocity (cm/μs)	5	5.3	5.0	6.7	4.3
Approx. value of $\omega\tau$	1.5	4.9	7	5.4	3.5
Performance (2)					
Point-measurement error radial tracks					
On pads: $\sigma_{r\varphi}$ (mm)	0.15 (c)	0.20 (c)	0.17 (c)	0.18 (c)	0.4
σ_z (mm) at 90°/30°	0.16/(c)	0.3/1.0 (c)	0.74/2 (c)	0.9 (c)	—
On wires: σ_z (mm)		0.3/1.0(c)	2 (b)	0.9 (c)	0.4–1
Double-track resolution					
On pads: $\Delta_{r\varphi}$ (cm)	2.4 (c)	2.5 (a)	1.5 (b)	1.5 (c)	—
Δ_z (cm) at 90°/30°		3/3 (a)	2/6 (b)	1.5 (c)	—
Δ_x (mrad)	25–40 (c)		35 (c)		
On wires: Δ_z (cm)		3 (a)	2 (b)	1.5 (c)	0.5–1.5
Accuracy of momentum measurement (3) $\delta p_t/p_t^2$ (GeV/c)$^{-1}$	0.009 (c)	0.015 (c)	0.0012 (c)	0.0015 (c)	—
Accuracy of momentum measurement (5) $\delta p_t/p_t^2$ (GeV/c)$^{-1}$	0.0065 (c)		0.0008 (c)		—

1. Numbers are given for each one of the 8 TPCs ('modules'), each having two halves with opposite drift directions.
2. The performance figures are based on (a) calculation and laboratory tests; (b) prototype and test-beam measurements; (c) measurements in the running experiment
3. High p_t, full track length, TPC alone
4. Zig-zag shape along $r\varphi$, depth ± 2 mm.
5. High p_t, full track length, TPC plus other detectors.

10.4. They are all very similar because their construction followed closely the principles established by the pioneers in this field, the designers of the PEP-4 TPC. These chambers are cylinders around the colliding beams; the drift direction is parallel to the magnetic field and to the direction of the incoming particles. The drift volume is divided in the middle by the high-voltage membrane which is set at a negative potential so that the circular end plates with their detection electrodes can be kept at earth potential (Fig. 10.31). The disposition of the various layers of electrodes in the end plates is seen in Fig. 10.32a, and Fig. 10.32b shows the arrangement of the pads in the cathode plane in the example of the ALEPH TPC. The inner and outer cylinder mantles that delimit the gas volume also carry the high-voltage electrode rings, which shape the axial, homogeneous electric drift field.

The four large TPCs differ among each other in two respects: the PEP-4 and TOPAZ TPCs are constructed to work at higher gas pressure and with straight pad rows; the ALEPH and DELPHI TPCs have ambient pressure and circular pad rows. The gas pressure is an important choice for any drift chamber (see the discussion in Sect. 11.5), but for TPCs one must consider in addition the magnetic anisotropy of diffusion – the transverse width of the diffusion cloud increases with the pressure. The circular pad rows in the two TPCs at CERN are an improvement over the (earlier) straight rows, since for stiff radial tracks they do not cause the measurement error that was called the 'angular pad effect' in Sect. 6.2.

Whereas the width of the pad determines the number of electronic channels and hence the cost of the project, the length of the pads (extent in the radial direction) is chosen as a balance between the largest possible signal and a loss of accuracy of lower-momentum tracks, owing to the angular pad effect (cf. Sect. 6.3.3). Also, long pads deteriorate the two-track separation for steep tracks (Sect. 10.7.1). The long pads in the ALEPH TPC reflect a choice in favour of a larger signal.

The problem of track overlap may be quantified as a loss of tracks rather than points specifying the minimum opening angle Δ_α measured between two

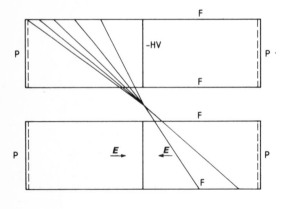

Fig. 10.31. Geometry of the Berkeley TPC and the three TPCs derived from it. F = field cage, P = proportional wire chambers, E = electric field

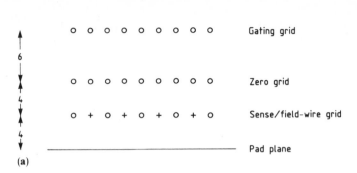

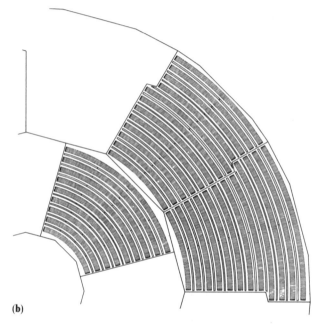

Fig. 10.32a, b. Disposition of the electrodes on the end plate of the ALEPH TPC (distances in mm).
(a) Wire grids; (b) pad plane

tracks at the vertex, below which losses are observed. The minimum opening
angle is smaller for pairs with opposite curvature because the magnetic field
bends them apart. Such loss studies have been performed for the PEP-4 and the
ALEPH TPCs and the corresponding values of Δ_α are also listed in Table 10.4.
They represent averages over the full solid angle.

In Table 10.4, besides the four large TPCs mentioned above, we have
included one other TPC built around a collider, the CDF vertex TPC system,
which consists of eight double chambers near the beam pipe and predominantly
serves the purpose of a vertex detector.

10.7.3 A Type 3 Chamber with a Radial Drift Field

The central tracking device for the ASTERIX experiment at CERN [AHM 90] is a cylindrical volume around a gas target. The inner cylinder mantle is a thin aluminized foil with a diameter of 16 cm, the outer mantle has a diameter of 30 cm and carries the detection electrodes: 90 axial sense wires and 270 field wires backed by helical cathode strips. The inner mantle is at a potential of − 10 kV; the circular end faces (1 m apart) incorporate field degraders down to the earth potential of the outer cylinder.

With the radial drift field at right angles to the axial magnetic field, which amounts to 0.8 T, the drift paths for the electrons take on the curved forms visible in Fig. 10.33; this gave rise to the name *'spiral projection chamber'* for this device [GAS 81]. Filled with a gas mixture of argon (50%) and ethane (50%), the chamber was primarily used for the detection of X-rays in the keV range, but also for the measurement of particle tracks. A stiff track crosses the drift cells that belong to 5 different sense wires, thus allowing 5 track coordinates to be measured. The number of measured points depends very much on the track curvature.

For the reconstruction of the primary vertex point, an accuracy was achieved which amounted to $\sigma = 0.4$ mm transverse to the beam and $\sigma = 2.1$ mm in the longitudinal direction.

10.7.4 A TPC for Heavy-Ion Experiments

The interaction of a heavy ion of several hundred GeV per nucleon with a fixed target often produces hundreds of secondary particles. TPCs have been built to study such high-multiplicity events downstream of a fixed target, and we discuss

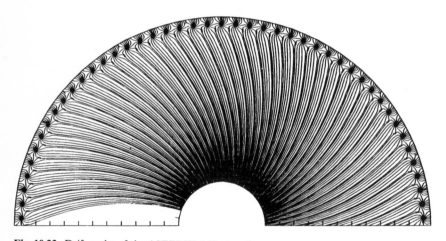

Fig. 10.33. Drift paths of the ASTERIX drift chamber

here the Brookhaven chamber as an example [ETK 89]. Like the first TPC, it operates in a magnetic field (0.5 T) parallel to the electric drift field (330 V/cm). However, the electronic signals are derived not from cathode pads but from short (≈ 1 cm) anode wires, which are spaced out by 2.54 mm along several parallel straight rows.

There are three thin-walled TPC modules, one behind the other, which are separate containers of the gas, a mixture of argon (79%), isobutane (16%) and dimethoxymethane (5%) at ambient pressure. In each module the sensitive surface is horizontal, and the vertical y coordinate is measured using the drift time. The x coordinate along each anode row is measured by the wire addresses and a suitable interpolation, but only the fact of a hit above some discriminator threshold is recorded. There are twelve anode rows per module, one every 3.8 cm along the z direction of the incoming beam, and altogether 3072 channels per module.

For the high-multiplicity events a good two-particle separation capability had to be the essential goal of the construction and of the pattern-recognition programme. The measured result was that two tracks could be separated in 50% of the cases if they lay inside the same area that measured 5×5 mm^2 in x–y. The effective r.m.s. point-measuring accuracy amounted to $\sigma_x = 0.9$ mm and $\sigma_y = 0.75$ mm for secondary tracks from Si interaction, whereas isolated beam tracks could be measured much better. The drift velocity was 2.3 cm/μs.

The beam projectiles are fully ionized, and since the gas ionization that they create is proportional to the square of their electric charge, there is a beam-rate limitation in the TPC owing to the build-up of primary positive charge. The authors observed that an intensity of 2×10^4 oxygen nuclei per AGS spill of 1 s distorted the tracks to the extent that their apparent sagitta was 0.5 mm in y–z.

In Fig. 10.34 we see the x–y projection of an event with 77 tracks pointing to the vertex. Whereas it must be very difficult to measure the track reconstruction efficiency, it is impressive to what extent the drift-chamber technique is capable of reconstructing such complex events.

10.7.5   A Type 3 Chamber as External Particle Identifier

A drift chamber specialized in particle identification, ISIS2, was built by Allison et al. [ALL 84] as part of the European Hybrid Spectrometer. Located behind a fixed target at the SPS machine at CERN, this spectrometer contained two magnets and numerous chambers for the measurement of particle tracks and momenta. The purpose of ISIS2 was to precisely record the ionization of the particle tracks, whereas their coordinates would only have to be measured with moderate accuracy. A schematic diagram of this huge and almost empty box is seen in Fig. 10.35. The useful volume is 4 m high, 2 m wide and 5.12 m long; it is divided into two drift spaces by a single horizontal wire plane of alternate anodes and cathodes at half the chamber height. The particles are supposed to enter the chamber through the front window, which is part of the field cage that

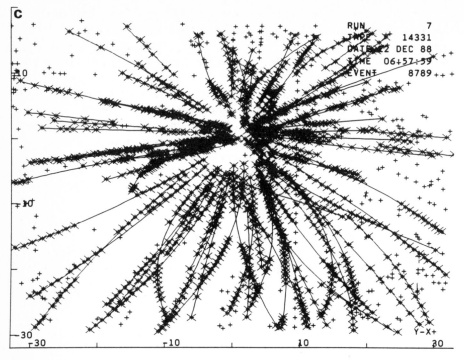

Fig. 10.34. Projection along the beam of a heavy-ion event (silicon beam on gold target) reconstructed in the Brookhaven TPC [ETK 89]

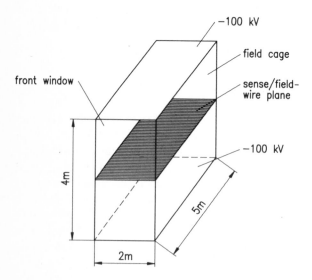

Fig. 10.35. Geometry of the ISIS2 chamber

shapes the vertical electric field of 500 V/cm. There is no magnetic field. The ionization electrons from each track drift in the uniform field and are amplified on 640 proportional wires connected in pairs to 320 channels of multihit electronics. The chamber is filled with a gas mixture of 80% Ar and 20% CO_2 at ambient pressure. The performance of this type 3 chamber was characterized in Chap. 9 in the context of particle identification.

10.7.6 A TPC for Muon-Decay Measurements

In a search for muon–electron conversion, the TRIUMF group employed a TPC which was to measure and identify low-momentum electrons and positrons in the region of 100 MeV/c [BRY 83, HAR 84]. The apparatus surrounded a target foil in which muons were degraded and stopped. The TPC was shaped as a hexagonal cylinder as shown in Fig. 10.36, approximately 70 cm long and equally large in diameter, with a central high-voltage plate; the two end plates carried the readout wires and were at earth potential. The drift field E was 250 V/cm to yield a saturated drift velocity of 7 cm/μs in an argon–methane gas mixture (80/20). A magnetic field of 0.9 T was oriented parallel to E. In order to

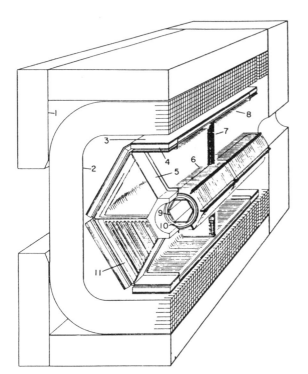

Fig. 10.36. A perspective view of the TRIUMF TPC [BRY 83]. (1) Magnet iron; (2) coil; (3) outer trigger scintillators; (4) outer trigger proportional counters; (5) end-cap support frame; (6, 8) field cage wires; (7) central high-voltage plane; (9) inner trigger scintillators; (10) inner trigger proportional wire chamber; (11) TPC end-cap proportional wire modules

protect the gas volume as much as possible from positive ions produced at the anode wires, a double layer of a monopolar gating grid (see Sect. 8.2.1) was mounted at the entrance of the amplification region. It was kept 'closed' until a trigger signal arrived.

10.8 Chambers with Extreme Accuracy

The accuracy of coordinate measurement is the crucial point of most drift chambers. As we have discussed throughout this book, the essential parameters are those itemized in Table 10.5. The choices open to the designer that correspond to these parameters are quoted in the right-hand column of the table. We notice that some of them represent conflicting requirements. For example, slow gas is not saturated at practical electric fields, and reduced sense-wire distances cause larger wire bowing owing to electrostatic forces that add to the gravitational sag. There are of course all the other important boundary conditions of the particle experiment – interaction rate, space available, radiation material

Table 10.5. Factors relevant for accuracy and corresponding choices of design parameters

Diffusion:	Gas with low diffusion
	Small drift length
	Elevated gas pressure
	Magnetic field
Drift-path variations:	Reduced sense-wire distance
	Focussing of drift trajectories
	Drift-path restrictions
Ionization clustering:	Elevated gas pressure
Time measurement:	Fast electronics
	Slow gaas
	Optimal estimator
Avalanche localization:	Increased number of channels
	Narrow avalanche
	Accurate pulse-height measurement
	Optimal estimator
Uniformity of drift velocity:	Homogeneous electric field
	No free charges in drift volume
	Small field-wire distances
	Mechanical accuracy
	Saturated drift velocity
Knowledge of wire positions:	Mechanical accuracy
	Short wires
	Controlled wire sag

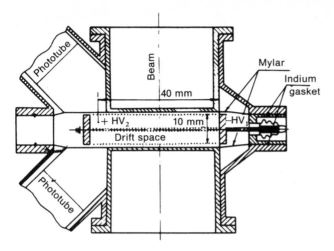

Fig. 10.37. Cross section of a drift chamber based on electroluminescence by Baskakov et al. [BAS 79], which had two sense wires

tolerable, and, last but not least, cost – which further restrict the choices to be made in the quest for the very best accuracy. Therefore, it is a much harder goal to achieve good accuracy in a drift chamber that is part of a particle experiment than in a test chamber built for a more specialized investigation. The problems connected with small wire distances and mechanical accuracy are obviously better solved in chambers with small dimensions. For these reasons, the chambers that have reached extreme values of accuracy are small and specialized.

A chamber containing xenon gas under a pressure of 20 atm was presented by Baskakov et al. and is depicted in Fig. 10.37 [BAS 79]. It is the most precise drift chamber known to us. Its two anode wires have no electrical amplifiers attached to them but are viewed by two photomultipliers, which register the light pulse emitted by the avalanche. Drift chambers based on electroluminescence had previously been studied by Charpak, Majewski and Sauli [CHA 75]; these are suitable for high particle rates because the emission of light in the high electric field of the anode wire is so strong (especially at high gas pressure) that only little or no charge amplification is necessary. However, large systems of such drift chambers have not yet been built.

With their chamber, Baskakov et al. were able to achieve an average accuracy of 18 µm on one wire over a sensitive drift length between 10 and 35 mm (16 µm at 20 mm). This accuracy comes mainly from the low drift velocity of 0.13 cm/µs and the small diffusion at this pressure. The sensitive area was nearly 4×4 cm^2.

A more conventional model is the precision chamber built by Belau et al. for the recording of particles behind a target [BEL 82]. It operates with a gas mixture of 75% propane and 25% ethylene at 4 atm pressure and presents

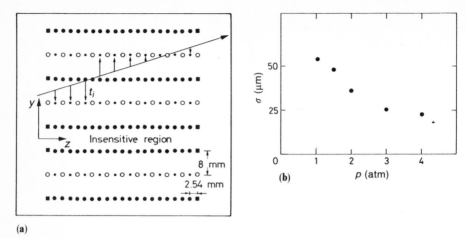

(a)

(b)

Fig. 10.38a, b. Precision drift chamber by Belau et al. [BEL 82]. (a) Wire configuration of three cells; (b) spatial resolution of a single wire as a function of gas pressure

a sensitive area of $10 \times 10 \text{ cm}^2$. The proportional wires (spaced 1.3 mm) are arranged in 6 horizontal planes, the cathode wires in parallel planes, creating 12 drift spaces 8 mm high and 50 mm deep, and one central insensitive space for a high-intensity beam (Fig. 10.38a). The achieved accuracy was 23 µm per wire after eliminating or calibrating the inhomogeneous drift regions near the wires (18% of the volume). This accuracy was obtained with a fast gas (drift velocity 5 cm/µs) and correspondingly fast electronics (rise time 0.8 ns) on 60 channels. The mechanical accuracy is based on ceramic spacers for the wires, ground to a precision of a few microns. The variation of the achieved measuring accuracy with gas pressure is seen in Fig. 10.38b.

References

[ABR 89]   G. Abrams, The MARK II detector for the SLC, *Nucl. Instrum. Methods Phys. Res.*
            A **287**, 55 (1989)
[ADE 90]   B. Adeva et al., The construction of the L3 experiment, *Nucl. Instrum. Methods Phys.*
            Res. A **289**, 35 (1990)
[ADL 89]   J. Adler et al., The MARK III vertex chamber, *Nucl. Instrum. Methods Phys. Res.*
            A **276**, 42 (1989)
[AHM 90]   S. Ahmad et al., The ASTERIX spectrometer at LEAR, *Nucl. Instrum. Methods*
            Phys. Res. A **286**, 76 (1990)
[ALB 89]   H. Albrecht et al., ARGUS – a universal detector at DORIS II, *Nucl. Instrum.*
            Methods Phys. Res. A **275**, 1 (1989)
[ALE 89]   J. P. Alexander et al., The Mark II vertex drift chamber, *Nucl. Instrum. Methods*
            Phys. Res. A **283**, 519 (1989)

[ALL 74]   W.W.M. Allison, C.B. Brooks, J.N. Bunch, J.H. Cobb, J.L. Lloyd, R.W. Fleming, The identification of secondary particles by ionisation sampling, *Nucl. Instrum. Methods* **119**, 499 (1974)

[ALL 82]   W.W.M. Allison, Relativistic particle identification by dE/dx: the fruits of experience with ISIS, in *Int. Conf. for Colliding Beam Physics, SLAC, 61* (1982)

[ALL 82a]   J. Allison, R.J. Barlow, C.K. Bowdery, I. Duerdoth, P.G. Rowe, An electrodeless drift chamber, *Nucl. Instrum. Methods* **201**, 341 (1982)

[ALL 84]   W.W.M. Allison, C.R. Brooks, P.D. Shield, M. Aguilar-Benitez, C. Willmott, J. Dumarchez, M. Schouten, Relativistic charged particle identification with ISIS2, *Nucl. Instrum. Methods Phys. Res.* **224**, 396 (1984)

[AND 86]   H. Anderhub et al., A time expansion chamber as a vertex detector for the experiment MARK J at DESY, *Nucl. Instrum. Methods Phys. Res. A* **252**, 357 (1986)

[AND 88a]   H. Anderhub et al., A time expansion chamber as a vertex detector, *Nucl. Instrum. Methods Phys. Res. A* **263**, 1 (1988)

[AND 88b]   H. Anderhub et al., Operating experience with the MARK-J time expansion chamber, *Nucl. Instrum. Methods Phys. Res. A* **215**, 50 (1988)

[ASH 87]   W.W. Ash et al., Design, construction, prototype tests and performance of a vertex chamber for the MAC detector, *Nucl. Instrum. Methods Phys. Res. A* **261**, 399 (1987)

[ATW 91]   W.B. Atwood et al., Performance of the ALEPH Time Projection Chamber, *Nucl. Instrum. Methods Phys. Res. A* **306**, 446 (1991)

[AVE 89]   R.E. Avery, Bose–Einstein Correlations of pions in e^+e^- annihilation at 29 GeV center-of-mass energy, PhD thesis, University of California, 1989, also LBL preprint 26593

[BAR 89]   G.J. Barber et al., Performance of the three-dimensional readout of the ALEPH inner tracking chamber, *Nucl. Instrum. Methods Phys. Res. A* **279**, 212 (1989)

[BAS 79]   V. I. Baskakov et al., The electroluminescenting drift chamber with spatial resolution 16 μm, *Nucl. Instrum. Methods* **158**, 129 (1979)

[BEC 82]   Ch. Becker, W. Weihs, G. Zech, Wireless drift tubes, electrodeless drift chambers and applications, *Nucl. Instrum. Methods* **200**, 335 (1982)

[BEC 89]   G.A. Beck et al., Radial wire drift chambers for the H1 forward track detector at HERA: Design, construction and performance, *Nucl. Instrum. Methods Phys. Res. A* **283**, 471 (1989)

[BED 88]   F. Bedeschi et al., Design and construction of the CDF central tracking chamber, *Nucl. Instrum. Methods Phys. Res. A* **268**, 50 (1988)

[BEL 82]   E.R. Belau, W. Blum, Z. Hajduk, T.W.L. Sanford, Construction and performance of a small drift chamber with 23 μm spatial resolution, *Nucl. Instrum. Methods* **192**, 217 (1982)

[BER 91]   J.P. Berge, Private communication, Sept. 1991

[BET 86]   D. Bettoni et al., Drift chambers with controlled charge collection geometry for the NA34/HELIOS experiment, *Nucl. Instrum Methods Phys. Res. A* **252**, 272 (1986)

[BOS 89]   F. Bosi et al., Performance of the UA2 Jet Vertex Detector at the CERN Collider, *Nucl. Instrum. Methods Phys. Res. A* **283**, 532 (1989)

[BRA 89]   C. Brand et al., The DELPHI time projection chamber, *Nucl. Instrum. Methods Phys. Res. A* **283**, 567 (1989)

[BRE 69]   T. Bressani, G. Charpak, D. Rahm and Č. Zupančič, Track localization by means of a drift chamber, in *Proc. of the International Seminar on 'Filmless Spark and Streamer Chambers'* (April 1969) (Joint Institut for Nuclear Research, Dubna, USSR, 1969) p. 275

[BRE 75]   A. Breskin, G. Charpak, F. Sauli, M. Atkinson, G. Schultz, Recent observations and measurements with high-accuracy drift chambers, *Nucl. Instrum. Methods* **124**, 189 (1975)

[BRE 78]   A. Breskin, G. Charpak, F. Sauli, A solution to the right–left ambiguity in drift chambers, *Nucl. Instrum. Methods* **151**, 473 (1978)

[BRY 83]   D. Bryman et al., The time projection chamber at TRIUMF, in *Proceedings of a Workshop on 'The Time Projection Chamber'*, ed. by J.A. Macdonald held at TRIUMF, Vancouver, Canada, June 1983. *AIP Conference Proceedings No. 108* (American Institute of Physics, New York 1984)

[CAR 90]   J.R. Carter et al., The OPAL vertex drift chamber, *Nucl. Instrum. Methods Part. Phys. A* **286**, 99 (1990), and private communication from J. Carter

[CHA 68]   G. Charpak, R. Bouclier, T. Bressani, J. Favier, Č. Zupančič, The use of multiwire proportional counters to select and localize charged particles, *Nucl. Instrum. Methods* **62**, 262 (1968)

[CHA 70]   G. Charpak, D. Rahm and H. Steiner, Some developments in the operation of multiwire proportional chambers, *Nucl. Instrum. Methods* **80**, 13 (1970) (received 13 November 1969)

[CHA 73]   G. Charpak, F. Sauli, W. Duinker, High accuracy drift chambers and their use in strong magnetic fields, *Nucl. Instrum. Methods* **108**, 413 (1973)

[CHA 75]   G. Charpak, S. Majewski and F. Sauli, The scintillating drift chamber: A new tool for high-accuracy, very-high-rate particle localization, *Nucl. Instrum. Methods* **126**, 381 (1975)

[CHA 84]   G. Charpak and F. Sauli, High-resolution electronic particle detectors, *Annu. Rev. Nucl. Part. Sci.* **34**, 285 (1984)

[CLA 76]   A.R. Clark et al. (Johns Hopkins University, Lawrence Berkeley Laboratory, University of California at Los Angeles, University of California at Riverside, Yale University), Proposal for a PEP facility based on the time projection chamber, SLAC-PUB-5012 (1976)

[DAU 86]   E. Daubie et al., Drift chambers with delay line readout, *Nucl. Instrum. Methods Phys. Res. A* **252**, 435 (1986)

[DAV 79]   W. Davies-White, G.E. Fischer, M.J. Lateur, R.H. Schindler, R.F. Schwitters, J.L. Siegrist, H. Taureg, H. Zaccone, D.L. Hartill, A large cylindrical drift chamber for the MARK II detector at SPEAR, *Nucl. Instrum. Methods* **160**, 227 (1979)

[DEC 90]   D. Decamp et al., ALEPH, A detector for electron–positron annihilation at LEP, *Nucl. Instrum. Methods Phys. Res. A* **294**, 121 (1990)

[DEL 91]   DELPHI Collaboration, The DELPHI detector at LEP, *Nucl. Instrum. Methods Phys. Res. A* **303**, 233 (1991)

[DOE 85]   R. Dörr, C. Grupen, A. Noll, Characteristics of a multiwire circular electrodeless drift chamber, *Nucl. Instrum. Methods Phys. Res. A* **238**, 238 (1985)

[DUR 90]   D. Durret et al., Calibration and performance of the MARK-II drift chamber vertex detector, SLAC-PUB 5259, LBL 29108 (1990)

[ETK 89]   A. Etkin et al., Modular TPC's for relativistic heavy-ion experiments, *Nucl. Instrum. Methods Phys. Res. A* **283**, 557 (1989)

[FAB 90]   C. Fabjan, private communication, April 1990

[FIS 89]   H.M. Fischer et al., The OPAL jet chamber, *Nucl. Instrum. Methods Phys. Res. A* **283**, 492 (1989)

[FLÜ 81]   G. Flügge, B. Koppitz, R. Kotthaus, H. Lierl, Review of contributed papers for experimentation at LEP, *Phys. Scr.* **23**, 499 (1981)

[FRA 82]   A. Franz and C. Grupen, Characteristics of a circular electrodeless drift chamber, *Nucl. Instrum. Methods* **200**, 331 (1982)

[GAS 81]   U. Gastaldi, The spiral projection chamber (SPC): A central detector with high resolution and granularity suitable for experiments at LEP, *Nucl. Instrum. Methods* **188**, 459 (1981)

[GRA 89]   H. Grässler et al., Simultaneous track reconstruction and electron identification in the H1 radial drift chambers, *Nucl. Instrum. Methods Phys. Res. A* **283**, 622 (1989)

[HAR 84]   G. Harder, Optimierung der Ortsauflösung der zylindrischen Driftkammer des Detektors ARGUS, Diploma Thesis, DESY Internal Report F15-84/01 (unpublished)

[HAY 88]   K.G. Hayes, Drift chamber vertex detectors for SLC/LEP, *Nucl. Instrum. Methods Phys. Res. A* **265**, 60 (1988)

[KAM 86]   T. Kamae et al., The TOPAZ time projection chamber, *Nucl. Instrum. Methods Phys. Res. A* **252**, 423 (1986)

[LYN 87]   G. Lynch, Performance of the PEP-4 TPC, Talk given in MPI Munich, 22 June 1987

[MAR 77]   G. Marel et al., Large planar drift chambers, *Nucl. Instrum. Methods* **141**, 43 (1977)

[MUL 73]   J. Mulvey, Comments on particle identification using the relativistic rise of ionization energy loss, in *Int. Conf. on Instrumentation for High Energy Physics*, ed. by Stipcich and Stanislao, held in Frascati, Italy (Lab. Naz. del Com. Naz. per l'Energia Nucleare, Frascati 1973), p. 259

[NEL 87]   H.N. Nelson, Design and construction of a vertex chamber, and measurement of the average B hadron life time, SLAC 322 (1987) (unpublished)

[NOR 90]   A. Norton, private communication, July 1990

[NYG 74]   D.R. Nygren, Proposal to investigate the feasibility of a novel concept in particle detection, LBL internal report, Berkeley, February 1974

[NYG 83]   D.R. Nygren, TPC workshop summary, in *Proceedings of a Workshop on 'The Time Projection Chamber'*, ed. by J.A. Macdonald held at TRIUMF, Vancouver, Canada, June 1983. *AIP Conference Proceedings No. 108* (American Institute of Physics, New York 1984)

[OPA 91]   OPAL Collaboration, The OPAL detector at LEP, *Nucl. Instrum. Methods Phys. Res. A* **305**, 275 (1991)

[SHI 88]   A. Shirahashi et al., Performance of the TOPAZ time projection chamber, *IEEE Trans.* **NS-35**, 414 (1988)

[SIE 90]   P. Siegrist, CERN, private communication, September 1990

[SLD 85]   SLD Design Report, SLAC Report 273 (revised 1985)

[SNI 88]   F. Snider et al., The CDF Time Projection Chamber system, *Nucl. Instrum. Methods Phys. Res. A* **268**, 75 (1988)

[TOK 90]   W.H. Toki, Review of straw chambers, in *Proceedings of the 5th Int. Conf. for Colliding Beam Physics*, Novosibirsk (USSR), March 1990; also as: SLAC preprint SLAC-PUB-5232 (1990)

[VIL 86]   F. Villa (ed.), *Vertex Detectors, Proceedings of a Workshop for the INFN Eloisatron Project*, held September 1986 (Plenum, New York London 1988)

[WAG 81]   A. Wagner, Central detectors, *Phys. Scr.* **23**, 446 (1981)

[WAL 71]   A.H. Walenta, J. Heintze and B. Schürlein, The multiwire drift chamber, a new type of multiwire proportional chamber, *Nucl. Instrum. Methods* **92**, 373 (1971) (received 27 November 1970)

[WAL 78]   A.H. Walenta, Left–right assignment in drift chambers and MWPC's using induced signals. *Nucl. Instrum. Methods* **151**, 461 (1978)

[WAL 79]   A.H. Walenta, The time expansion chamber and single ionization cluster measurement, *IEEE Trans. Nucl. Sc.* **NS 26**, 73 (1979)

[WIL 86]   H.H. Williams, Design principles of detectors at colloiding beams, *Annu. Rev. Nucl. Part. Sc.* **36**, 361 (1986)

[YOU 86]   C.C. Young et al., Performance of the SLD central driftchamber prototype, *IEEE Transactions Nucl. Sc.* **33**, 176 (1986)

11. Drift-Chamber Gases

Gas is the medium in which the processes of ionization, drift and amplification develop. We have discussed these processes in previous parts of this book. Here we want to present some additional practical information relevant to the choice of gas mixtures and their pressure, and relevant to the impurities that can be tolerated in the gas.

11.1 General Considerations
Concerning the Choice of Drift-Chamber Gases

There are many gas mixtures that are successfully used in drift chambers. The tables in Chap. 10 contain a representative collection. Often we find the noble gas argon mixed with some quenching organic molecular gas. But purely organic mixtures are also in use. These gases fulfil the basic requirement that the electron lifetime is sufficiently long and that a stable amplification process exists.

An important choice has to be made with regard to the drift velocity and its dependence on the electric field. A highly preferable situation arises if the drift velocity varies little with the field, for then the coordinate measurement is less dependent on the unavoidable field gradients, and on the pressure and temperature variations of the gas. This happens at the maximum of the velocity–field relation ('saturated drift velocity'). Figure 11.1 shows the drift velocity of argon–isobutane mixtures, which have this desirable property over a relatively long range of E field.

At these high velocities – typically 5 cm/μs – a fast time measurement is a necessity. If one wants to relax this requirement, one can reduce the drift field, and the drift velocity is no longer saturated, with the consequence that it depends on space charge and other changes of the drift field, thus requiring very careful calibrations.

The 'dilemma of the Lorentz angle' comes into sight here (see Sect. 10.3.2). If the electric and magnetic fields are orthogonal one may wish to achieve a small drift velocity at a high E field in order to keep the Lorentz angle small (2.7, 11).

A high E field at small electron energy must be the goal in the interests of a low diffusion width; cf. the discussion following (2.63). The last two requirements combine well in the 'cold' gas mixtures based on carbon dioxide or dimethylether. In these, the random electron velocities and the drift velocity are

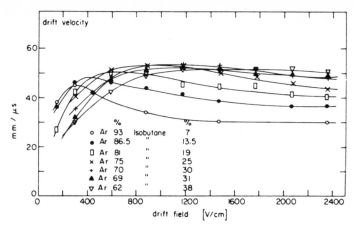

Fig. 11.1. Drift velocity as a function of the drift field, for various argon–isobutane mixtures [BRE 74]

much lower than in the 'hot' gases at the same field strength. The cold gases have the disadvantage that an operating point must be found in the unsaturated situation. High E fields are more difficult to handle, since spurious discharges are a threat to the reliability of a chamber. High E fields also cause electrostatic wire displacements; these are larger for greater wire lengths.

In order to give an overview of achievable diffusion limits we reproduce in Fig. 11.2 the characteristic electron energies calculated by Paladino and Sadoulet [PAL 75] for the pure gases Ar, CH_4 and CO_2, and for various argon–isobutane mixtures. The width of the diffusion after 1 cm of drift, calculated according to (2.63), is shown in Fig. 11.3.

In TPCs, where the E and B fields are parallel, there is an advantage in having a high value of $\omega\tau$, or a high product of the B field and the mobility (cf. Sect. 2.1). One reason is the enormous suppression factor for the transverse diffusion constant (2.72), another has to do with the track distortions caused by imperfections in the uniformity of the magnetic field. Since in a good solenoid the inhomogeneities are predominantly radial (not azimuthal), the distortions of the curvature of radial tracks are suppressed by one power of $\omega\tau$; see (2.6).

Argon–methane mixtures with small CH_4 concentrations can be used in the saturated situation of the drift velocity, and there they exhibit large values of τ. It can be seen in Table 10.4 that such gas mixtures have actually been chosen for all the large TPCs.

11.2 Inflammable Gas Mixtures

Most of the quenching gases used in drift chambers can burn in air. In large systems they may therefore represent a security risk. We would like to know under what circumstances and in which concentrations they are dangerous.

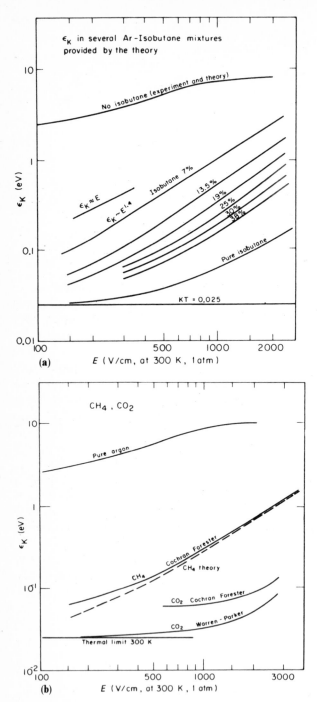

Fig. 11.2a, b. Characteristic electron energies calculated and reported by [PAL 75]. (a) for various mixtures of argon–isobutane; (b) argon, carbon dioxide and methane

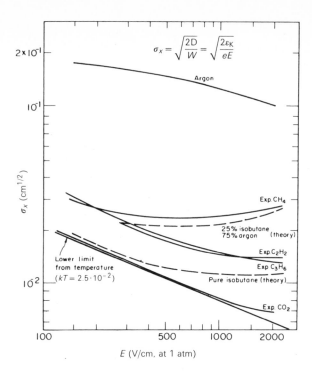

$$\sigma_x = \sqrt{\frac{2D}{W}} = \sqrt{\frac{2\varepsilon_K}{eE}}$$

Fig. 11.3. Diffusion width after 1 cm of drift, according to (2.63), using calculated electron energies [PAL 75]

A gas mixture that contains some oxygen is 'inflammable' if a flame can exist and propagate in it. (Confusingly enough, this same mixture is also called 'flammable'.) The fact is established in a standardized vertical tube with ignition from below. Whether a flame develops or not depends on the nature of the gas components, their temperature and pressure, and in particular on their concentrations. By varying the concentrations of a gas mixture, one finds non-inflammable combinations whose content of combustible or oxygen is not enough to make a flame. Therefore there are lower and upper limits of concentration which mark the region of inflammability.

The simplest mixture is a two-component gas, for example methane in air. At normal temperature and pressure it is inflammable between 5 and 15% CH_4 by volume in the total, whilst the stoichiometric ratio for complete combustion is 9.5%. There is no simple principle from which to derive such limits: they depend on details of the flame dynamics. Let us note that these limits are markedly different when the flame proceeds with the ignition on top rather than on the bottom. The monograph by Lewis and von Elbe [LEW 61] may be consulted concerning this subject. In Table 11.1 we present inflammability limits for typical quenching gases in air.

When an inert gas is added to the previous mixture, we have a three-component mixture, for example methane, argon and air. Now the inflammability limits depend also on the argon concentration. If there is too much argon, then there will be no flame, irrespective of the methane/air ratio. The three-

Table 11.1. Limits of inflammability of gases and vapours in air [LEW 61]

Compound	Formula	Limits of inflammability	
		Lower	Upper
		(vol.% of total mixture)	
Methane	CH_4	5.3	15.0
Ethane	C_2H_6	3.0	12.5
Propane	C_3H_8	2.2	9.5
Butane	C_4H_{10}	1.9	8.5
Isobutane	$i\text{-}C_4H_{10}$	1.8	8.4
Pentane	C_5H_{12}	1.4	7.8
Isopentane	$i\text{-}C_5H_{12}$	1.4	7.6
2,2-dimethylpropane	C_5H_{12}	1.4	7.5
Hexane	C_6H_{14}	1.2	7.5
Ethylene	C_2H_4	3.1	32.0
Propylene	C_3H_6	2.4	10.3
Methylalcohol	CH_3OH	7.3	36.0
Isopropylalcohol	C_3H_7OH	2.0	12.0
Diethylether	$C_4H_{10}O$	1.9	48.0

Table 11.2. Standard limiting safe mixtures in air. A mixture of a combustible gas (component 1) and an inert gas (component 2) is considered inflammable in air if, with a suitably chosen admixture of air (component 3), the three-component mixture can be ignited. The table quotes concentrations c_1 of component 1 at and below which no admixture of air can make a flame. The corresponding concentrations c_2 are also given

Combustible gas	c_1 (%)	Inert gas	c_2 (%)	Reference[a]
CH_4	13/9	He	87/91	[BUR 48, ZAB 65]
$i\text{-}C_4H_{10}$	2.7	Ne	97.3	[SCH 92]
CH_4	9	Ar	91	[BUR 48]
C_2H_6	2	Ar	98	[SCH 92]
$n\text{-}C_5H_{12}$	2	Ar	98	[SCH 92]
CH_4	23	CO_2	77	[BUR 48, ZAB 65]
C_2H_6	12	CO_2	88	[BUR48, ZAB 65]
C_3H_8	11	CO_2	89	[BUR 48, ZAB 65]
$n\text{-}C_4H_{10}$	9.7	CO_2	90.3	[BUR 48, ZAB 65]
$i\text{-}C_4H_{10}$	9.3	CO_2	90.7	[BUR 48]
$n\text{-}C_5H_{12}$	7.5	CO_2	92.5	[BUR 48, ZAB 65]
$n\text{-}C_6H_{14}$	6.6	CO_2	93.4	[BUR 48, ZAB 65]

[a] Where two references are given they are in agreement, except for CH_4/He.

component gas mixtures are conveniently described by the diagrams in Fig. 11.4: a particular mixture is given by the three volume concentrations c_1, c_2 and c_3 in the total mixture, which add up to unity; they are represented by a point in the unilateral triangle whose three coordinates are measured parallel

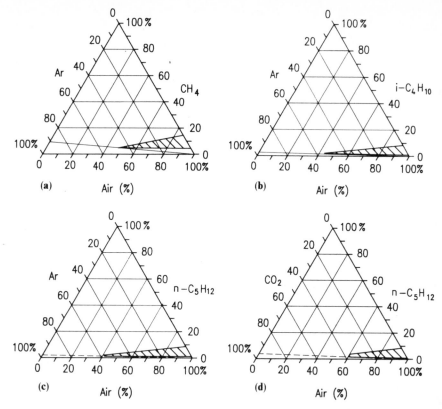

Fig. 11.4a–d. Diagrams to describe three-component gas mixtures by their volume concentrations c_1, c_2 and c_3 (see text). Limits of inflammability for four different mixtures containing air

to its sides (the sum of the three coordinates is always equal to the length of one side).

Results of flame tests are now recorded in a diagram that delineates the region of inflammability. Examples from measurements by the Physikalisch-Technische Bundesanstalt, made at the request of CERN [SCH 92], are shown in Fig. 11.4.

Chamber gas does not contain oxygen, and inflammable mixtures will come about by unintentional contact with air. A safe chamber gas is one that produces a non-inflammable mixture with any amount of air. Looking, for example at Fig. 11.4a, all the combinations of methane–argon–air concentrations that have this property lie below the tangent that connects the region of inflammability with the point P ($c_1 = 0, c_2 = 0, c_3 = 1$). All mixtures with a fixed concentration ratio c_1/c_2 lie on a straight line through P, which intersects the opposite side at the value c_2 – the methane–argon mixture is safe in this sense if the methane concentration is not larger than 9%. Some safe concentrations of combustible gases are collected in Table 11.2.

A special point of interest is the tip of the region of inflammability that belongs to the smallest air concentration. One might expect that it always represents the stoichiometric concentrations. But this is not so – it is often displaced from the stoichiometric ratio towards lower concentrations of the component that has the higher diffusion [LEW 61]. Here is one more hint that no simple theory will predict the safe concentrations.

Inflammable gas mixtures are not easily avoidable for limited-streamer chambers, because most mixtures with sufficient absorption power for the ultraviolet light are also inflammable. Special studies of this problem have been undertaken by [BEN 89] and several other groups quoted there.

11.3   Gas Purity, and Some Practical Measurements of Electron Attachment

Together with the gases intended to serve in a chamber, there will always be some contamination, by impurities in the bottles delivered from industry, from outgassing construction materials in the gas stream or produced in operation, or arriving from the outside world through membranes that are not entirely tight.

Even minor contamination may have adverse effects on the chamber operation. It can change the drift velocity (a drastic example is reported in Sect. 11.3.3), or they may remove drifting electrons by attachment (see Sects. 11.3.1 and 11.3.2). They are a cause of chamber ageing (Sect. 11.6). But they can also have a beneficial effect: ionization tracks produced by pulsed UV lasers often depend on some unknown molecular compound that happens to be in the chamber gas (cf. Sect. 11.4).

We have described in Chap. 2 various mechanisms of negative-ion formation which may lead to a loss of electrons drifting through a chamber. It is chiefly by varying the partial pressures of two clean gases that the two-body and three-body processes can be experimentally distinguished. The values reported in Figs. 2.11 and 2.12 were obtained in this way, mostly by workers engaged in atomic-physics research.

For the purposes of drift chambers it is not always necessary to isolate the exact process or to know the electron energy. On the other hand one wants to have electron attachment rates for the multi-component gas mixtures and for drift fields typical for applications in particle experiments. We collect here some measurements of this kind, keeping in mind that the detailed interpretation of the attachment process is not always required.

11.3.1   Three-Body Attachment to O_2, Mediated by CH_4, i-C_4H_{10} and H_2O

Electron loss rates were measured in various drift-chamber gases with additions of oxygen, water or methanol, by Huk, Igo-Kemenes and Wagner [HUK 88].

Working initially at 4 bar with Ar (88%) + CH_4 (10%) + i-C_4H_{10} (2%) and with O_2 concentrations up to 440 ppm, they varied the total gas pressure and the hydrocarbon and O_2 concentrations. They were able to show that the measured electron attenuation rate R_{tot} associated with oxygen could be understood as a superposition of rates according to (2.77):

$$R_{tot} = \sum R_i \,,$$

$$R_i = k_1^{(i)} N(O_2) N(X_i) \,,$$

where $N(O_2)$ is the O_2 concentration and the $N(X_i)$ are the concentrations of the other gas components. (The rate proportional to the square of $N(O_2)$ is negligible at these low O_2 contaminations.) The various contributions could be separated, and the three-body rate coefficients of CH_4, i-C_4H_{10} and Ar were determined for a practical range of drift fields. The contribution of Ar was zero inside the errors, as expected. The results are quoted in Tables 11.3 and 11.4. The coefficient $C(O_2, X)$, measured in units of $(\mu s^{-1} bar^{-2})$, is related to k_1 of (2.77), measured in units of $(cm^6 s^{-1})$, by the relation

$$C(O_2, X) = \left(\frac{6.02 \times 10^{20}}{22.4} \right)^2 \left(\frac{bar^{-2}}{cm^6} \right) 10^{-6} k_1$$

at N.T.P.

It should be noted that the three-body rate coefficients $C(O_2, CH_4)$ and $C(O_2, i\text{-}C_4H_{10})$ given in Tables 11.3 and 11.4 for specific values of the reduced electric field represent averages over the electron energies in the range of the gas mixtures used. For details the reader is referred to the article by Huk et al. [HUK 88], which is also an excellent entry point to the recent literature.

The authors then investigated the influence of water and methanol. In the absence of oxygen, neither 0.3% H_2O nor 0.1% CH_3OH in Ar(90%) + CH_4(10%) caused any electron absorption outside the errors. But the situation changed when a small amount (200 ppm) of O_2 was added to the argon–methane mixture. It turned out that 0.1% of H_2O is as efficient as 10% of

Table 11.3. Three-body electron attachment coefficients for O_2 and third body CH_4 determined by [HUK 88] at fixed reduced drift fields E/P in drift-chamber gas composed of Ar(90%) + CH_4(10%) + O_2 (200 ppm) and Ar(80%) + CH_4(20%) + O_2 (200 ppm)

E/P (V/cm bar)	$C(O_2, CH_4)$ ($\mu s^{-1} bar^{-2}$)
100	167 ± 32
138	191 ± 30
163	176 ± 20
200	149 ± 27
250	146 ± 26

Table 11.4. Three-body electron attachment coefficients as in Table 11.3, but for O_2 and i-C_4H_{10} in gas composed of Ar(90%) + CH_4(10%) + O_2 (200 ppm) + i-C_4H_{10}(0 to 4%)

E/P (V/cm bar)	$C(O_2, i\text{-}C_4H_{10})$ ($\mu s^{-1} bar^{-2}$)
100	3240 ± 350
138	2819 ± 300
163	2490 ± 300
200	1970 ± 300
250	1570 ± 250

CH_4 to mediate electron attachment to oxygen under the operating conditions investigated. 500 ppm of methanol did not change the attachment rate to oxygen inside the errors.

11.3.2 'Poisoning' of the Gas by Construction Materials

A long lifetime of electrons, which is essential for the functioning of drift chambers with long drift paths, is not yet guaranteed when the chamber has been filled with extremely clean gas. The very material of the chamber itself and of its gas distribution system may contaminate the gas with electronegative (i.e. electron-absorbing) impurities which emanate from the surface or the volume of these materials. As we discussed in Sect. 2.2.7, the two-body attachment rates of some halogen-containing compounds are so large that even at concentrations as low as 10^{-8} they will absorb a sizeable fraction of the electrons that have to drift over a distance of one metre.

In practice, every piece of material that will come into contact with the gas stream of a long-drift chamber has to be controlled with respect to outgassing before it is built in. Test facilities for this purpose have been in use at Berkeley, CERN and probably other laboratories. Basically, a long drift tube is connected to a gas distribution system which comprises a receptacle for the material in question. In Berkeley the 'poisioning time' T_p was measured for which the test piece has to be in contact with the $6\,m^3$ of the TPC volume before the attenuation of electrons over 1 m of drift increases by 1%. At CERN the results of the measurements were expressed in terms of the percentage electron loss A over 1 m of drift after the test piece had been in contact with $1\,m^3$ of gas for 24 h.

The drift tubes were operated under the conditions of the large chamber that was being built in the respective laboratory (PEP-4 TPC at Berkeley: Ar(80%) + CH_4(20%) at 10 bar; ALEPH TPC at CERN: Ar(90%) + CH_4(10%) at 1 bar; cf. Table 10.4). The results of these technical measurements cannot be directly compared because the various contributing

2-body and 3-body attachment rates behave differently with respect to gas composition, pressure and electric field.

The results of these tests are documented in two lists [BRO 79] and [LEH 89] containing about a hundred items each. We present in Tables 11.5

Table 11.5. List of poisoning times T_p of some selected materials measured in the Berkeley Materials Test Chamber. T_p is the time it takes for one unit of the indicated material to increase the attenuation rate of electrons travelling over a drift distance of 1 m by 1%, when in contact with 6 m^3 of the gas mixture, which was Ar(80%) + CH$_4$(20%) at 10 bar; the drift field was 1.5 kV/cm [BRO 79]

Material	Unit	T_p (best estimate) (h)	T_p (lower limit) (h)
Epoxy Versamid 140 and Epon 826	1 m^2	360	310
Teflon 'TFE'	1 m^2	(1)	210
Etched copper-clad Kapton	1 m^2	(1)	1500
Mylar	1 m^2	(1)	680
G-10 (Westinghouse)	1 m^2	660	600
Tufftane polyuretane film	1 m^2	(1)	910
Nylon connectors (3M)	1 piece	750	740
Glass-filled polyester connectors (Amp)	1 piece	160	75
Neoprene seal (Victaulic)	1 piece	100	99
White Nitrile seal (Victaulic)	1 piece	7.1	7
Tygon tubing (Xorton plastics R 3603)	1 ft	28	26
Spirex paint	1 m^2	87	80

1. Data consistent with infinite T_p.

Table 11.6. List of electron attenuations A caused by some selected materials in the ALEPH Materials Test Facility. A is the percentage electron loss over a drift distance of 1 m, caused by 1 m^2 of the indicated material in contact for 24 h with 1 m^3 of the gas mixture, which was Ar(90%) + CH$_4$(10%) at 1 bar; the drift field was 110 V/cm. A has been scaled linearly from smaller test pieces [LEH 89]

Material	A (%/m^2)	$\pm \Delta A$ (%/m^2)
Aluminium plate	0.0	0.02
Mylar foil, 25 mm thick	0.00	0.005
Delrin (polyacetal/polyoxymethylen) ('POM')	0.25	0.10
Plexiglas (PMMA)	0.7	0.1
Stesalit with copper back plane, cleaned with isopropyl-alcohol	0.3	0.2
Same, cleaned in ultrasonic bath containing freon	~40	
Same as above, 4 months later	20	5
HT anti-corona coating SL 1300 (Peters, Kempten)	40	15
HT anti-corona coating Plastic 70 (Kontakt-Chemie)	0.1	0.1
Natural rubber (cleaned in hot water, 70 °C, 2 h)	25	5
Viton vacuum seals (uncleaned)	70	20
Araldite AW 106 glue (1-week-old)	0.21	0.06
Same as above, but 5-weeks-old	0.06	0.03
Same as above, but 14-weeks-old	0.01	0.02

and 11.6 some examples in order to show that many common construction materials do not poison the gas, whereas a number of other materials, especially certain cleaning agents, paints and seals, have been found to be disastrous.

11.3.3   The Effect of Minor H_2O Contamination on the Drift Velocity

We present in Fig. 11.5 drift velocities for two argon–methane mixtures, typical TPC gases, which in addition contain some very low concentrations of water vapour. These measurements from the DELPHI group are basically in agreement with calculations of Biagi [BIA 89] and of Schmidt [SCH 86]. Let us note that for some, especially low, values of the electric field, one tenth of a per cent of water vapour changes the drift velocity almost by a factor of 2. This drastic behaviour is due to the properties of the water molecule with its static electric dipole moment, which causes the inelastic scattering cross-section for low-energy electrons to be exceedingly large, thus reducing the drift velocity according to (2.19).

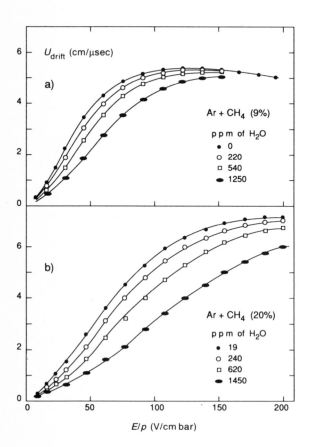

Fig. 11.5a, b. Drift velocities measured by [CAT 89] for two argon–methane mixtures containing a small additional amount of water vapour. (**a**) $Ar(91\%) + CH_4(9\%)$; (**b**) $Ar(80\%) + CH_4(20\%)$. The measurement errors are estimated to be 1% or less; the lines are drawn to connect the measured points

11.4   Chemical Compounds Used for Laser Ionization

The most often employed technique of creating ionization tracks in the gas of a drift chamber relies on the presence of molecules in the chamber gas that can be ionized in a two-step process as discussed in Sect. 1.3. In many chambers the gas has sufficient impurities, owing to the outgassing of materials or as remnants from the production process. If the concentration is not large enough for the available laser power, or if it is not stable, a suitable compound has to be added to the chamber gas. We can look for candidates among the organic vapours with ionization potentials below twice the laser photon energy. A good number of them are known to produce ionizable tracks, but definitive measurements of the second-order cross-section equivalent (see Sect. 1.3.1) do not yet exist.

A typical measurement is shown in Fig. 11.6 [HUB 85]. The ionization density is seen to follow the quadratic dependence of the laser flux. In Tables 11.7 and 11.8 we present linear ionization densities achieved with single laser shots in various vapours. Laser tracks imitating particle tracks can obviously be created with vapour partial pressures in the range of 10^{-3} to 10^{-6} Torr and with laser power densities of 1 $\mu J/mm^2$. The reader is referred to a review article by Hilke [HIL 86] for further discussion.

As we know from Sect. 1.3.1 and (1.93), the yield of one shot at a given energy depends on the space and time structure of the laser pulse. Once these are determined together with the ionization, definitive measurements of the second-

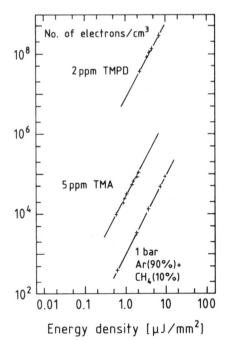

Fig. 11.6. Measured ionization density as a function of the energy density of the laser at $\lambda = 266$ nm in Ar + CH$_4$ (uncleaned), doped with 5 ppm TMA, and with 2 ppm TMPD. The slopes of the curves are 1.99 ± 0.06 (Ar + CH$_4$), 1.9 ± 0.1 (TMA) and 1.9 ± 0.2 (TMPD)

Table 11.7. Linear ionization densities achieved in various organic vapours with one shot of the N_2 laser ($\lambda = 337$ nm, $E_\gamma = 3.66$ eV). The results are rescaled to a beam cross-section of 1 mm^2, an energy of 1 µJ, and a partial pressure of 10^{-3} Torr

Substance	Ionization potential (eV)	Vapour pressure at 20 °C (Torr)	Linear ionization density (el./cm)	Reference
α-naphthylamine α-$C_{10}H_7NH_2$	7.3	2.7×10^{-4}	4×10^3	[GUS 84]
N,N-dimethylaniline (DMA) $C_6H_5N(CH_3)_2$	7.14	0.25	0.05	[LED 84]
N,N-dipropylaniline (DPA)$C_6H_5N(C_3H_7)_2$	7.1	1.2×10^{-2}	130	[GUS 84]
Diethyl ferrocene	6.6	5.4×10^{-2}	10	[GUS 84]
N,N,N',N'-tetramethyl-p-phenylenediamine (TMPD)	6.18	2.3×10^{-3}	7×10^4	[GUS 84]

Table 11.8. Linear ionization densities achieved in various organic vapours with one shot of the frequency-quadrupled Nd:YAG laser ($\lambda = 266$ nm, $E_\gamma = 4.64$ eV). The results are rescaled to a beam cross-section of 1 mm^2, an energy of 1 µJ, and a partial pressure of 10^{-3} Torr

Substance	Ionization potential (eV)	Vapour pressure at 20 °C (Torr)	Linear ionization density (el./cm)	Reference
Toluene C_7H_8	8.82	22	7×10^3	[DRY 86]
Trimethylamine (TMA) $(CH_3)_3N$	8.5; 7.82	—	60 70	[LED 85] [HUB 85]
Phenol C_6H_5OH	8.51; 8.3	—	> 250	[HUB 86]
Naphthalene $C_{10}H_8$	8.12	—	500	[HUB 86]
Triethylamine (TEA) $(CH_3CH_2)_3N$	7.5	50	200	[LED 85]
N,N-dimethylaniline (DMA) $C_6H_5N(CH_3)_2$	7.14	0.25	10^4	[LED 85]
N,N,N',N'-tetramethyl-p-phenylenediamine (TMPD)	6.18	2.3×10^{-3}	4×10^4	[HUB 85]

order cross-section will become possible. Not all ionizable substances are suitable for doping the gas of a chamber. For example, the vapour tetrakis(dimethylamine)ethylene (TMAE) is a highly ionizable agent because of its low ionization potential of 5.5 eV. But it quickly deposits on metal surfaces, rendering them photosensitive, and it is difficult to remove [LED 85]. The authors quoted in Tables 11.7 and 11.8 have devoted much work to this problem of vapour sticking to the surfaces of their chambers.

Another aspect of doping is the observation of deposits on the anode or cathode of wires. Even modest concentrations of ionizable organic vapours may accelerate the process of 'ageing'. Such studies, made with the UA1 detector, are reported by Beingessner et al. [BEI 88a]. The effects of ageing are also discussed in Sect. 11.6.

Finally, we mention that under conditions of heavy irradiation, Beingessner et al. [BEI 88b] observe that tetramethyl-p-phenylenediamine (TMPD) vapour disappears from the detector, presumably owing to charge transfer from the avalanche-produced heavy ions that are in the drift region; when they have ionized and cracked the TMPD molecules, the ionized TMPD molecules and crack products drift away to the cathode.

11.5 Choice of the Gas Pressure

In the design of a drift chamber, one of the first important decisions must be taken on the gas pressure, because it determines to a large extent the over-all construction. The technical inconvenience of a pressure vessel and the increase of matter in the path of the particles are balanced to a certain degree by some advantages in the measuring accuracy, and we want to review them here. We note in passing that the wall thickness of a vessel must increase in proportion to the over-pressure it has to hold.

We have stated in (2.86) that for the mobility tensor the electric and magnetic fields scale with the gas density. One would therefore like to discuss the gas density always in connection with the corresponding change of the electric and magnetic field strengths, thus keeping the drift-velocity vector constant. But in practice an increase of the relevant field strength is easy in the electric and difficult in the magnetic case. Therefore, we proceed by specifying the magnetic field first, the gas density is discussed next, and then the electric fields are adjusted accordingly. The gas density is varied by changing the pressure, since one usually works at room temperature.

Let us consider the various consequences of a change of the gas pressure by a factor $p > 1$, say from 1 bar to p bar. Table 11.9 contains a summary of the most important effects.

11.5.1 Point-Measuring Accuracy

In measurement accuracy, insofar as it is limited by diffusion, one gains by increasing the pressure – unless the limitation is in the diffusion transverse to the magnetic field, and $\omega\tau \gg 1$. Therefore, we distinguish between the following:

Table 11.9. Influence of the gas pressure on various parameters relevant for drift chambers. If the gas pressure is changed by a factor p, then, in a first approximation, the particular parameter will change by the factor indicated in the last column

Mean free path between collisions	l_0	$1/p$
Mean time between collisions (at constant electron energy)	τ	$1/p$
Electron attachment:		
2-body rate	R_2	p
3-body rate	R_3	p^2
Diffusion constant:		
without magnetic field	$D(0)$	$1/p$
parallel to B	$D_L(\omega)$	$1/p$
orthogonal to B ($\omega\tau \gg 1$)	$D_T(\omega)$	p
orthogonal to B ($\omega\tau \ll 1$)	$D_T(\omega)$	$1/p$
Electric and magnetic fields:		
(adjustment for constant electron energy	E	p
and drift field)	B	p
Synchrotron radiation background:		
rate of charge directly produced by the radiation	B_s	p
amount of charge from the amplification process in the volume at any given time	Q	1
ratio of disturbing to ordinary drift field	E_s/E	$1/p$
Ionization:		
mean distance between clusters	λ	$1/p$
total ionization	N_{tot}	p
ionization effective for coordinate measurement	N_{eff}	$\sim \sqrt{p}$
most probable ionization	I_{mp}	$p f(p)$[a]
variance of most probable value	ΔI_{mp}	$p^{-0.32}$[b]
ratio between minimum and maximum	I_{Fermi}/I_{min}	decreasing[c]
velocity saturation point	γ^*	$1/\sqrt{p}$
Radiation effects (Coulomb scattering, pair production, bremsstrahlung) radiation length	X_{rad}	$1/p$

[a] The increase does not follow a simple law because the atomic structure is involved, but it is essentially proportional to p times a logarithmic term of p. A curve for argon is shown in Fig. 1.11.

[b] [ALL 80].

[c] See discussion in Chap. 9.

(a) Longitudinal diffusion or $\omega\tau \ll 1$: For the width of the diffusion cloud one gains a factor $1/\sqrt{p}$ (2.61, 63), and in the accuracy to find the centre of the cloud one gains another factor $1/\sqrt{p}$, because the statistical fluctuations vary with $1/\sqrt{N_{tot}}$, where N_{tot} is the total number of electrons in the diffusion cloud. Therefore, the over-all gain is a factor $1/p$.

(b) Transverse diffusion with $\omega\tau \gg 1$: For the width of the diffusion cloud one *loses* a factor $\sqrt{p}$ according to (2.61, 63, 72). The statistical fluctuations compensate this; therefore, the over-all factor is 1.

The contributions to the measurement accuracy that arise from the drift-path variations have statistical fluctuations that vary with $1/\sqrt{N_{\text{eff}}}$ (see (1.73) and Sect. 6.2.3). As the pressure is increased, N_{eff} increases as well, but only slowly. For the purpose of the present estimates we take N_{eff} proportional to $\sqrt{p}$ (Figs. 1.22 and 1.23). The situation would change where declustering occurred in any important measure. The drift-path variations are partly due to the wire geometry, i.e. constant when p varies, but partly they have a pressure dependence of their own. This is the case for the wire $E \times B$ effect. Here we refer to the angle at which the electrons approach the wires in their immediate neighbourhood (Sect. 6.3.1). The tangent of the effective Lorentz angle ψ will decrease roughly proportional to $1/p$, as p increases. A smaller angle ψ will reduce the drift-path variations. Depending on how important the wire $E \times B$ effect is in comparison to the other drift-path variations, we may say in summary that the contributions to the measurement accuracy that arise from the drift-path variations decrease as a function of p which is somewhere between $1/p$ and $1/p^{0.25}$.

11.5.2   Lorentz Angle

Here we refer to the direction Ψ under which the electrons travel in the main drift space. For a drift chamber of type 2 it is imperative to keep Ψ low. One way to achieve this is to increase the gas pressure. For a first orientation it is enough to work in the approximation of one electron energy. Referring to Sect. 2.2,

$$\tan \Psi = \omega\tau \ .$$

Since $\omega\tau = (e/m)B$, at a given magnetic field one seeks a small τ. The mean time between collisions is related to the density N and the collision cross-section σ by

$$\tau = \frac{1}{N\sigma c} \ .$$

An increase of the density N by a factor p will reduce τ to τ/p if the electric field E is also increased to Ep, because the electron velocity c is a function of E/N and then stays the same. In the presence of a constant and strong magnetic field B, this is not strictly true because c is also a function of B/N, but the dependence is not very strong, as can be seen from (2.49). In conclusion we may say that an increase of the gas pressure and the electric field by a factor p reduces the tangent of the Lorentz angle approximately by the same factor.

11.5.3   Drift-Field Distortions from Space Charge

In drift chambers with long drift distances like TPCs one always has to beware of the adverse effect of any space charge that may exist in the drift volume. Such a charge may arise from some background such as synchrotron radiation in e^+e^- colliders. It will distort the uniform drift field. The effect of an increase of the gas pressure in such a situation is the following.

The background B_s increases proportional to p, and since the wire gain G is assumed to be adjusted to produce the same signal level as before, we have $B_s G = $ constant. If the drift potential has been increased in proportion to the pressure, the ion drift velocity remains the same, and the amount of positive ion charge that flows back into the drift volume (assuming constant effectiveness of any ion shutter) produces the same amount of total charge Q inside the drift volume as before. The adverse electric field it creates there is the same, but relative to the increased drift field it is *smaller* by the factor by which the pressure has been increased.

11.6   Deterioration of Chamber Performance with Usage ('Ageing')

Drift and proportional chambers that have been in use for some time have a tendency to mal-function sooner or later – an increase of the dark current, a lowering of the gain and a loss of pulse-height resolution are the typical symptoms. Once it has started, the disease seems to become worse and to spread from a few wires to many, and finally the chamber may not hold the operating voltage any more.

This behaviour is intimately connected with the gas mixture in the chamber and with certain contaminants. But also the material properties of the anodes and cathodes as well as their size play a role in this area, which is far from being clearly understood. Given the practical importance of the subject and that the new accelerators will produce extremely high radiation levels, efforts towards better understanding are under way. A workshop held in Berkeley in 1986 [WOR 86] summarized the experience by collecting reports and by recording a very lively discussion.

11.6.1   General Observations in Particle Experiments

The wires of degraded chambers carry deposits of various kinds, either as spots or as complete coatings of anode or of cathode wires, smooth or hairy, white or black or oily, in the region of the strongest exposure. One quantifies the lifetime of a wire by adding up the charge that it has collected during its life; it is calculated as the total charge that has drifted towards it times the avalanche gain. Lifetimes at which performance losses have been reported are in the range 10^{-4}–1 Coulomb per cm of wire length. It appears that the lower limit of 10^{-4} C/cm has become rather uncommon since the problems with the particular gas mixture consisting of Ar (75%), isobutane (24.5%) and freon (0.5%) have

Table 11.10. Drift and proportional chambers exposed to high radiation doses – reports of some accelerator experiments with high densities of accumulated charge

Chamber	Reference	Basic gas mixture (percentages)	Additional measures (concentrations in %)	Accumulated charge density (C/cm)	Observed effects[a]
TASSO Central Detector	[BIN 86]	$Ar(50)$ $+C_2H_6$ (50)	C_2H_5OH (1.6) +oil in ethane removed gas filters inserted	0.2–0.4	None
TASSO Vertex Chamber	[BIN 86]	$Ar(95)+CO_2(5)$ $+C_2H_5OH$ (0.12) (3 bar)	H_2O (1)	0.25	Some
Split field Magnet Chambers	[ULL 86]	$Ar(53)$ $+i\text{-}C_4H_{10}(40)$ $+(CH_2OH)_2$ CH_2 (7)	—	0.3	Anode wire coating, thickness 1 µm
ACCMOR Drift Chambers	[TUR 86]	Ar (50) $+C_2H_6$ (50)	$(CH_3)_2$ $CHOH$ (0.2) $Ar(60)+C_2H_6$ (40)	0.02	None
ARGUS Drift Chamber	[DAN 89]	$C_3H_8(97)$ $+(CH_2OH)_2$ CH_2 (3)	H_2O (0.2)	0.005	None
BNL Hyper-nuclear Spec-trometer DC	[PIL 86]	$Ar(50)$ $+C_2H_6(50)$	covering soft urethane adhesive	0.2	None
FNAL Tagged Photon Spec-rometer	[EST 86]	$Ar(50)$ $+C_2H_6$ (50)	C_2H_5OH wire cleaning	0.1–0.2	Efficiency loss
UA1 Central Chamber	[YVE 86]	Ar (40) $+C_2H_6$ (60)	—	0.01	None
Argonne ZGS Beam Chambers	[SPI 86]	Ar (65) $+CO_2$ (35) $+CBrF_3$ (0.5)	new bubbler oil, new plastic tubes, new wires	0.1–1	Efficiency loss, deposits (Si) on anodes, cathodes
EMC Muon Chambers	[HIL 86]	$Ar(79.5)$ $+CH_4$ (19.5)	—	0.15	Anode deposits, efficiency loss 30%
				after 0.25:	End of operation
J-Spectrometer, BNL (1974), front proportional chambers	[BEC 92]	$[Ar^+$ $(CH_2OH)_2$ $CH_2]$ $(0°C)^b$	—	0.1–1	Deposits on flat cathode, chambers still efficient

[a] After the additional measures   [b] Argon saturated with methylal at 0 °C

become understood [SAU 86]. (This mixture was employed in the very first proportional chamber systems and, among experts, went under the name of 'magic gas'; it was used for very high avalanche gain factors beyond the proportional mode.) Most often, lifetimes are reported to lie above 10^{-2} C/cm before deterioration sets in. With special precautions such lifetimes may be extended by an order of magnitude or more. Table 11.10 contains a collection of particle experiments where detailed reports about the behaviour of ageing chambers and of some special precautions were given.

11.6.2 Dark Currents

The build up of a dark current drawn by the anode wire of a deteriorated chamber even in the absence of radiation can be explained by the 'thin-field emission effect', or Malter effect [MAL 36]. Malter has shown in his experiment that a positive surface charge deposited on a thin insulating film that covers a cathode can provoke the emission of electrons from the cathode through the film. A very high electric field may be created in the film between the deposited charge and the counter-charges accumulated on the metallic cathode on the opposite side. Electrons are extracted from the metal through field emission and may find their way through the film into the space between anode and cathode. In Malter's experiment the positive surface charges were produced by an external beam of particles; when this was removed, the field emission was not immediately interrupted but decayed slowly, because the surface charge took an appreciable time to leak away.

In a wire chamber, some thin insulating coating on the cathodes is charged up by the positive ions produced in the avalanches and drifting to the cathodes. The surface charge density depends on the balance between the neutralization rate and the ion collection rate, the latter being proportional to the rate of charge multiplication on the anode. If the surface charge density increases above some critical value, secondary electron emission from the cathode into the gas via the Malter effect becomes possible. The secondary electrons reach the anode and produce avalanches, thus increasing the positive-ion production rate. If the secondary-electron emission rate is sufficiently high, the process becomes self-sustained and the dark current remains, even if the source of the primary radiation is removed from the chamber.

If one stops the avalanches by turning down the anode voltage, the dark current is obviously interrupted, but it does not immediately reappear when the anode voltage is later turned up again. A certain intensity of radiation on the area in question is required to reestablish the current.

An efficient way to prevent the development of cathode deposits is to keep the electric field at the cathode surfaces low. Plane cathodes and wires with larger diameters show smaller deposits.

In chambers with appreciable standing currents, the addition of a small amount of water in the gas has often been found to suppress the currents

[WOR 86, DAN 89]. An explanation is that the cathode deposits are made conductive so as to increase the ion leakage rate through the insulating film. It is also general knowledge that small amounts of alcohol have a tendency to prevent or delay the appearance of dark currents.

Another cause of standing currents and of the breakdown of chambers are the so-called whiskers, fine strands of material growing from the cathode towards the anode. Glow discharges may appear owing to the large fields they create.

The growth of whiskers is primarily a question of the gas composition. For example, a change from the gas composition 1 $(Ar(89\%) + CO_2(10\%) + CH_4(1\%))$ to the similar gas composition 2 $(Ar(90\%) + CO_2(5\%) + CH_4(5\%))$ produced whiskers which grew from the cathode wires in the field direction, according to a test by the JADE group [KAD 86]. When going back to composition 1, the whiskers shrank and finally disappeared. In another test [FOS 86] it was shown that whisker growth on cathode wires is supported by gas mixtures containing the lower alkanes, which presumably form $(CH_2)_n$-type polymers. On the other hand, no mixtures of carbon dioxide and argon support whisker growth; rather they would make existing whiskers retreat and disappear. Table 11.11 summarises these tests which were done in the extreme conditions of glow discharge. Although such extreme conditions do not usually apply to well-working chambers, they may develop temporarily at some imperfect spot of a chamber.

Let us add to this list of causes of dark currents the photo-effect discussed in Sect. 4.1. Depending on the photo-absorption properties of the gas, photons

Table 11.11. Whisker growth measured in glow discharges produced on a cathode wire [FOS 86]

Gas mixture (percentages)	Whisker growth allowed?	Comments
Argon (50) + ethane (50)	Yes	Whiskers grow down to concentrations of ethane of 5% or more
Argon (48) + ethane (48) + ethanol (4)	Yes	Growth is suppressed by adding ethanol
Argon (40) + ethane (40) + carbon dioxide (20)	Yes	
Argon + carbon dioxide	No	No mixture of Ar and CO_2 supports whisker growth
Argon (70) + carbon dioxide (25) + ethanol (5)	No	
Argon (90) + methane (10)	Yes	
Argon (95) + methane (5)	Yes	
Argon (50) + isobutane (50)	Yes	Threshold for whisker growth about 2 times higher than argon–ethane mixtures
Argon (60) + isobutane (40)		
Argon (92) + isobutane (8)		

emitted in an avalanche may create electrons in the gas or on the cathode which produce new avalanches. Whether this feed-back situation can lead to limited dark currents, depends on the presence of some damping mechanism.

11.6.3 Ageing Tests

The process of ageing has been investigated in numerous laboratory experiments. With the help of intense radioactive sources, the ageing process is forced to occur in shorter intervals of time. Although it is established that the measured fatal dose is not independent of this intensity (it comes out larger at too high intensities [VAV 86]) the method is useful for the comparison of different factors that influence the ageing process. A list complete up to 1985 is contained in [WOR 86]. Later work may be followed consulting [ATA 87] and [KAD 90].

A first line of investigation concerns the nature of the quenching agent, which, either alone or in combination with a noble gas (typically Ar), would serve as the counting gas. Perhaps some quenching agents age later than others? Figure 11.7 shows the results of Turala and co-workers [DWU 83]; the relative pulse height as well as the peak of the Fe-55 signal began to deteriorate at

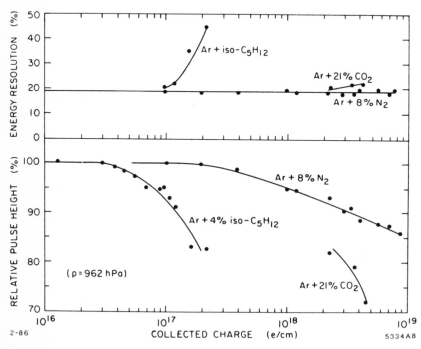

Fig. 11.7. Energy resolution and peak position of the ^{55}Fe signal in various gas mixtures as functions of the total collected charge, measured by [TUR 86]

lifetimes that were different by an order of magnitude or more, as isopentane was exchanged for carbon dioxide or nitrogen.

It is generally reported that the addition of a small percentage of some alcohol (CH_3CH_2OH, $(CH_2OH)_2CH_2$, $(CH_3)_2CHOH$) to the counting gas extends significantly the lifetime of chambers even if it does not prevent the development of deposits.

Another line of enquiry starts from the observation that chemical elements found in the wire deposits are occasionally foreign to the chemical compounds of the gas or the wires. Silicon, a dominant signal in X-ray fluorescence or Auger-electron-spectrometry analyses of some wire deposits, could only have been contained in the gas at such low concentrations as are permitted by the tight tolerances of the gas-purity specifications. In addition there may have been some parts of the chamber installation where silicon-containing compounds reached the gas at very low vapour pressure. In one instance, some silicon oil present in the counting gas with a vapour pressure of 10^{-9} Torr was observed to cause rapid growth of polymers on the anode [HIL 86]. This implies a surprisingly high silicon collection efficiency of the avalanche process. Could it be that other substances 'poison' the gas in a similar manner? And if so, under what circumstances? In Fig. 11.8, the surprising results of a test performed by

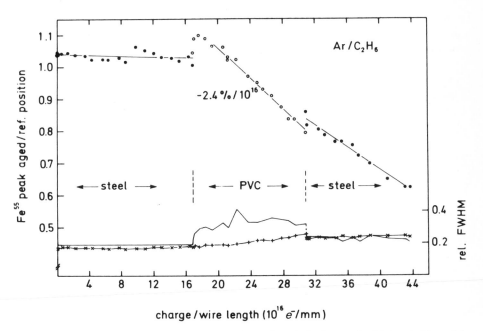

Fig. 11.8. Effect of a soft, 10 m long, PVC gas hose. *Dots* (referring to the left-hand side scale) indicate the position of the ^{55}Fe peak relative to the (unaged) reference position. The *crosses* (referring to the right-hand side scale) show the width of the ^{55}Fe distribution relative to its peak position in the radiation damaged wire region. The lower drawn-out curve (referring to the right-hand scale) represents the same quantity, but measured at the (unaged) reference position. The period of time in which the PVC hose was put into the input gas stream is also indicated [KOT 86]

Kotthaus are depicted. The input gas stream of argon–ethane was conducted through clean steel tubes; when the integrated collected charge had reached a certain value marked in the figure, a 10-m-long hose of soft PVC was introduced into the input gas stream, resulting in an immediate deterioration of the test-tube performance, as measured by the pulse height and the resolution of the Fe-55 spectrum. When this hose was removed again, the performance did not return to its original values, but the loss of gain persisted. It is also remarkable that the resolution changed quite strongly on a piece of wire had not aged (outside the area of exposure), but only during the presence of the PVC hose. Such observations support the idea that the process of polymerization, which later continues on its own, is started by certain dangerous substances, in this case perhaps one of the organic softening agents that are mixed into the PVC giving it its characteristic odour. In his summary [VAV 86], Va'vra makes a list of contaminations that, according to experience, should be avoided in order to achieve extended chamber lifetimes; it includes halogens, oil traces (also from bubblers), rubber (especially silicon rubber), polyurethane adhesive, PVC and teflon tubing, soft epoxies and adhesives, aggressive solder, any unknown organic materials, and large amounts of G10 (glass-fibre-enforced material containing Si). Cold traps have a beneficial effect [HIL 86].

A thorough understanding of the ageing process must involve the chemical processes of polymerization in the plasma of the avalanches. There are certainly several competing mechanisms at work; this is apparent from the wide variety of manifestations of the deposits and also from the fact that drastically different element abundances are measured in different spots along the wires [KOT 86].

At present we do not even know whether a better understanding of the plasma polymerization on our wires will give us the means to substantially prolong the lifetime of the chambers. For the time being we are therefore left with a few rules – to keep the gas and the wires clean, to add some alcohol or water if necessary, and to keep the cathode wires thick. Since it is the total collected charge which destroys the wire, it is useful to keep the amplification low.

References

[ATA 87]   M. Atac, Wire chamber aging and wire material, *IEEE Trans.* **NS 34**, 476 (1987)
[BEC 92]   U. Becker, private communication (1992)
[BEI 88a]   S.P. Beingessner, T.C. Meyer and M. Yvert, Influence of chemical trace additives on the future ageing of the UA1 central detector, *Nucl. Instrum. Methods A* **272**, 669 (1988)
[BEI 88b]   S.P. Beingessner, T.C. Meyer, A. Norton, M. Pimiä, C. Rubbia and V. Vuillemin, First running experience with the laser calibration system of the UA1 central detector, in *Proc. Int. Conf. on Advanced Technology and Particle Physics*, Villa Olmo (Como), Italy, 1988

[BEN 89]   A.C. Benvenuti et al., A nonflammable gas mixture for plastic limited streamer tubes. *Nucl. Instrum. Methods Phys. Res. A* **284**, 339 (1989)

[BIA 89]   S.F. Biagi, A multiterm Boltzmann analysis of drift velocity, diffusion, gain and magnetic effects in argon-methane-water-vapour mixtures, *Nucl. Instrum. Methods Phys. Res. A* **283**, 716 (1989)

[BIN 86]   D.M. Binnie, Experience with the TASSO chambers, in [WOR 86], p. 213

[BRE 74]   A. Breskin, G. Charpak, B. Gabioud, F. Sauli, N. Trautner, W. Duinker, G. Schultz, Further results on the operation of high-accuracy drift chambers, *Nucl. Instrum. Methods* **119**, 9 (1974)

[BRO 79]   F. Brown, N. Hadley, P. Miller, M. Pettersen, D. Phan, G. Przybylski and P. Rohrisch, List of poisoning times for materials tested in the materials test chamber, LBL Berkeley note TPC-LBL-79-8 (unpublished)

[BUR 48]   J.H. Burgoyne and G. Williams-Leir, Limits of inflammability of gases in the presence of diluants, a critical review, *Fuel* **27**, 118 (1948)

[CAT 89]   A. Cattai, H.G. Fischer, A. Morelli, Drift velocity in Ar + CH_4 + water vapours, DELPHI internal note 89-63, 28 July 1989, and H.G. Fischer, private communication

[DAN 89]   M. Danilov, V. Nagovitsin, V. Shibaev, I. Tichomirov, E. Michel, W. Schmidt-Parzefall, A study of drift chamber ageing with propane, *Nucl. Instrum. Methods Phys. Res. A* **274**, 189 (1989)

[DRY 86]   S.L.T. Drysdale, K.W.D. Ledingham, C. Raine, K.M. Smith, M.H.C. Smyth, D.T. Stewart, M. Towrie and C.M. Houston, Detection of toluene in a proportional counter gas by resonant two-photon ionization spectroscopy, *Nucl. Instrum. Methods A* **252**,521 (1986)

[EST 86]   P. Estabrooks, Aging effects in a large drift chamber in the Fermilab Tagged-Photon Spectrometer, in [WOR 86], p. 231

[FOS 86]   B. Foster, Whisker growth in test cells, in [WOR 86], p. 227, and private communication

[GUS 84]   E.M. Gushchin, A.N. Lebedev and S.V. Somov, Tracks produced by an N_2 laser in a streamer chamber, *Instrum. and Exp. Tech.* (Translation of *Prib. i. Tekh. Eksp.*) **27**, 546 (1984)

[HIL 86]   H.J. Hilke, Detector calibration with lasers, a review, *Nucl. Instrum. Methods A* **252**, 169 (1986)

[HIL 86a]   H.J. Hilke, Summary of ageing studies in wire chambers by AFS, DELPHI and EMC groups, in [WOR 86], p. 153

[HUB 85]   G. Hubricht, K. Kleinknecht, E. Müller, D. Pollmann and E. Teupe, Ionization of counting gases and ionizable gaseous additives in proportional chambers by UV lasers, *Nucl. Instrum. Methods A* **228**, 327 (1985)

[HUB 86]   G. Hubricht, K. Kleinknecht, E. Müller, D. Pollmann, K. Schmitz and C. Stürzl, Investigation of UV laser ionisation in naphtalene and phenol vapours added to proportional chamber gases, *Nucl. Instrum. Methods A* **243**, 495 (1986)

[HUK 88]   M. Huk, P. Igo-Kemenes and A. Wagner, Electron attachment to oxygen, water and methanol in various drift chamber gas mixtures, *Nucl. Instrum. Methods Phys. Res. A* **267**, 107 (1988)

[KAD 86]   H. Kado, Performance of the JADE Vertex Detector, in [WOR 86], p. 207

[KAD 90]   J. Kadyk, J. Wise, D. Hess and M. Williams, Anode wire aging tests with selected gases, *IEEE Trans.* **NS 37**, 478 (1990)

[KOT 86]   R. Kotthaus, A laboratory study of radiation damage to drift chambers, in [WOR 86], p. 161

[LED 84]   K.W.D. Ledingham, C. Raine, K.M. Smith, A.M. Campbell, M. Towrie, C. Trager and C.M. Houston, Laser induced ionization in proportional counters seeded with low-ionization-potential vapours, *Nucl. Instrum. Methods A* **225**, 319 (1984)

[LED 85]   K.W.D. Ledingham, C. Raine, K.M. Smith, M.H.C. Smyth, D.T. Steward, M. Towrie and C.M. Houston, Wavelength dependence of laser-induced ionization in proportional counters, *Nucl. Instrum. Methods A* **241**, 441 (1985)

[LEH 89]   I. Lehraus, R. Matthewson and W. Tejessy, Materials test facility for the ALEPH TPC, CERN Note EF 89-8, also as ALEPH 89-89 (unpublished)

[LEW 61]   B. Lewis and G. von Elbe, *Combustion, Flames and Explosions of Gases*, 2nd edn (Academic, New York London 1961)

[MAL 36]   L. Malter, Thin film field emission, *Phys. Rev.* **50**, 48 (1936)

[PAL 75]   V. Palladino and B. Sadoulet, Application of classical theory of electrons in gases to drift proportional chambers, *Nucl. Instrum. Methods* **128**, 323 (1975)

[PIL 86]   P.H. Pile, Radiation damage control in the BNL Hypernuclear Spectrometer drift chamber system, in [WOR 86], p. 219

[RIE 72]   F.F. Rieke and W. Prepejchal, Ionization cross-sections of various gaseous atoms and molecules for high-energy electrons and positrons, *Phys. Rev. A* **6**, 1507 (1972)

[SAU 86]   F. Sauli, When everything was clear, in [WOR 86], p. 1

[SCH 86]   B. Schmidt, Drift und Diffusion von Elektronen in Methan und Methan-Edelgas-Gemischen, Dissertation, Univ. Heidelberg (1986)

[SCH 92]   Private communication J. Schmid, CERN (1992)

[SPI 86]   W. Haberichter, H. Spinka, Wire chamber degradation at the Argonne ZGS, presented by H. Spinka, in [WOR 86], p. 99

[TOW 86]   M. Towrie, J.W. Cahill, K.W.D. Ledingham, C. Raine, K.M. Smith, M.H.C. Smyth, D.T.S. Stewart and C.M. Houston, Detection of phenol in proportional-counter gas by two-photon ionization spectroscopy, *J. Phys. A (At. Mol. Phys.)* **19**, 1989 (1986)

[TUR 86]   A. Dwurazny, Z. Hajduk, M. Turala, Ageing effects in gaseous detectors, and search for remedies, presented by M. Turala, in [WOR 86], p. 113

[ULL 86]   OMEGA and SFM Collaborations, The OMEGA and SFM experience in intense beams, presented by O. Ullaland, in [WOR 86], p. 107

[WOR 86]   Proceedings of the Workshop on Radiation Damage to Wire Chambers, held at Berkeley, January 1986 (J. Kadyk workshop organizer), Lawrence Berkeley Laboratory, University of California, Berkeley, LBL 21 170 (1986) (unpublished)

[VAV 86]   J. Va'vra, Review of wire chamber ageing, in [WOR 86], p. 263

[YVE 86]   S. Beingessner, T. Meyer, V. Vuillemin, M. Yvert, Our aging experience with the UA1 central detector, presented by M. Yvert, in [WOR 86], p. 67

[ZAB 65]   M. Zabetakis, Flammability characteristics of combustible gases and vapours, Bulletin 627 of the Bureau of Mines, United States Department of the Interior 1965. The curves are generally in agreement with the numbers of [BUR 48], except for CH_4/He

Subject Index

Springer-Verlag
and the Environment

We at Springer-Verlag firmly believe that an international science publisher has a special obligation to the environment, and our corporate policies consistently reflect this conviction.

We also expect our business partners – paper mills, printers, packaging manufacturers, etc. – to commit themselves to using environmentally friendly materials and production processes.

The paper in this book is made from low- or no-chlorine pulp and is acid free, in conformance with international standards for paper permanency.